Advances in
ORGANOMETALLIC CHEMISTRY

VOLUME 40

Advances in Organometallic Chemistry

EDITED BY

F. GORDON A. STONE

DEPARTMENT OF CHEMISTRY
BAYLOR UNIVERSITY
WACO, TEXAS

ROBERT WEST

DEPARTMENT OF CHEMISTRY
UNIVERSITY OF WISCONSIN
MADISON, WISCONSIN

VOLUME 40

ACADEMIC PRESS
San Diego   London   Boston   New York
Sydney   Tokyo   Toronto

Copyright © 1996 by ACADEMIC PRESS

Academic Press, Inc.
525 B Street, Suite 1900, San Diego, California 92101-4495, USA
http://www.apnet.com

Academic Press Limited
24-28 Oval Road, London NW1 7DX, UK
http://www.hbuk.co.uk/ap/

International Standard Serial Number: 0065-3055

International Standard Book Number: 0-12-031140-2

PRINTED IN THE UNITED STATES OF AMERICA
96   97   98   99   00   01   QW   9   8   7   6   5   4   3   2   1

Contents

Silylhydrazines: Lithium Derivatives, Isomerism, and Rings

KATRIN BODE and UWE KLINGEBIEL

The Organometallic Chemistry of Halocarbonyl Complexes of Molybdenum(II) and Tungsten(II)

PAUL K. BAKER

Substituent Effects as Probes of Structure and Bonding in Mononuclear Metallocenes

MELANIE L. HAYS and TIMOTHY P. HANUSA

Reactions of 17- and 19-Electron Organometallic Complexes

SHOUHENG SUN and DWIGHT A. SWEIGART

A Review of Group 2 (Ca, Sr, Ba) Metal-Organic Compounds as Precursors for Chemical Vapor Deposition

WILLIAM A. WOJTCZAK, PATRICK F. FLEIG, and MARK J. HAMPDEN-SMITH

Contributors

Numbers in parentheses indicate the pages on which the authors' contributions begin.

PAUL K. BAKER (45), Department of Chemistry, University of Wales, Bangor, Gwynedd LL57 2UW, Wales, United Kingdom

KATRIN BODE (1), Institute of Inorganic Chemistry, University of Goettingen, D-37077 Goettingen, Germany

PATRICK F. FLEIG (215), Department of Chemistry and The Center for Micro-Engineered Materials, University of New Mexico, Albuquerque, New Mexico 87131

MARK J. HAMPDEN-SMITH (215), Department of Chemistry and The Center for Micro-Engineered Materials, University of New Mexico, Albuquerque, New Mexico 87131

TIMOTHY P. HANUSA (117), Department of Chemistry, Vanderbilt University, Nashville, Tennessee 37235

MELANIE L. HAYS (117), Department of Chemistry, Vanderbilt University, Nashville, Tennessee 37235

UWE KLINGEBIEL (1), Institute of Inorganic Chemistry, University of Goettingen, D-37077 Goettingen, Germany

SHOUHENG SUN (171), Department of Chemistry, Brown University, Providence, Rhode Island 02912

DWIGHT A. SWEIGART (171), Department of Chemistry, Brown University, Providence, Rhode Island 02912

WILLIAM A. WOJTCZAK (215), Department of Chemistry and The Center for Micro-Engineered Materials, University of New Mexico, Albuquerque, New Mexico 87131

ADVANCES IN ORGANOMETALLIC CHEMISTRY, VOL. 40

Silylhydrazines: Lithium Derivatives, Isomerism, and Rings

KATRIN BODE and UWE KLINGEBIEL

Institute of Inorganic Chemistry
University of Goettingen
D-37077 Goettingen, Germany

I

INTRODUCTION

The versatility of nitrogen in its compounds depends in large measure on the existence of a range of oxidation states between -3 and $+5$. In its combination with silicon, systems are known in which N has an oxidation state of -3 as in the derivatives of silylamines, of -2 as in the derivatives of silylhydrazines, and of -1 as in the derivatives of silylazenes:

$$\overset{|}{\underset{|}{-Si}}-\overset{-3}{N}\diagdown \quad , \quad \overset{|}{\underset{|}{-Si}}-\overset{-2}{\underset{|}{N}}-\overset{|}{\underset{|}{N}}-\overset{|}{\underset{|}{Si}}- \quad , \quad \overset{|}{\underset{|}{-Si}}-\overset{-1}{N}=N-\overset{|}{\underset{|}{Si}}-$$

Whereas in these compounds silicon is always tetrahedral, nitrogen has a changing coordination geometry: in silylamines it is planar, but in silylazenes it is linear.

This article reviews silylhydrazines, their preparation, properties, and reactions. Other reviews on Si–N compounds have appeared elsewhere.[1-8] However, as the chemistry of silylhydrazines has developed very fast, especially during the past 5 years, it seems appropriate to write an article that deals only with silylhydrazines.

1

II

PREPARATION

A. *Synthesis and Properties of Silylhydrazines*

The synthesis of the first cyclic and acyclic silylhydrazines were reported by Aylett[9] and Wannagat *et al.*[10,11] in 1956–1958. Two main methods of preparation were developed.

1. The first is the treatment of a hydrazine with a halosilane. The silylhydrazines are formed by intermolecular cleavage of a hydrogen halid [Eq. (1)].

$$\tag{1}$$

This is the most common method for chloro-, bromo-, or iodosilanes.[6] Often, auxilary bases such as triethylamine or pyridine are added. But when hydrazine is treated with fluorosilanes or silanes, no condensation is observed because of the reduced reactivity. Fluorosilanes only form adducts with hydrazines, so that the reaction is stopped at step (a). Because of the extremely strong Si—F bond energy, no cleavage of HF or N_2H_4-condensation is observed.[3,6] In this case another preparation method must be chosen.

2. The second method of preparation is the treatment of lithiated hydrazine with halosilanes, especially fluorosilanes, with elimination of lithium halide [Eq. (2)]:

$$R_3Si—Hal + LiHN—NH_2 \longrightarrow R_3Si—NH—NH_2 + LiHal \tag{2}$$

The formation of the silylhydrazines depends on the reactivity and the bulkiness of the halosilanes. The condensation increases with increasing number of the hydrogen atoms. While iodosilane or bromo(methyl)-silane yield the tetrakis(silyl)hydrazines,[9,12] the reaction of hydrazine with chlorodimethylsilane gives only the tris(silyl)hydrazine[12]:

$$\begin{array}{cc} Me_2HSi & SiHMe_2 \\ & N-N \\ Me_2HSi & H \end{array}$$

When three methyl groups are bound to the silicon, isomeric bis(trimethylsilyl)hydrazines are formed.[7,10] Wannagat et al. succeeded in separating the isomeric N,N- and N,N'-bis(trimethylsilyl)hydrazines[11]:

$$\begin{array}{cc} Me_3Si & SiMe_3 \\ & N-N \\ H & H \end{array} \quad \text{and} \quad \begin{array}{cc} Me_3Si & H \\ & N-N \\ Me_3Si & H \end{array}$$

A monosilylhydrazine can only be obtained when bulky substituents such as triphenylsilyl groups stabilize the hdyrazine[10]:

$$\begin{array}{cc} Ph_3Si & H \\ & N-N \\ H & H \end{array}$$

B. Mono(silyl)hydrazines

1. Synthesis and Properties

The only mono(silyl)hydrazine known until 1993 was the triphenylsilyl-hydrazine 1, described by Wannagat and Liehr in 1958[10] [Eq. (3)].

$$Ph_3SiCl \ + \ 2\,N_2H_4 \ \longrightarrow \ \begin{array}{cc} Ph_3Si & H \\ & N-N \\ H & H \end{array} \ + \ N_2H_5Cl \qquad (3)$$

$$\mathbf{1}$$

Primary silylhydrazines with R = Me, Et, or Pr cannot be isolated, as they immediately undergo further condensation to bis(silyl)hydrazines with elimination of hydrazine.[6,10] For the triphenylsilylhydrazine this condensation only occurs under more drastic conditions at 90°C [Eq. (4)].

$$2 \ \begin{array}{cc} Ph_3Si & H \\ & N-N \\ H & H \end{array} \ \xrightarrow[-N_2H_4]{90°C} \ \begin{array}{cc} Ph_3Si & H \\ & N-N \\ H & SiPh_3 \end{array} \qquad (4)$$

$$\mathbf{2}$$

Some other mono(silyl)hydrazines have been synthesized in the reaction of fluorosilanes with lithiated hydrazine:

$$R_3SiF \; + \quad \underset{Li}{\overset{H}{\diagdown}}N-N\overset{H}{\underset{H}{\diagup}} \quad \longrightarrow \quad \underset{R_3Si}{\overset{H}{\diagdown}}N-N\overset{H}{\underset{H}{\diagup}} \quad + \; LiF$$

$$\textbf{3-8}$$

(5)

$\textbf{3}^{13}$	$\textbf{4}^{14}$	$\textbf{5}^{16}$	$\textbf{6}^{16}$	$\textbf{7}^{15}$	$\textbf{8}^{17}$
SiR_3	$SitBu_2F$	$SitBu_2Me$	$SitBu_2Ph$	$SitBu(iPr)_2$	$Si(NiPr_2)_2F$	$SiN(SiMe_3)_2MeF$

This class of compounds is kinetically stabilized by bulky *tert*-butyl groups (**3–6**) or amine groups (**7,8**). In **3**, **7**, and **8**, HF elimination should be possible, but it is hindered by the very strong SiF bond energy. These mono(silyl)hydrazines are very stable molecules and show no tendency to undergo condensation at room temperature.

Their stability allows a directed synthesis of asymmetrical bis(silyl)hydrazines by the reactions of lithiated mono(silyl)hydrazines with halosilanes (Section B,2). These reactions often lead to the formation of isomeric products.

2. *Lithium Derivatives of Mono(silyl)hydrazines*

Isomeric products are formed in the reaction of lithiated di-*tert*-butylmethylsilylhydrazine with fluorosilanes. The formation of these isomers requires prior coordination of the Li^+ ion with the two N atoms of the hydrazine unit. The crystal structure of the lithiated di-*tert*-butylmethylsilylhydrazine **4** (**9**)[14] exhibits two different silylhydrazide units I and II, which are bound by six Li^+ ions to form a hexameric entity.

$$\left[-\underset{|}{\overset{|}{Si}}-\bar{N}-N\overset{H}{\underset{H}{\diagup}}\right]^{\ominus} \quad ; \quad \left[-\underset{|}{\overset{|}{Si}}-\overset{H}{\underset{}{N}}-N\overset{H}{\underset{}{\diagup}}\right]^{\ominus}$$

I   II

The Li^+ ions are bound to three different structural units: Li1 is coordinated with one N atom and two NH groups; Li2 with one N atom, one N_2, and one NH unit; and Li3 with one NH, one N_2, and one NH_2 unit. Thus, four Li^+ ions are bound side-on. (See Fig. 1 and Table I.)

Unlike unit II, in the solid state I does not coordinate Li^+ ions side-on. The side-on bond lengths for the atoms Li2 and Li3 were determined to be 192 and 188 pm, respectively. The angles at N5 and N6, which bind Li3 side-on, are nearly identical; the angles at N1 and N2, which coordinate Li2, deviate by 7°. Thus, Li3 is positioned centrosymmetrically above the N—N bond.

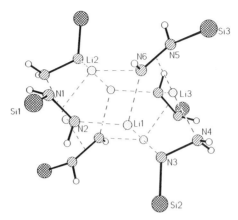

FIG. 1.   Crystal structure of **9**.

Table II shows a comparison of the data for *ab initio* calculations of the lithium hydrazide system $(NHNH_2)Li^+$ (A)[18,19] with that for the side-on units N1–N2–Li2 and N5–N6–Li3 (B) in **9**, derived from the X-ray structure determination:

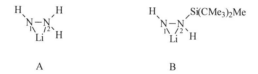

A   B

TABLE I

SELECTED BOND LENGTHS [pm] AND ANGLES
[°] OF **9**

Bond lengths (pm)		Bond angles (deg)	
Li1—N1	199.5	N2–Li2–N1	42.5
Li1—N3	201.6	N2–N1–Li2	63.5
Li1—N6	201.8	N1–N2–Li2	72.2
Li2—N2	200.8	N4–N3–Si2	115.4
Li2—N6	203.6	N4–N3–Li1	108.4
Li2—N1	210.4	Si2–N3–Li1	112.7
N1—N2	149.3	N5–N6–Li3	68.6
Si1—N1	174.4	N6–N5–Li3	68.5
Si2—N3	173.0		
N3—N4	149.3		
N5—N6	147.7		
Si3—N5	174.4		

TABLE II

CALCULATED AND MEASURED PARAMETERS IN A
AND B, RESPECTIVELY

Bond lengths (pm)	A	B
N—N	145	149.3; 147.7
N^1—Li	161	200.8; 202.3
N^2—Li	189	210.4; 202.4
Bond angles (deg)	A	B
N–Li–N	56	42.5; 42.8
Li–N^1–N^2	49	72.2; 68.6
Li–N^2–N^1	76	63.5; 68.5

The structural analysis confirms the *ab initio* calculations, according to which Li^+ ions in hydrazines can be coordinated equally by both N atoms of the hydrazine. This phenomenon also accounts for the isomerizations during the secondary substitutions.

C. Bis(silyl)hydrazines

1. Symmetrical Bis(silyl)hydrazines

When mono(silyl)hydrazines are not stable at room temperature, they condensate to give the symmetrical bis(silyl)hydrazines **10–15**:

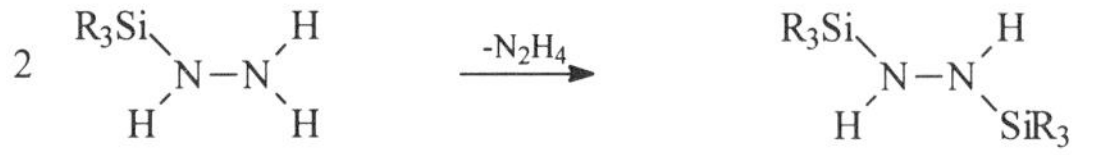

$$2 \quad \underset{H}{\overset{R_3Si}{\diagdown}} N-N \underset{H}{\overset{H}{\diagup}} \quad \xrightarrow{-N_2H_4} \quad \underset{H}{\overset{R_3Si}{\diagdown}} N-N \underset{SiR_3}{\overset{H}{\diagup}} \tag{6}$$

10-15

	10^{10}	11^{10}	12^7	13^{10}	14^{10}	15^{17}
SiR_3	$SiMe_3$	$SiEtMe_2$	$SitBuMe_2$	$SiEt_3$	$SiPr_3$	$SitBu_2H$

These condensation reactions often lead to the formation of isomers. Structural isomerism of bis(silyl)hydrazines was first observed in 1964.[19–24] In the absence of strong steric or electronic constraints, the bis(silyl)hydrazines such as bis(trimethylsilyl)hydrazine give in a thermoneutral reaction essentially equal amounts of the *N,N*- and *N,N'*-isomers at equilibrium.[7,23] Wannagat *et al.* found that the reaction of hydrazine with trimethylchlorosilane at room temperature results only in the formation of *N,N'*-bis(trimethylsilyl)hydrazine, whereas the same reaction in boiling solvents leads to a mixture of *N,N*- and *N,N'*-bis(trimethylsilyl)hydrazine. Both could be separated by preparative gas chromatography. Their struc-

TABLE III

PHYSICAL DATA OF THE BIS(TRIMETHYLSILYL)HYDRAZINES[11]

	R_3Si—NH—NH—SiR_3 (Ia)	$\begin{array}{c} R_3Si \\ \diagdown \\ N\text{—}NH_2 \text{ (Ib)} \\ \diagup \\ R_3Si \end{array}$
Melting point	$-64 \pm 1°$	$-64 \pm 1°$
Boiling point	149° (731 torr)	158° (733 torr)
Vapor pressure curve log p [torr]	$7.93-2128/T$	$7.72-2083/T$
Evaporation enthalpy	9.73 kcal/mol	9.53 kcal/mol
Trouton constant	23.1 Cl	22.1 Cl
Refractive index	1.4205	1.4295
Density	0.8117	0.8348
Molar refraction		
MR_L found	55.06	54.54
calc'd	55.41	55.41
MR_E found	250.6	252.2
calc'd	253.6	253.6
τ (SiCH$_3$) in ppm	9.99	9.93

tures may be derived from infrared and Raman spectra. The properties are shown in Tables III–V.[11]

Silatropy was also observed in bis(silyl)hydrazines with bulkier silyl substituents.[20–22,25–29] Both steric and electronic effects are seen to influence the position of the equilibrium. Silicon shows an extremely high mobility in anionic rearrangements compared to that of phenyl or methyl groups. In contrast to carbon-substituents for which a 1,2-sigmatropic shift is disallowed by the Woodward–Hoffmann rules,[30,31] in organosilicon group migration rearrangements can occur without these symmetry restrictions because pentacoordinate transition states are possible.[7]

TABLE IV

VAPOR PRESSURE CURVE OF THE BIS(TRIMETHYLSILYL)HYDRAZINES[11]

R_3Si—NH—NH—SiR_3				$(R_3Si)_2N$—NH_2			
t (°C)	p (torr)	t (°C)	p (torr)	t (°C)	p (torr)	t (°C)	p (torr)
17	4	65	43	17	3,5	65	37
23	5	80	79	23	4,5	80	64
35	10	100	171	35	9	100	140
55	28	149	731	55	24	158	733

TABLE V

VIBRATION SPECTRA OF THE BIS(TRIMETHYLSILYL)HYDRAZINES[11]

	$(CH_3)_3Si—NH—NH—Si(CH_3)_3$ (Ia)		$(CH_3)_3Si$, $(CH_3)_3Si$ $N—NH_2$ (Ib)	
	IR	Raman	IR	Raman
ν NH	3348 s	3345 mp		
ν_{as} NH$_2$			3350 s	3352 s
ν_s NH$_2$			3270 ss	3275 mp
			3205 ss	
ν_{as} CH$_3$	2962 st	2960 sst	2958 st	2962 sst
ν_s CH$_3$	2905 s	2900 sstp	2901 m	2905 sstp
δ NH$_2$			1575 m	1580 ss
	1440 ss		1440 ss	
δ_{as} CH$_3$	1395 m	1410 m	1395 s	1417 m
δ_s CH$_3$	1248 sst	1255 s	1248 sst	1252 s
ν NN	1060 st		1060 st	1075 ss
ν_{as} SiNSi			1005 sst	1010 ss
ν SiN	875 sst	890 ss		
			890 sst	
ρ_{as} CH$_3$	840 sst	845 s	838 sst	838 s
		780 ss	760 st	
ρ_s CH$_3$	750 m	750 ss	752 st	750 ss
ν_{as} SiC$_3$	685 m	690 st	680 m	689 st
ν_s SiC$_3$	615 s		647 ss	655 sstp
ν_s SiC$_3$			617 m	615 ss
ν SiN	606 s	611 sstp		
ν_s SiNSi			478 m	490 sstp
ρ SiC$_3$	330 s	328 ss	345 st	340 ss
			315 st	
δ_{as} SiC$_3$	250 m		245 m	
δ_s SiC$_3$		215 st		225 st
δ SiNSi				190 st
δ SiNN		192 st		
		130 s		

Another possibility for the formation of symmetrical bis(silyl)hydrazines is the condensation of the corresponding mono(silyl)hydrazines under more drastic conditions, as shown for the formation of the triphenylsilylhydrazine 2 by heating 1 above its melting point at 90°C. In order to synthesize the compounds 16–19, the mono(silyl)hydrazines have to be heated for seven days at 220°C [Eq. (7)].

$$2 \quad \underset{\substack{16^{13} \\ }}{\overset{R_3Si}{\underset{H}{\diagdown}}N-N\overset{H}{\underset{H}{\diagup}}} \quad \xrightarrow[-N_2H_4]{220°C} \quad \underset{\substack{16\text{-}19 \\ 19^{16}}}{\overset{R_3Si}{\underset{H}{\diagdown}}N-N\overset{H}{\underset{SiR_3}{\diagup}}}$$

(7)

SiR_3	$SitBu_2F$	$SitBu_2Me$	$SitBu_2Ph$	$SitBu(iPr)_2$

When the silyl groups are too bulky, as in **7** and **8**, no tendency to condense is observed even under the preceding conditions.

2. *Asymmetrical Bis(silyl)hydrazines*

The first asymmetric bis(silyl)hydrazines **20–22** were obtained in the 1950s by treating hydrazine with a mixture of different halosilanes [Eq. (8)].

$$N_2H_4 + R_3SiHal + R'_3SiHal \xrightarrow{-2\ HHal} \underset{\mathbf{20\text{-}22}}{\overset{R_3Si}{\underset{H}{\diagdown}}N-N\overset{H}{\underset{SiR'_3}{\diagup}}}$$

(8)

$\mathbf{20}^{11}$: $R_3Si = SiMe_3$, $SiR'_3 = SiEt_3$

$\mathbf{21}^{7}$: $R_3Si = SiMe_3$, $SiR'_3 = SiMe_2Et$

$\mathbf{22}^{7}$: $R_3Si = SiMe_3$, $SiR'_3 = SiMe_2tBu$

Because of the stability of the mono(silyl)hydrazines **3–8**, their lithium derivatives can be used to synthesize asymmetric N,N'-bis(silyl)hydrazines **(23–40)**, as shown in Eq. (9).

$$\underset{}{\overset{R_3Si}{\underset{H}{\diagdown}}N-N\overset{H}{\underset{Li}{\diagup}}} + R_3'SiF \xrightarrow{-LiF} \underset{\mathbf{23\text{-}40}}{\overset{R_3Si}{\underset{H}{\diagdown}}N-N\overset{H}{\underset{SiR'_3}{\diagup}}}$$

(9)

	SiR_3	SiR'_3	References
23	$SitBu_2F$	$SiMe_3$	13
24	$SitBu_2F$	$SitBuMe_2$	13
25	$SiMe_3$	$SiFMe[N(HSiMe_3)]$	17
26	$SiMe_3$	$SiFMe[N(SiMe_3)_2]$	17
27	$SitBu_2Me$	$SitBuMe_2$	15
28	$SitBu_2Me$	$Si\ i\text{-}Pr_2F$	15
29	$SitBu_2Me$	$SiPh_2F$	17

	SiR_3	SiR'_3	References
30	$SitBu_2Me$	$SiMes_2F$	17
31	$SitBu_2Me$	$SiF_2[N(SiMe_2tBu)_2]$	17
32	$SitBu_2Me$	$Si\ i\text{-}Pr_2H$	37
33	$SiF(NiPr_2)_2$	$SitBu_2Me$	17
34	$SiF(NiPr_2)_2$	$SitBu_2F$	17
35	$SiF(NiPr_2)_2$	$Si\ i\text{-}Pr_2F$	17
36	$SiF(NiPr_2)_2$	$SiMe_3$	17
37	$SitBu_2Ph$	$SitBu_2F$	16
38	$SitBu_2Ph$	$Si\ i\text{-}Pr_2F$	16
39	$SitBu(iPr)_2$	$Si\ i\text{-}Pr_2F$	16
40	$SitBu(iPr)_2$	$SitBu_2F$	16

Depending on the steric effects, either N,N-bis(silyl)hydrazines **41** or isomeric products **42a/b–45a/b** are formed [Eq. (10)].

$$
\begin{array}{c}
R_3Si\text{-}N(H)\text{-}N(H)\text{-Li} + R_3'SiF \xrightarrow{-\ LiF}
\begin{cases}
\tfrac{1}{2} \quad
\begin{matrix} R_3Si \\ R'_3Si \end{matrix}\!\!N\text{-}N\!\!\begin{matrix} H \\ H \end{matrix} \\[4pt]
\mathbf{41,\ 42a\text{-}45a} \\[8pt]
\tfrac{1}{2} \quad
\begin{matrix} R_3Si \\ H \end{matrix}\!\!N\text{-}N\!\!\begin{matrix} H \\ SiR'_3 \end{matrix} \\[4pt]
\mathbf{42b\text{-}45b}
\end{cases}
\end{array}
\tag{10}
$$

	SiR_3	SiR'_3	References
41	$SitBu_2F$	$SitBuF_2$	13
42	$SitBu_2F$	$SiF_2[N(SiMe_3)_2]$	13
43	$SitBu_2F$	$SiF_2[N(tBuSiMe_3)]$	13
44	$SitBu_2Me$	$SiF_2[N(SiMe_3)_2]$	14
45	$SitBu_2Me$	$SitBuF_2$	17

3. Crystal Structures of Lithiated Bis(silyl)hydrazines

Four crystal structures of lithiated bis(silyl)hydrazines are described.[16,32–35] The Li centers exhibit tri- and tetracoordination, but because of Li–C interactions, the Li atoms may also adopt higher coordination numbers. All lithiated bis(silyl)hydrazines show side-on and end-on coordinated lithium.

     a. *Monolithiated Bis(silyl)hydrazines.*   The isomeric N,N- and N,N'-bis(trimethylsilyl)hydrazines react with *n*-BuLi in *n*-hexane to give an insoluble lithium derivative **46a**. On addition of small amounts of thf, **46a** is dissolved exothermally and crystallizes as **46b** on cooling. At the same time, N,N'-bis(trimethylsilyl)hydrazine is formed. **46b** crystallizes as a dimeric structure. As shown in Eq. (11), the monomers consist of two different hydrazide units: one N-lithium-N',N'-bis(trimethylsilyl)hydrazide and one N,N'-dilithium-N,N'-bis(trimethylsilyl)hydrazide unit.[31,33]

$$\begin{array}{c} H \\ 3\ Me_3SiN_2SiMe_3 \\ Li \end{array}$$

46a

$$\frac{1}{2}\left[\begin{array}{c} Me_3Si \diagdown \qquad \diagup SiMe_3 \\ \qquad N-N \\ Li \diagup \qquad \diagdown Li \end{array}\right. + \begin{array}{c} H \diagdown \qquad \diagup SiMe_3 \\ \qquad N-N \\ Li \diagup \qquad \diagdown SiMe_3 \end{array}\Bigg]_2 \qquad (11)$$

$$\xrightarrow[\ \ \ -(Me_3SiNH)_2\ \ \ ]{+\ thf}$$

46b

The structure of **46b** shown in Fig. 2 is a tetramer, which is held together by Li clusters. The Li$^+$ ions are bound to different units (Table VI): Li1 connects the two units belonging to the monomer and is bound end-on to N3 and end-on to N2. By being bound side-on to N3a and N4a, the hydrazine unit of the second monomer, the coordination sphere of Li1 is saturated.

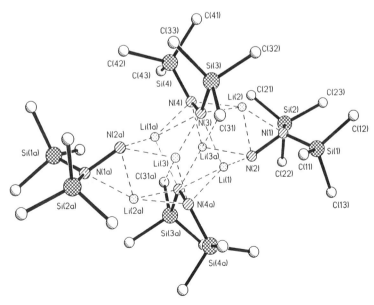

FIG. 2.   Structure of **46b**.

TABLE VI

SELECTED BOND LENGTHS [pm] AND ANGLES
[°] OF **46b**

Bond lengths (pm)		Bond angles (deg)	
Li1—N2	216.3	N4–Li2–N2	108.5
Li2—N2	212.1	N2–Li2–N3	111.9
Li2—N1	229.5	N2–N1–Si1	110.5
Li2—N4	200.3	N4–N3–Li2	61.2
Li2—N3	223.3	N3–N4–Li2	77.6
Li3—N4	215.5	N4a–Li1–N3	99.4
Li3—N3	221.1	N4–Li2–N3	41.2
N1—N2	155.3	N2–Li2–N1	40.1
N3—N4	150.5	N2–N1–Li2	63.5
N1—Si2	178.4	Li2–N2–Li1	69.0

Li2 is also tetracoordinate: it is bound side-on to the nitrogen atoms of the monomeric unit N1/N2 and N3/N4 so that in contrast to Li1 and Li3, Li2 does not contribute to form the dimer. Li3, the only lithium ion showing no side-on coordination, is found to be bound to N2 and N4, connecting the monomeric units, and to the nitrogen atom N3a belonging to the second monomer.

The crystal structure shows that dilithiated products are formed in the reaction with halosilanes. The formation of these dilithiated by products can be avoided when a nonpolar solvent is used instead of thf.

The monolithiated bis(*tert*-butyldimethylsilyl)hydrazine **47**[16] is obtained in the reaction of the *N,N'*-bis(*tert*-butyldimethylsilyl)hydrazine with *n*-BuLi in thf in a molar ratio 1 : 1.

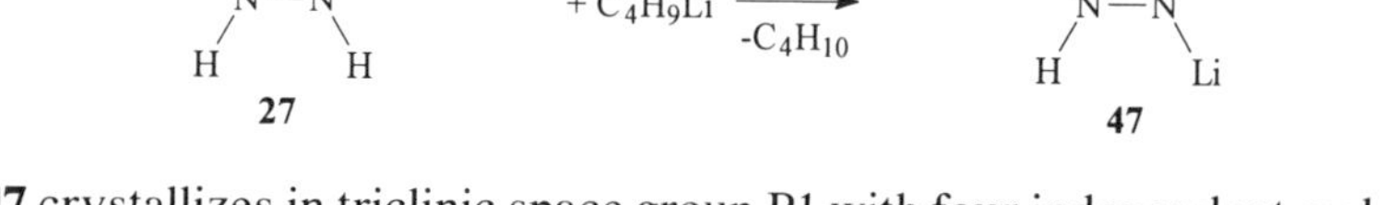

$$\underset{\mathbf{27}}{\overset{\displaystyle CMe_3Me_2Si}{\underset{H}{\diagdown}}N\!-\!N\!\underset{H}{\overset{SiMe_2CMe_3}{\diagup}}} + C_4H_9Li \xrightarrow[-C_4H_{10}]{} \underset{\mathbf{47}}{\overset{\displaystyle CMe_3Me_2Si}{\underset{H}{\diagdown}}N\!-\!N\!\underset{Li}{\overset{SiMe_2CMe_3}{\diagup}}} \quad (12)$$

47 crystallizes in triclinic space group P1 with four independent molecules in one unit cell. In the solid state, the monomers are connected to dimeric units.

Compound **47** (Fig. 3, Table VII) is centrosymmetric and shows a tricyclic ring system consisting of one planar (Li–N)$_2$ four-membered ring and two LiN$_2$ three-membered rings forming a ladder structure via the Li—N bond. The four molecules of the unit cell are not identical: three molecules exhibit a disorder with respect to the nitrogen atoms, but the fourth does not.

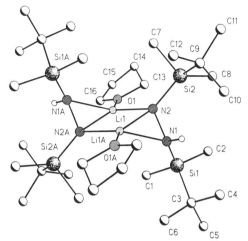

FIG. 3. Structure of **47**.

TABLE VII

SELECTED BOND LENGTHS [pm] AND ANGLES [°] OF **47**

Bond lengths (pm)		Bond angles (deg)	
1. Dimer			
Si1—N1	169.5	N1a–Li1–N2a	43.4
N1—N2	151.4	N2a–Li1–N2	108.8
N1a–Li1	211.8	Li1a–N2–Li1	71.2
N2–Li1	204.1	N1–N2–Li1	116.1
Li1–O1	194.3	N2–N1–Li1A	62.9
Si2—N2	180.6	N1–N2–Si2	112.4
		Si2–N2–Li1	130.1
		N2–N1–Si1	118.7
		O1–Li1–N2	115.6
		O1–Li1–N2a	134.3
2. Dimer			
Si4—N4	176.2	N3–N4–Li2	119.0
		Si4–N4–Li2	125.6
3. Dimer			
Si6—N6	175.9	N5–N6–Si6	115.3
N6—Li3a	199.7	Si6–N6–Li3	125.0
4. Dimer			
Si8—N8	169.8	N7–N8–Li4	120.9
N7a—Li4	207.1	N8–N7–Li4a	67.0
N8—Li4a	203.0	N7–N8–Si8	116.7
		Si8–N8–Li4a	140.4
		Si8–N8–Li4	121.6
		N8–N7–Si7	121.2

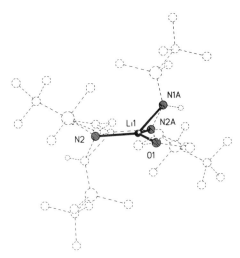

FIG. 4.   Structure of **47**.

The lithium ions Li1 and Li1a are identically coordinated. They are bound side-on to the nitrogen atoms of the hdyrazide unit and are coordinated to N2 or N2a in the four-membered ring. Their coordination sphere is saturated by thf coordination. The lithium ions exhibit an interesting geometry: if the relatively weak Li–N contact in the three-membered ring (Li1–N1a) is ignored, the angle sum of lithium is found to be 358.0°, which means that lithium shows nearly a planar coordination geometry, illustrated in Fig. 4.

In contrast to **46**, no formation of dilithiated hydrazide units is observed in **47**, which is the only purely monolithiated bis(silyl)hydrazine. Similar to this structure, a monolithiated *N-tert*-butyl-*N*′-trimethylsilylhydrazine has been described.[16]

b. *Dilithiated Bis(silyl)hydrazines.* N,N'-Bis(trimethylsilyl)hydrazine is converted into the dilithium derivative $[(Me_3Si)_2N_2Li_2] \cdot 2$ thf (**48**) by *n*-BuLi in a 1:2 molar ratio[33,34] [Eq. (13)]. Compound **48** forms a tetramer that lies on a crystallographic inversion center.

$$(Me_3SiNH)_2 \quad \xrightarrow[\text{$-2\,n$ C}_4\text{H}_{10}]{\overset{+2\,n\text{-BuLi}}{\overset{+2\text{ thf}}{}}} \quad 1/4 \left[\begin{array}{c} Me_3Si \diagdown \qquad \diagup SiMe_3 \\ N{-}N \\ Li \diagup \qquad \diagdown Li \end{array} \right]_4 \quad * \; 2 \text{ thf} \qquad (13)$$

48

As shown in Fig. 5, eight Li^+ ions connect the tetramer. They are bound to two different structural units: Li1 is coordinated to one nitrogen atom,

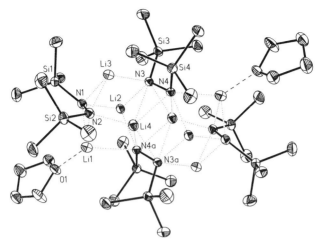

FIG. 5.   Structure of **48**.

to one N_2 unit, and to thf; Li2 is bound to three nitrogen atoms; Li3 is bound to one nitrogen atom and one N_2 unit; and Li4 is coordinated to one nitrogen atom and two N_2 units. (See Table VIII.) Thus, three Li^+ ions are found to be bound side-on.[34] The molecular structure of **48** remains intact at elevated temperatures. Actually, the Li_8N_8 "cluster" does not even break down with thf coordination.[33,34]

As this compound contains exclusively N,N'-dilithium-N,N'-bis(silyl)-hydrazide units, it is understandable that no isomerism is observed in the

TABLE VIII

SELECTED BOND LENGTHS [pm] AND ANGLES [°] OF **48**

Bond lengths (pm)		Bond angles (deg)	
Li1—N1	195.3	N2–N1–Si1	124.9
Li1—N2	203.0	N2–N1–Li2	99.9
Li1—O1	203.4	N2–N1–Li1	69.8
N1—N2	154.9	N2–N1–Li3	63.9
N1—Li2	198.3	N1–N2–Li3	70.1
Li2—N3	206.3	N1–N2–Li1	64.5
Li3—N2	193.4	N4–N3–Li4	63.3
Li3—N1	202.5	N3–N4–Si4	116.6
Li4—N3	238.8	N3–N4–Li4	77.1
Li4—N4	219.0		
N3—N4	156.1		

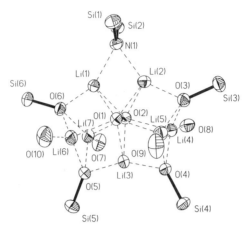

FIG. 6.   Structure of **49**.

reactions with fluorosilanes. Instead, SiN_2 three-membered rings, $(SiN_2)_2$ six-membered rings, or tetrakis(silyl)hydrazines are formed.

The hydrolysis of **48** in air leads to the formation of $[(Me_3SiOLi)_4 \cdot Li_2O_2 \cdot LiN(SiMe_3)_2] \cdot 2thf$ **49**, which contains one peroxide and four oxide ions (Fig. 6, Table IX). The O_2^{2-} unit lying in the center is side-on coordinated by Li1, Li2, and Li3. Li1 and Li2 saturate their coordination sphere by binding to one Me_3SiO^- and to one $(Me_3Si)_2N^-$ unit. Li3 is bound also to two Me_3SiO^- groups. Li4, 5, 6, and 7 connect the Me_3SiO^- units and are bound to one thf molecule and end-on to the O_2^{2-} unit. That means that all Li^+ ions are tetracoordinate.

The Li—O bond lengths of Li1 and Li2, the Li^+ ions that are coordinated side-on, were determined to be 200.9 and 202.9 or 203.0 pm, and those

TABLE IX
SELECTED BOND LENGTHS [pm] AND ANGLES [°] OF **49**

Bond lengths (pm)		Bond angles (deg)	
N1—Li1	201.7	Si2–N1–Si1	123.7
N1—Li2	204.0	Li1–N1–Li2	68.9
O1—O2	155.7	O2–O1–Li3	66.6
O1—Li2	200.9	O2–O1–Li1	66.6
O1—Li1	202.9	O2–O1–Li2	68.0
O2—Li1	200.9	O1–O2–Li1	68.0
O2—Li2	203.0		

TABLE X

Selected Bond Lengths [pm] and Angles [°] of 50

Bond lengths (pm)		Bond angles (deg)	
N—Si	172	N2–N1–Li1	67
Li—N	193–200	N1–N2–Li2	66
N—N	156		

of Li3, 196.8 and 196.5 pm. The angle O2–Li3–O1 is 46.7°. Thus, Li3 in particular is positioned centrosymmetrically above the O1—O2 bond.

The dilithiated *tert*-butyldimethylsilylhydrazine 50[35] is present as a trimer. The monomeric unit shows both Li$^+$ bound side-on. (See Table X.)

$$\left[\begin{array}{c} \text{Li(1)} \\ \text{Me}_2\text{CMe}_3\text{Si}\diagdown\diagup\diagdown\diagup\text{SiMe}_2\text{CMe}_3 \\ \text{N(1)}\!-\!\text{N(2)} \\ \diagdown\diagup \\ \text{Li(2)} \end{array} \right]_3$$

50

Three monomeric units associate to give a Li$_6$N$_6$ framework, which exhibits pentacoordinated nitrogen atoms and tetracoordinated Li centers. In addition, the Li$^+$ ions show two Li–H contacts. There are three different Li—N—Li angles: 70°, 95°, and 108°. The molecule shows a crystallographically imposed C$_2$ symmetry.

D. Tris(silyl)hydrazines

1. Synthesis of Tris(silyl)hydrazines

Tris(silyl)hydrazines are obtained by different methods: they can be obtained in the reaction of chlorosilanes with hydrazines in the presence of Et$_3$N as in Eq. (14)[12]:

$$\text{Me}_2\text{HSiCl} + 3\,\text{NEt}_3 + \text{N}_2\text{H}_4 \xrightarrow{-3\,[\text{HNEt}_3]\text{Cl}} (\text{Me}_2\text{HSi})\text{HN}-\text{N(SiHMe}_2)_2 \qquad (14)$$

53

In 1959, lithiated silylhydrazines were first used in the preparation of tris(silyl)hydrazines by Wannagat et al.[11]:

$$\underset{\underset{\text{H}}{\diagup}\text{N}-\text{N}\underset{\underset{\text{H}}{\diagdown}}{}\overset{\overset{\text{Me}_3\text{Si}}{\diagdown}}{}\overset{\overset{\text{SiMe}_3}{\diagup}}{} \xrightarrow[\substack{-\text{C}_4\text{H}_{10} \\ -\text{LiCl}}]{\substack{+\text{C}_4\text{H}_9\text{Li} \\ +\text{Me}_3\text{SiCl}}} \underset{\underset{\text{Me}_3\text{Si}}{\diagup}\text{N}-\text{N}\underset{\underset{\text{H}}{\diagdown}}{}\overset{\overset{\text{Me}_3\text{Si}}{\diagdown}}{}\overset{\overset{\text{SiMe}_3}{\diagup}}{} \qquad (15)$$

51

By the second method, tris(silyl)hydrazines can be substituted not only with two, but also with three different silyl groups (**72–74**).

The following tris(silyl)hydrazines have been synthesized to date:

$$\underset{H}{\overset{R_3Si}{\diagdown}}N-N\underset{SiR''_3}{\overset{SiR'_3}{\diagup}}$$

	SiR$_3$	SiR$'_3$	SiR$''_3$	References
51	SiMe$_3$	SiMe$_3$	SiMe$_3$	36
52	SiMeH$_2$	SiMeH$_2$	SiMeH$_2$	12
53	SiMe$_2$H	SiMe$_2$H	SiMe$_2$H	12
54	SiPhH$_2$	SiPhH$_2$	SiPhH$_2$	12
55	SiMe$_3$	SiMe$_3$	SiEt$_3$	36
56a	SiMe$_3$	SiEt$_3$	SiEt$_3$	36
56b	SiEt$_3$	SiEt$_3$	SiMe$_3$	36
57	SiMe$_3$	SiMe$_2$	SiMe$_2$tBu	37
58	SitBuMe$_2$	SitBuMe$_2$	SitBuF$_2$	32
59	SitBuMe$_2$	SitBuMe$_2$	SiMe$_2$Cl	32
60	SitBuMe$_2$	SitBuMe$_2$	SiF$_3$	16
61	SitBuMe$_2$	SitBuMe$_2$	SiPhF$_2$	16
62	SitBu$_2$Me	SitBu$_2$Me	SiPhF$_2$	32
63	SitBu$_2$Me	SitBu$_2$Me	SiMeF$_2$	32
64	SiMe$_3$	SiMe$_3$	SisBuPhF	38
65	SiMe$_3$	SiMe$_3$	SitBuiBuF	38
66	SiMe$_3$	SiMe$_3$	SiMePhF	38

Isomerism is observed also in the formation of tris(silyl)hydrazines (**56a/b**). The reaction of **46** with fluorosilanes leads to the formation of the isomeric products **67** and **68** or **69** and **70** [Eq. (16)].

$$\overset{\textbf{46}}{\underset{}{}}$$

$$+ \ F_2Si\overset{R}{\underset{Me}{\diagdown}} \quad \bigg\downarrow \ - \ LiF$$

$$1/2 \ \underset{MeFRSi}{\overset{Me_3Si}{\diagdown}}N-N\underset{H}{\overset{SiMe_3}{\diagup}} \quad + \quad 1/2 \ \underset{Me_3Si}{\overset{Me_3Si}{\diagdown}}N-N\underset{SiFRMe}{\overset{H}{\diagup}} \qquad (16)$$

67: R = Me[39]   **68:** R = Me[39]
69: R = Ph[39]   **70:** R = Ph[39]

In the reaction of **47** with iPr_2SiF_2, shown in Eq. (17), the isomeric products **71** and **72**[32] are obtained according to the side-on coordination of lithium in **47**.

$$
\begin{array}{l}
\text{Me}_2\text{CMe}_3\text{Si} \quad \text{SiCMe}_3\text{Me}_2 \\
\qquad\qquad \text{N}-\text{N} \\
\qquad\qquad \text{H} \qquad \text{H}
\end{array}
\xrightarrow[\substack{-\text{C}_4\text{H}_{10} \\ -\text{LiF}}]{\substack{+\text{C}_4\text{H}_9\text{Li} \\ +(\text{CHMe}_2)_2\text{SiF}_2}}
$$

1/2
$$
\begin{array}{l}
\text{Me}_2\text{CMe}_3\text{Si} \quad \text{SiCMe}_3\text{Me}_2 \\
\qquad\qquad \text{N}-\text{N} \\
\qquad\qquad \text{H} \qquad \text{Si}(\text{CHMe}_2)_2\text{F}
\end{array}
$$
71

1/2
$$
\begin{array}{l}
\text{Me}_2\text{CMe}_3\text{Si} \qquad \text{H} \\
\qquad\qquad \text{N}-\text{N} \\
\text{Me}_2\text{CMe}_3\text{Si} \qquad \text{Si}(\text{CHMe}_2)_2\text{F}
\end{array}
$$
72

(17)

Similarly, lithiated **27** reacts with $PhSiF_3$ to give the isomeric products **73** and **74** in a 1 : 3 ratio[16] [Eq. (18)].

$$
\begin{array}{l}
\text{Me}_2\text{CMe}_3\text{Si} \quad \text{Si}(\text{CMe}_3)_2\text{Me} \\
\qquad\qquad \text{N}-\text{N} \\
\qquad\qquad \text{H} \qquad \text{H}
\end{array}
\xrightarrow[\substack{-\text{C}_4\text{H}_{10} \\ -\text{LiF}}]{\substack{+\text{C}_4\text{H}_9\text{Li} \\ +\text{PhSiF}_3}}
$$

1/2
$$
\begin{array}{l}
\text{Me}_2\text{CMe}_3\text{Si} \quad \text{Si}(\text{CMe}_3)_2\text{Me} \\
\qquad\qquad \text{N}-\text{N} \\
\qquad\qquad \text{H} \qquad \text{SiPhF}_2
\end{array}
$$
73

1/2
$$
\begin{array}{l}
\text{Me}_2\text{CMe}_3\text{Si} \quad \text{Si}(\text{CMe}_3)_2\text{Me} \\
\qquad\qquad \text{N}-\text{N} \\
\text{F}_2\text{PhSi} \qquad \text{H}
\end{array}
$$
74

(18)

The condition for the formation of isomers is a side-on coordination of the lithium in **47**, the lithium derivative of **27**. Depending on the bulkiness of the fluorosilane, this side-on coordination makes possible a substitution that is suitable for the system. In comparison, the reaction of lithiated **27** with iPr_2SiF_2 leads only to the formation of the tris(silyl)hydrazine **75**,[16] so that kinetic control of the reaction must be supposed:

$$
\begin{array}{l}
\text{Me}_2\text{CMe}_3\text{Si} \quad \text{Si}(\text{CMe}_3)_2\text{Me} \\
\qquad\qquad \text{N}-\text{N} \\
\qquad\qquad \text{H} \qquad \text{H}
\end{array}
\xrightarrow[\substack{-\text{C}_4\text{H}_{10} \\ -\text{LiF}}]{\substack{+\text{C}_4\text{H}_9\text{Li} \\ +(\text{CHMe}_2)_2\text{SiF}_2}}
\begin{array}{l}
\text{Me}_2\text{CMe}_3\text{Si} \quad \text{Si}(\text{CMe}_3)_2\text{Me} \\
\qquad\qquad \text{N}-\text{N} \\
(\text{CHMe}_2)_2\text{FSi} \qquad \text{H}
\end{array}
$$
75

(19)

2. *Crystal Structure of a Lithium–Tris(silyl)hydrazide*

Only the lithium salt of a tris(trimethylsilyl)hydrazine **76** is described.[35,40] It crystallizes as a dimer via a Li—N bond in a ladder structure, forming a tricyclic system with one $(LiN)_2$ four-membered ring and two LiN_2 three-

TABLE XI

SELECTED BOND LENGTHS [pm] AND ANGLES [°] OF **76**

Bond lengths (pm)		Bond angles (deg)	
Li1—N1	191.7	N1–Li1–N2	43.7
Li1—N1a	197.0	N1–Li1–N1a	107.8
Li1—N2	213.5	N2–N1–Li1	75.8
Li1—C	271–273	Si1–N1–Li1	149.0
N1—N2	152.0	Li1–N1–Li1a	72.2
		Si1–N1–N2	124.6
		N1–N2–Li1	60.5
		N1–N2–Si3	115.3

membered rings. The nitrogen atom N1 is tetrahedral. Because of the short Li1–N1–Li1a angle (72.2°), the N1–Li1–N1a angle is extended to 107.8°. The Li1—N1 bond length is 191.7 pm and is very short compared to the calculated Li—N bond length. The Li—N1a bond, connecting the monomers to give a dimeric unit, has a length of 197.0 pm.

Both Li^+ ions Li1 and Li1a are coordinated identically: they are bound side-on to the nitrogen atoms of the hydrazine unit and end-on to the ring nitrogen atom N1. Their coordination sphere is saturated by two Li–C contacts. Computational positioning of H-atoms with a rigid C—H bond distance of 108 pm and tetrahedral geometry results in Li–H contacts like those cited or evidenced for binding interactions.[41] The dimer exists in the gas phase, as well. The mass spectrum shows the molecular ion **76** (Table XI).

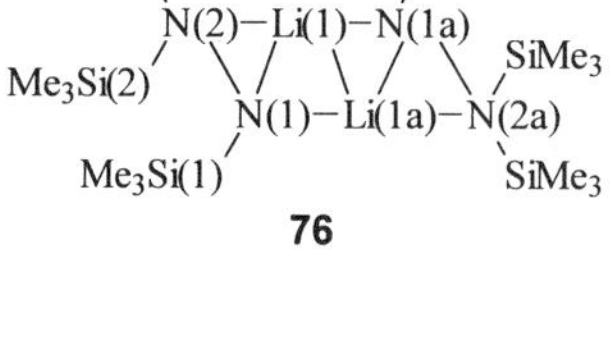

Me₃Si(3)\   SiMe₃
       N(2)—Li(1)—N(1a)
Me₃Si(2)/ \ / \ / \ SiMe₃
       N(1)—Li(1a)—N(2a)
Me₃Si(1)/   SiMe₃

76

E. Tetrakis(silyl)hydrazines

1. Synthesis of Tetrakis(silyl)hydrazines

Four different methods have been employed for the synthesis of tetrakis (silyl)hydrazines.[9,12,16,40,42–45]

(a) The first tetrakis(silyl)hydrazine, the tetrasilylhydrazine **77**, was obtained in the gas-phase reaction of hydrazine with excess silyl iodide,[9] as shown in Eq. (20).

$$4 \, SiH_3I \; + \; 5 \, N_2H_4 \quad \longrightarrow \quad (H_3Si)_2N\!-\!N(SiH_3)_2 \; + \; 4 \, N_2H_5I \qquad (20)$$

$$\mathbf{77}$$

The compound has also been obtained by blowing anhydrous hydrazine vapor with a stream of dry nitrogen stepwise into a 5-liter bulb containing an excess of SiH_3Br.[46] The compound, which is colorless and nonvolatile at $-64°$, is purified by repeated low-temperature vacuum distillation.[9,46] Because the compound may explode if it comes into contact with air, it must be prepared and handled in a Stock-type high-vacuum line.

(b) The decomposition of bis(trimethylsilyl)diimine[47] above 35°C leads to the formation of $\mathbf{78}$[48] [Eq. (21)].

$$2 \; Me_3Si\!-\!N\!=\!N\!-\!SiMe_3 \quad \xrightarrow{\;-\,35°C\;} \quad \begin{array}{c} Me_3Si \diagdown \qquad \diagup SiMe_3 \\ N\!-\!N \\ Me_3Si \diagup \qquad \diagdown SiMe_3 \end{array} \quad + \quad N_2 \qquad (21)$$

$$\mathbf{78}$$

(c) Another method [Eq. (22)] is the treatment of hydrazine with halosilane in the presence of Et_3N[12,42]:

$$4 \, PhH_2SiCl \; + \; 4 \, NEt_3 \; + \; N_2H_4 \quad \xrightarrow{\;-\,4\,[HNEt_3]Cl\;} \quad \begin{array}{c} PhH_2Si \diagdown \qquad \diagup SiH_2Ph \\ N\!-\!N \\ PhH_2Si \diagup \qquad \diagdown SiH_2Ph \end{array} \qquad (22)$$

$$\mathbf{79}^{12}$$

The symmetrical tetrakis(silyl)hydrazines $(SiR_3)_4N_2$ $\mathbf{77}$–$\mathbf{81}$ have been synthesized by this method.

	$\mathbf{77}^{9}$	$\mathbf{78}^{42,44,48}$	$\mathbf{79}^{12}$	$\mathbf{80}^{12}$	$\mathbf{81}^{12}$
SiR_3	SiH_3	$SiMe_3$	SiH_2Ph	$SiHMe_2$	SiH_2Me

(d) Tetrakis(silyl)hydrazines can also be obtained in the reaction of lithiated tris(silyl)- or dilithiated bis(silyl)hydrazines with halosilanes.[16,40,43,44] Examples are given in Eqs. (23) and (24).

$$\begin{array}{c} Me_3Si \diagdown \qquad \diagup SiMe_3 \\ N\!-\!N \\ Me_3Si \diagup \qquad \diagdown Li \end{array} \; + \; SiF_4 \quad \xrightarrow{\;-\,LiF\;} \quad \begin{array}{c} Me_3Si \diagdown \qquad \diagup SiMe_3 \\ N\!-\!N \\ Me_3Si \diagup \qquad \diagdown SiF_3 \end{array} \qquad (23)$$

$$\mathbf{82}$$

$$\underset{\text{Li}}{\overset{\text{Me}_3\text{Si}}{\diagdown}}\text{N}-\text{N}\underset{\text{Li}}{\overset{\diagup\text{SiMe}_3}{}} \quad + \quad 2\text{ PhSiF}_3 \quad \xrightarrow{-2\text{ LiF}} \quad \underset{\text{PhF}_2\text{Si}}{\overset{\text{Me}_3\text{Si}}{\diagdown}}\text{N}-\text{N}\underset{\text{SiF}_2\text{Ph}}{\overset{\diagup\text{SiMe}_3}{}} \quad (24)$$

$$\textbf{83}$$

According to method (d), unsymmetrical tetrakis(silyl)hydrazines **82–92** were synthesized:

$$\underset{\text{R}''_3\text{Si}}{\overset{\text{R}_3\text{Si}}{\diagdown}}\text{N}-\text{N}\underset{\text{SiR}'''_3}{\overset{\diagup\text{SiR}'_3}{}}$$

	SiR$_3$	SiR$'_3$	SiR$''_3$	SiR$'''_3$	References
82	SiMe$_3$	SiMe$_3$	SiMe$_3$	SiF$_3$	40
83	SiMe$_3$	SiMe$_3$	SiF$_2$Ph	SiF$_2$Ph	16
84	SiHMe$_2$	SiHMe$_2$	SiHMe$_2$	SiH$_2$Ph	12
85	SiMe$_3$	SiMe$_3$	SiMe$_3$	SiF$_2$Ph	40
86	SiMe$_3$	SiMe$_3$	SiF$_2$[N(SiMe$_3$)$_2$]	SiF$_2$[N(SiMe$_3$)$_2$]	37
87	SiMe$_3$	SiMe$_3$	SiMe$_2$F	SiMe$_2$F	38
88	SiMe$_3$	SiMe$_3$	SisBuF$_2$	SisBuF$_2$	38
89	SiMe$_3$	SiMe$_3$	SitBuF$_2$	SitBuF$_2$	39
90	SiMe$_3$	SiMe$_3$	SiMe$_2$CMe$_3$	SiF$_3$	37
91	SiMe$_2$CMe$_3$	SiMe$_2$CMe$_3$	SiF$_3$	SiF$_3$	43
92	SiMe$_2$CMe$_3$	SiMe(CMe$_3$)$_2$	SiF$_3$	SiF$_3$	43

2. *Structures of Tetrakis(silyl)hydrazines*

In 1970, the molecular structure of tetrasilylhydrazine (SiH$_3$)$_4$N$_2$ **77**[9] was determined by the method of electron diffraction.[46] The data were consistent with planar Si$_2$NN groups and a dihedral angle of 82.5°. The deviation of this angle from 90° was explained by torsional effects. The Si—N bond length (Table XII) is similar to those in (SiH$_3$)$_2$NH (172.5 pm) and in (SiH$_3$)$_2$NCH$_3$ (173.7 pm). The N—N bond distance may be compared with that of N$_2$H$_4$ (144.9 pm).

TABLE XII
Selected Bond Lengths [pm] and Angles [°] of **77**

Bond lengths (pm)		Bond angles (deg)	
Si—N	173.1	Si–N–Si	129.5
Si—H	148.7	N–Si–H	109.0
N—N	145.7		

TABLE XIII
SELECTED BOND LENGTHS [pm] AND ANGLES [°] OF **79**

Bond lengths (pm)		Bond angles (deg)	
N1—N2	48.2	Si1–N1–Si2	103.3
Si1—N1	172.6	Si1–N1–N2	114.1
C—C	137–139		

The first crystal structure of a tetrakis(silyl)hydrazine was described in 1992 by Schmidbaur and co-workers.[12] Hydrazine reacts with chloro-(phenyl)silane, PhH_2SiCl, in the presence of NEt_3 to give tetrakis(phenyl-silyl)hydrazine **79** [Eq. (20)]. The crystal structure of **79** shows that both N atoms exhibit almost completely trigonal planar coordination geometry. This geometry is comparable to the Si_2NNSi_2 skeleton of tetrasilylhydra-zine.[46] The dihedral angle of 87° between the two Si_2N planes is close to orthogonal, separating the nitrogen lone pairs perpendicular to the planes. The N—N bond length of 148.2 pm is only slightly longer than the corre-sponding distances found in tetrasilylhydrazine (Table XIII).

$$PhH_2Si(1) \diagdown \qquad \diagup Si(3)H_2Ph$$
$$N(1)-N(2)$$
$$PhH_2Si(2) \diagup \qquad \diagdown Si(4)H_2Ph$$

The first unsymmetrically substituted tetrakis(silyl)hydrazine that was determined by a crystal structure is the N,N'-[bis(difluorophenylsilyl)]-N,N'-[bis(trimethylsilyl)]hydrazine **83** [see Eq. (22)].[16] Like those of **77** and **79**, the N atoms of **83** show a planar coordination geometry, as the angle sum is found to be 360°. Because of the electron withdrawing effect of the fluorine atoms, these Si—N bonds are shortened compared to the other Si—N bonds of this molecule (Table XIV). The fluorine atoms of

TABLE XIV
SELECTED BOND LENGTHS [pm] AND ANGLES [°] OF **83**

Bond lengths (pm)		Bond angles (deg)	
N1—N2	149.2	Si1–N1–Si4	130.1
Si1—N1	175.8	Si1–N1–N2	117.4
Si4—N1	169.5	Si4–N1–N2	112.5
Si2—N2	176.5		
Si3—N2	169.3		

the SiF_2 group are nonequivalent in the ^{19}F-NMR spectrum. This is explained by hindered SiN bond rotation.[16] The N—N bond length of 149.2 pm is extended because of the bulky silyl groups bound to the N atoms of the hydrazine unit. This bulkiness also forces the silyl groups to adopt a staggered position in the molecule.

$$Me_3Si(2) \diagdown \diagup Si(1)Me_3$$
$$N(2)-N(1)$$
$$PhF_2Si(3) \diagup \diagdown Si(4)PhF_2$$

F. Bis- and Tris(hydrazino)silanes

1. Synthesis and Properties

The first bis(hydrazino)silanes were obtained in the early 1960s.[49–51] They contain organic groups such as Me or Ph bound to the nitrogen and silicon atoms, which protect the sensitive hydrazine unit. The bis (hydrazino)silanes were obtained in the reaction of diorganodichlorosilanes with the organo derivatives of hydrazine [Eq. (25)].

$$RR'SiCl_2 + 4\ H_2N\text{-}NR''R''' \longrightarrow 2\ [R''R'''N\text{-}NH_3]Cl\ + \begin{array}{c} R'' \\ \diagdown \\ R'' \diagup \end{array} N-N-\underset{R'}{\overset{R}{Si}}-N-N \begin{array}{c} \diagup R'' \\ \diagdown R''' \end{array} \quad (25)$$

$$\underset{H\quad R'\quad H}{}$$

93-104

	R	R'	R''	R'''	References
93	Ph	Ph	Me	Me	49
94	Ph	Ph	Ph	H	49
95	Ph	Ph	Ph	Ph	50
96	Ph	H	Me	Me	50
97	Ph	H	Ph	Ph	50
98	Me	H	Me	Me	50
99	Me	H	Ph	Ph	50
100	Me	Me	Me	Me	50
101	Me	Me	Ph	Ph	50
102	Me	Ph	Me	Me	50
103	Me	Ph	Ph	Ph	50
104	Ph	t-Bu	Me	Me	51

These compounds were described as highly sensitive to moisture and transformed into disiloxanes by water. They tend to condense forming six-membered rings, whereas chains of $(—R_2Si—NHNR'—)_x$ could only be obtained in a slightly impure form owing to hydrolysis and to oxidative splitting of N_2. In the reaction of lithiated **99** with trifluoro-*tert*-butylsilane, the bis(hydrazino)silane **105**[51] was obtained:

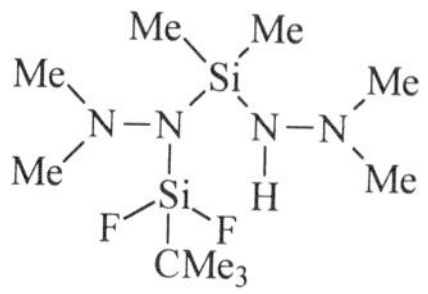

105

Bis(hydrazino)silanes are better stabilized with silyl groups bound to the nitrogen atoms. Although Zn-, Cd-, and Hg- connected silylhydrazines were described in 1970,[52] silyl-substituted bis(hydrazino)silanes were not known until later. The first bis(hydrazino)silane that was silyl substituted at the terminal nitrogen atom and that showed no tendency to condensation was the *tert*-butyl-bis[*N*,*N*'-bis(trimethylsilyl)hydrazino]phenylsilane **106**, described in 1981.[53] A decade later the synthesis of other stable bis(hydrazino)silanes was reported via different preparative methods using (a) mono-, (b) bis-, (c) tris-, and (d) tetrakis(silyl)hydrazines as precursors.

(a) As shown in Eq. (26), in the reaction of two equivalents lithiated mono(silyl)hydrazine **4** and **6** with di-*tert*-butyldifluorosilane, the bis (hydrazino)silanes **107**[17] and **108**[16] are obtained:

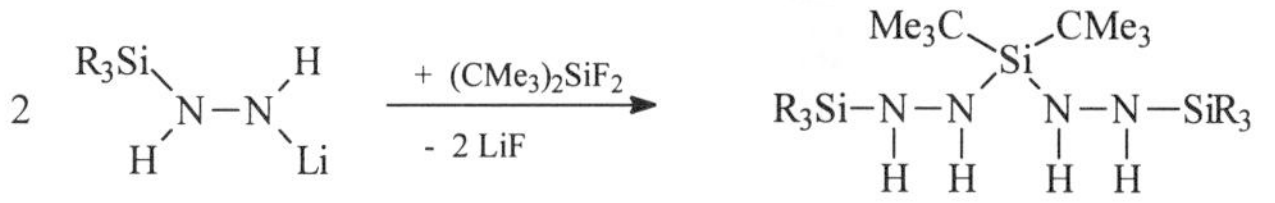

(26)

107: $SiR_3=Si(CMe_3)_2Me$

108: $SiR_3=SiCMe_3(CHMe_2)_2$

The reaction of two equivalents of **7** with dichlorodimethylsilane in the presence of Et_3N leads to the formation of **109**[37] [Eq. (27)].

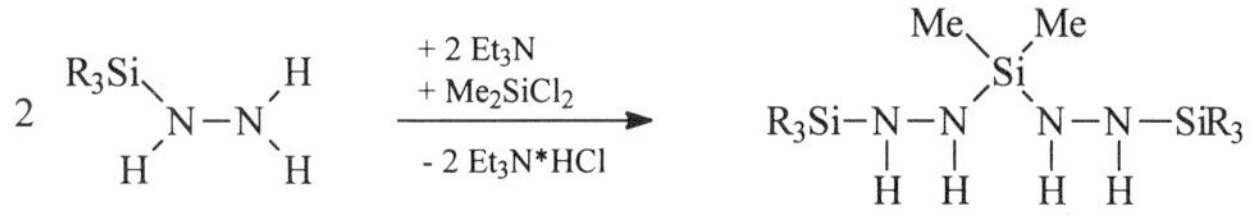

(27)

109: $SiR_3= Si[N(CHMe_2)]_2F$

(b) Lithiated bis(trimethylsilyl)hydrazine **46** reacts with tetrafluorosilane to give the difluorobis[*N*,*N*'-bis(trimethylsilyl)hydrazino]silane **110**[32] [Eq. (28)].

$$\mathbf{46} + 1/2 \; SiF_4 \xrightarrow{- LiF} \begin{array}{c} \overset{F}{\underset{Me_3Si}{\diagdown}} \overset{F}{\underset{\diagup}{Si}} \\ Me_3Si{\diagdown}\underset{N-N}{}{\diagup}Si{\diagdown}\underset{N-N}{}{\diagdown} SiMe_3 \\ \overset{\diagup}{H} \quad \underset{Me_3Si}{} \quad \underset{SiMe_3}{} \quad \overset{\diagdown}{H} \end{array} \quad (28)$$

$$\mathbf{110}$$

(c) Because of its fluoro functionality, the tris(silyl)hydrazine **74** is an appropriate precursor for bis(hydrazino)silanes. In its reaction with N,N-dimethylhydrazine [Eq. (29)], the unsymmetrical bis (hydrazino)silane **111**[16] is obtained via elimination of lithium fluoride.

$$\begin{array}{c} (CMe_3)_2MeSi{\diagdown}{\diagup}SiMe_2CMe_3 \\ \underset{H}{}N-N{} \\ \overset{\diagup}{H} \quad \overset{\diagdown}{SiPhF_2} \end{array} + \begin{array}{c} \overset{H}{\underset{\diagdown}{}} \overset{Me}{\underset{\diagup}{}} \\ N-N \\ \overset{\diagup}{Li} \quad \overset{\diagdown}{Me} \end{array} \xrightarrow{-LiF} \begin{array}{c} \overset{F}{\underset{\diagdown}{}}\overset{Ph}{\underset{\diagup}{}} \\ \quad Si \\ (CMe_3)_2MeSi-N-N{\diagdown}NH-NMe_2 \\ \overset{\diagup}{H} \quad SiMe_2CMe_3 \end{array} \quad (29)$$

$$\mathbf{111}$$

(d) In the same way, trifluorosilyl[tris(trimethylsilyl)]hydrazine **82** reacts with lithiated N,N-dimethylhydrazine or with lithiated tris(trimethylsilyl)hydrazine to give **112**[16] or **113**,[40] respectively:

$$\begin{array}{c} \overset{F}{\underset{\diagdown}{}}\overset{F}{\underset{\diagup}{}} \\ Me_3Si{\diagdown}\quad Si \\ N-N{\diagdown}NH-NMe_2 \\ Me_3Si{\diagup}\quad \overset{|}{SiMe_3} \end{array} \quad or \quad \begin{array}{c} \overset{F}{\underset{\diagdown}{}}\overset{F}{\underset{\diagup}{}} \\ Me_3Si{\diagdown}\quad Si \quad {\diagup}SiMe_3 \\ N-N{\diagdown}\quad N-N \\ Me_3Si{\diagup}\quad \overset{|}{Me_3Si}\quad \overset{|}{SiMe_3}{\diagdown}SiMe_3 \end{array}$$

$$\mathbf{112} \qquad\qquad\qquad \mathbf{113}$$

Similar to compound **113**, bis(hydrazino)fluoroboranes such as

$$\begin{array}{c} \overset{F}{\underset{|}{}} \\ Me_3Si{\diagdown}\quad \quad {\diagup}SiMe_3 \\ N-N-B-N-N \\ Me_3Si{\diagup}\quad \overset{|}{SiMe_3}\;\overset{|}{SiMe_3}\quad {\diagdown}SiMe_3 \end{array}$$

can also be synthesized in the reaction of silylhydrazines with BF_3 etherate.[40,54]

The tris(hydrazino)silane **114** was formed by treating **112** with lithiated N,N-dimethylhydrazine according to Eq. (30):

$$\mathbf{112} + \begin{array}{c} \overset{H}{\underset{\diagdown}{}} \overset{Me}{\underset{\diagup}{}} \\ N-N \\ \overset{\diagup}{Li} \quad \overset{\diagdown}{Me} \end{array} \xrightarrow{-LiF} \begin{array}{c} \overset{F}{\underset{|}{}}\;\overset{H}{\underset{|}{}} \\ Me_3Si{\diagdown}\quad \quad {\diagup}Me \\ N-N-Si-N-N \\ Me_3Si{\diagup}\quad \overset{|}{N-H}\quad {\diagdown}Me \\ \overset{|}{N} \\ \overset{\diagup}{Me}\;\overset{\diagdown}{Me} \end{array} \quad (30)$$

$$\mathbf{114}$$

All silyl-substituted bis- and tris(hydrazino)silanes are stable compounds. These are resin-like or crystalline products with no tendency to condensation and no sensitivity to moisture.

2. Crystal Structure of a Bis(hydrazino)Silane

Figure 7 shows the crystal structure of the difluorobis[tris(trimethylsilyl)hydrazino]silane **113**.[16] The nitrogen atom N1 shows a planar coordination geometry, whereas N2 is not exactly planar (sum of angles = 358.1°). The Si1–N2–Si3 angle is extended to 135°. Because of the electron withdrawing effect of the fluorine atoms, the Si1—N2 and Si1–N2a bonds, with lengths of 170.8 pm, are shortened compared to the Si2–N1 and Si3–N2 bonds (176.2 and 176.0 pm) (Table XV).

G. Cyclic Silylhydrazines

1. Classification and Preparation of Cyclic Silylhydrazines

The last section of this chapter deals with cyclic silylhydrazines. The synthesis of the first cyclic silylhydrazines were reported as early as 1958.[55] Since then, a great diversity of such ring systems containing the Si—N—N unit have been synthesized. These can be classified as follows.

The smallest rings are diazasilacyclopropanes presented in Section II,G,2. Four-membered rings are shown in Section II,G,3. Five- (Section

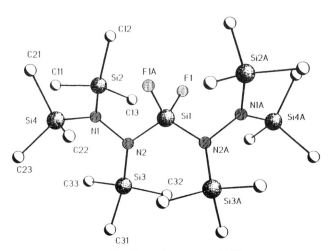

FIG. 7. Crystal structure of **113**.

TABLE XV

SELECTED BOND LENGTHS [pm] AND ANGLES [°] OF **113**

Bond lengths (pm)		Bond angles (deg)	
Si1—F1	157.8	F1–Si1–F1a	101.1
Si1—N2	170.8	F1–Si1–N2	109.4
Si2—N1	176.2	N2–Si1–N2a	120.8
Si3—N2	176.0	N2–N1–Si4	116.8
N1—N2	149.9	Si4–N1–Si2	125.2
		Si4–N1–Si2	125.2
		N2–N1–Si2	118.0
		N1–N2–Si3	114.4
		N1–N2–Si1	108.6
		Si1–N2–Si3	135.1

II,G,4), six- (Section II,G,5), and even seven-membered rings (Section II,G,6) are described; and finally the syntheses of coupled ring systems (Section II,G,7). Two main methods of preparation are described: one is the intermolecular cleavage of hydrogen chloride from hydrazine and dialkyl- or aryldichlorosilanes,[55] another is based on the cyclization of acyclic fluorosilylhydrazines.[15,38,39,56,57]

2. A Diazasilacyclopropane

The only known diazasilacyclopropane was synthesized in 1982.[56] As shown in Scheme 1, the bulky di-*tert*-butyl(fluoro)silyl group of the hydrazine derivative A, which is formed on reaction of monolithiated *N,N'*-bis(trimethylsilyl)hydrazine **46** with di-*tert*-butyldifluorosilane, reacts with excess **46** with formation of the diazasilylcyclopropane **115**.

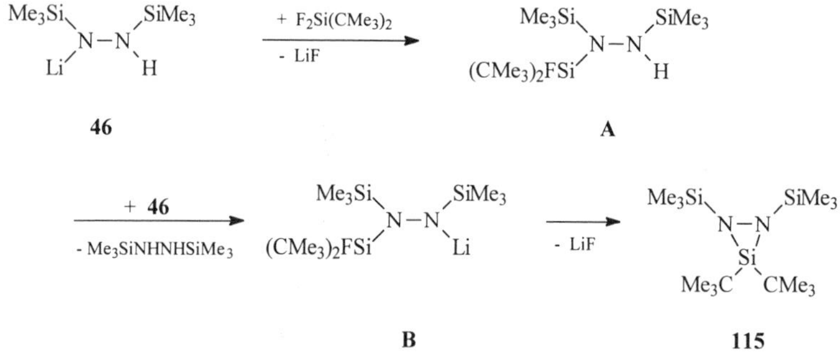

SCHEME 1

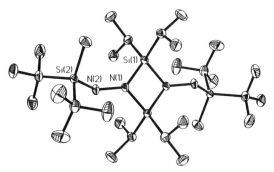

FIG. 8.   Structure of **116**.

The intermediate A was not isolated; it reacts with **46**, undergoing transmetalation via the lithium salt B to give the product **115**. Above +80°C, and on attempted distillation or gas-chromatographic purification of **115**, di-*tert*-butylbis(trimethylsilylamino)silane (Me$_3$SiNH)$_2$-Si(CMe$_3$)$_2$ was formed. The diazasilacyclopropane **115** can be purified by repeated recrystallization from *n*-hexane at −15°C. The upfield shift of the ^{29}Si NMR signal of the silicon containing small ring compounds is also observable in the case of **115** (in C$_6$F$_6$, TMS int.: δ ^{29}Si = −30, 73).

3. A Silylhydrazine Four-Membered Ring

Four-membered rings prepared from dimethylhydrazine were already described in the 1970s.[51] A cyclization with formation of a four-membered silylhydrazine ring was not achieved until 1994 because HHal and LiHal elimination from lithiated fluorosilylhydrazines always led to the formation of disilatetrazanes (see Section II,G,5). Klingebiel and co-workers[15] prepared a lithiated chlorosilylhydrazine using some lithiated fluorosilylhydrazines that react with Me$_3$SiCl not by substitution at the N-atom, but by halogen exchange.[58,59] A quantitative fluorine–chorine exchange occurs in the reaction of the lithium derivative of the N-(di-*tert*-butylmethylsilyl)-N′-(fluorodiisopropylsilyl)hydrazine **28** with Me$_3$SiCl at 0°C, and Me$_3$SiF and LiCl are formed. The resulting silahydrazone was isolated as the (2+2) cycloaddition product **116** (Scheme 2). Consequently, Me$_3$SiCl as a fluorine–chlorine exchange reagent affords chlorination of the lithium derivative of **28** with the formation of Me$_3$SiF.

The structure of the first silylhydrazine four-membered ring has been confirmed by X-ray analysis. Crystals suitable for X-ray analysis were obtained after a 2-week crystallization process by slowly condensing the *n*″-hexane solvent in an ice-cooled trap. Figure 8 shows the result of the X-ray analysis of **116**.

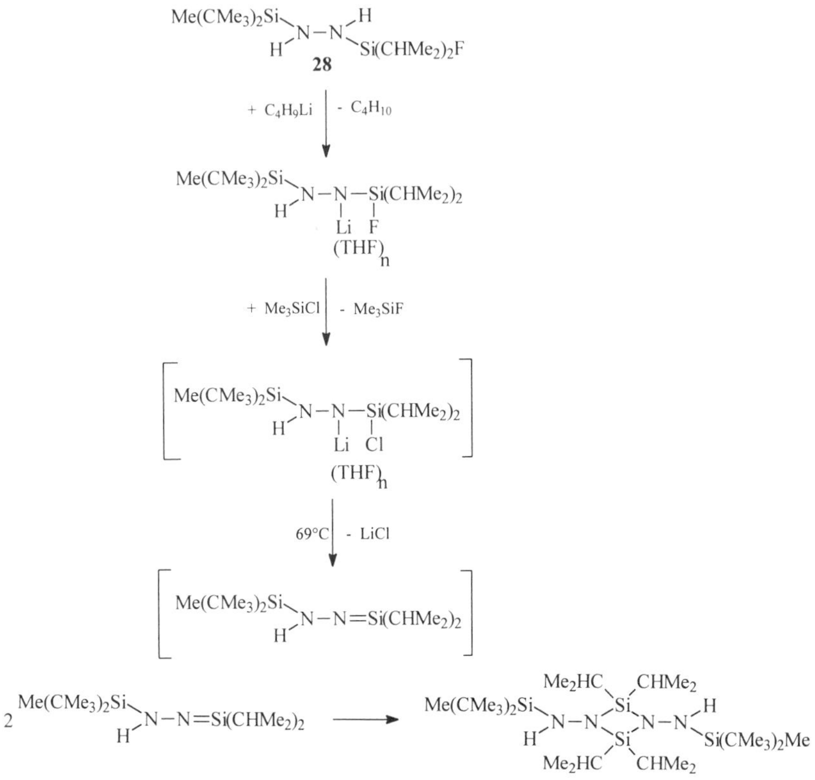

$$Me(CMe_3)_2Si{\diagdown}_{N-N}{\diagup}^H_{Si(CHMe_2)_2F}$$

28

$$+ C_4H_9Li \quad \Big| \quad - C_4H_{10}$$

$$Me(CMe_3)_2Si{\diagdown}_{N-N-Si(CHMe_2)_2}$$
$$\underset{Li \quad F}{|\quad|}$$
$$(THF)_n$$

$$+ Me_3SiCl \quad \Big| \quad - Me_3SiF$$

$$\left[Me(CMe_3)_2Si{\diagdown}_{N-N-Si(CHMe_2)_2} \atop \underset{Li \quad Cl}{|\quad|} \right]$$
$$(THF)_n$$

$$69°C \quad \Big| \quad - LiCl$$

$$\left[Me(CMe_3)_2Si{\diagdown}_{N-N=Si(CHMe_2)_2} \right]$$

$$2 \; Me(CMe_3)_2Si{\diagdown}_{N-N=Si(CHMe_2)_2} \longrightarrow$$

$$Me(CMe_3)_2Si{\diagdown}_{N-N}{\overset{Me_2HC\;\diagdown\;{\diagup}CHMe_2}{\underset{Me_2HC\;\diagup\;{\diagdown}CHMe_2}{Si}}_{N-N}{\diagup}^H_{Si(CMe_3)_2Me}}$$

116

SCHEME 2

The central four-membered ring is strictly planar, and the molecule possesses an inversion center. The atom N2 is 63 pm below the ring, so the ring N atom is not planar in contrast to that in the cyclodisilazanes. The sum of the angles in this case is 350.1°. (See Table XVI.)

TABLE XVI

SELECTED BOND LENGTHS [pm] AND ANGLES [°] OF **116**

Bond lengths (pm)		Bond angles (deg)	
Si1—N1	175.5	N1–Si1–N1i	84.3
N1—N2	144.4	N2–N1–Si1i	125.8
N2—Si2	173.0	N1–N2–Si2	125.5
		N2–N1–Si1	128.6
		Si1–N1–Si1i	95.7

4. Disilatriazoles

Cyclic five-membered silylhydrazine systems are obtained via different methods of preparation. One possibility is the reaction of bis(fluorosilyl)-hydrazines with lithiated amines as shown in Eq. (31)[38]:

$$\text{(31)}$$

or with lithiated hydrazines as in Eq. (32)[39]:

$$\text{(32)}$$

When tris(silyl)hydrazines contain an NH and an SiF group in the molecule, disilatriazoles can be obtained in the reaction of these silylhydrazines with metalorganic bases. When **67** and **68** are treated with *n*-butyllithium, the five-membered ring **119** is isolated as a cross dimer of the two isomers[39] [Eq. (33)].

$$\text{(33)}$$

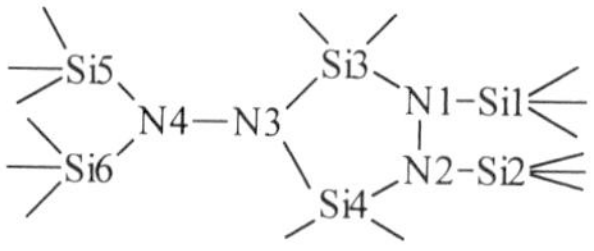

FIG. 9.   Structure of **120**.

The same reaction of **69** and **70** with n-butyllithium leads also to the formation of the disilatriazole **120**.[39] **120** is crystalline, and its structure was determined by X-ray analysis.[39]

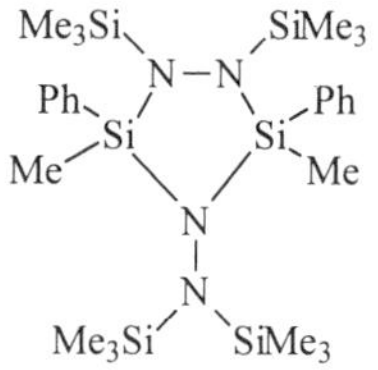

120

120 shows two symmetry-independent molecules. One of them exhibits a disorder, the second does not. The geometry of the second molecule is shown in Fig. 9. The Si_2N_3 ring is not planar. The Si–N–N–Si torsion angle is found to be 42°; the Si atoms Si3 and Si4 are found to be 55 pm above or 64 pm below the N1–N2–N3 plane (See Table XVII.)

Mono(silyl)hydrazines can also be used as precursors for the synthesis of five-membered silylhydrazine ring systems. In the reaction of the mono(silyl)hydrazine **7** with n-butyllithium in a 1 : 1 molar ratio, colorless crystals with a melting point of 184°C were obtained from n-hexane at 0°C. The reaction product did not contain the expected six-membered ring, but, as proved by NMR and IR investigations, consisted of the disilatriazole **121**,[15] as shown in Eq. (34).

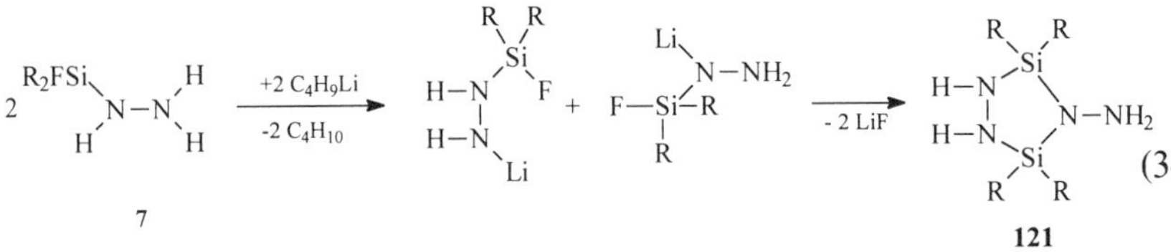

$$R = N(CHMe_2)_2$$

The ^{1}H-NMR spectrum shows two signals in the NH range with the same intensity, and the IR spectrum also shows NH and NH_2 vibrations.

TABLE XVII

SELECTED BOND LENGTHS [pm] AND ANGLES [°] OF **120**

Bond lengths (pm)		Bond angles (deg)	
N1—N2	152	Si3–N1–N2	108
N1—Si3	175	Si3–N1–Si1	129
N1—Si1	174	Si1–N1–N2	119
N2—Si4	175	N1–N2–Si4	108
Si3—N3	171	N3–Si3–N1	98
N3—N4	151	Si3–N3–Si4	113
		N4–N3–Si4	121

The mechanism of formation of **121** becomes clear with the result of the X-ray analysis of **9**. As mentioned earlier, two tautomers, $R_3SiNLiNH_2$ and $R_3SiNHLiNH$, exist in the solid state here. LiF elimination from fluorofunctional units in this case leads to the formation of a five-membered ring.[15]

5. Cyclodisilatetrazanes

Cyclo-3,6-disila-1,2,4,5-tetrazanes are six-membered rings containing two N—N bonds. In 1958, Wannagat et al.[55] reported the synthesis of a cyclodisilyltetrazane in the reaction of hydrazine with diphenyldichlorosilane [Eq. (35)].

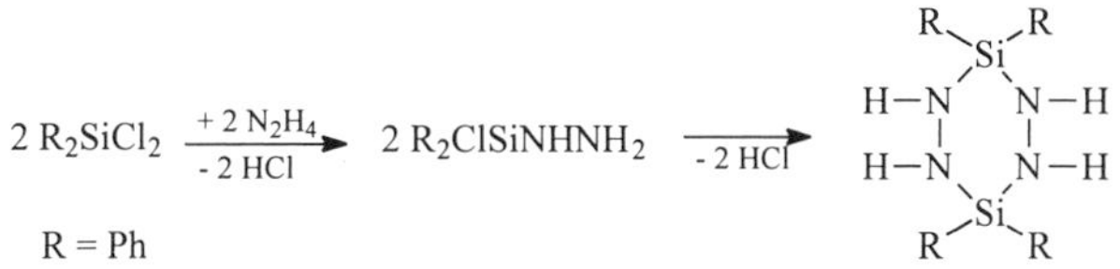

$$2\ R_2SiCl_2 \xrightarrow[-2\ HCl]{+2\ N_2H_4} 2\ R_2ClSiNHNH_2 \xrightarrow{-2\ HCl} \qquad (35)$$

R = Ph

122

In the reaction of difluorodiisopropylsilane with lithiated hydrazine, the six-membered ring **123** is formed.[32] No formation of HF is observed, which means that the reaction proceeds via the mono(silyl)hydrazine that is then lithiated as shown in Eq. (36).

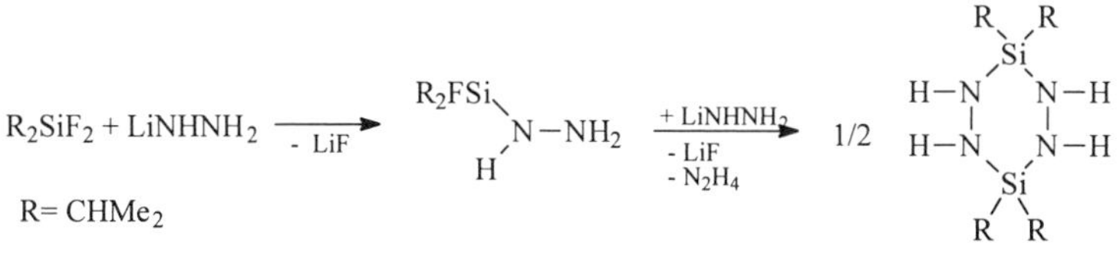

$$R_2SiF_2 + LiHNNH_2 \xrightarrow{-LiF} \overset{R_2FSi}{\underset{H}{\diagdown}}N-NH_2 \xrightarrow[-LiF\ -N_2H_4]{+LiHNNH_2} 1/2 \qquad (36)$$

R= CHMe₂

123

Mono(silyl)hydrazines can also be used as precursors for the synthesis of six-membered silylhydrazine ring systems. When 3 is lithiated it cyclizes above 5°C to give the six-membered ring 124[13] [Eq. (37)].

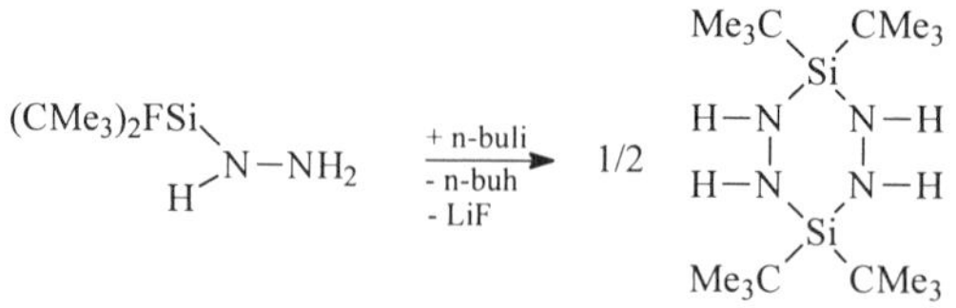

$$
(CMe_3)_2FSi \diagdown \quad \underset{\substack{+ \text{ n-buli} \\ - \text{ n-buh} \\ - \text{ LiF}}}{\xrightarrow{\hspace{1.5cm}}} \quad 1/2
$$

$$
\begin{array}{c}
Me_3C \diagdown \diagup CMe_3 \\
Si \\
H-N \diagup \quad \diagdown N-H \\
| \qquad | \\
H-N \diagdown \quad \diagup N-H \\
Si \\
Me_3C \diagup \diagdown CMe_3
\end{array}
$$

(37)

124

Another possibility for the preparation of cyclodisilatetrazanes is salt elimination from lithiated N-silyl-N'-fluorosilylhydrazines. The LiF elimination from the lithium derivates of the asymmetrical bis(silyl)hydrazines **23**, **24**, and **28** leads to the formation of the compounds **125–127**[15,32] [Eq. (38)].

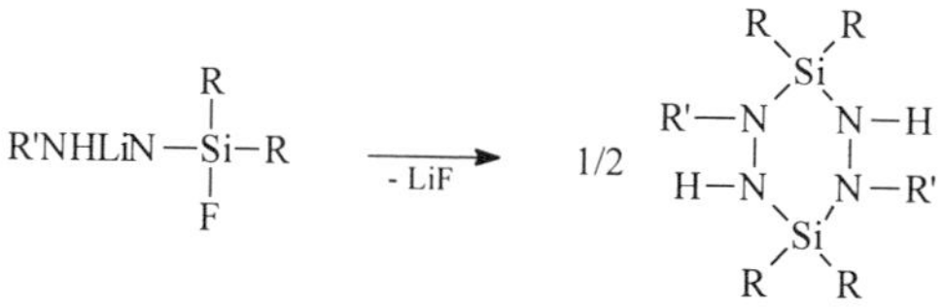

$$
\begin{array}{c}
R \\
| \\
R'NHLiN-Si-R \\
| \\
F
\end{array}
\quad \xrightarrow{- \text{ LiF}} \quad 1/2
\quad
\begin{array}{c}
R \diagdown \diagup R \\
Si \\
R'-N \diagup \quad \diagdown N-H \\
| \qquad | \\
H-N \diagdown \quad \diagup N-R' \\
Si \\
R \diagup \diagdown R
\end{array}
$$

(38)

125-127

	R'	R
125	Si(CMe$_3$)$_2$Me	CHMe$_2$
126	SiMe$_2$CMe$_3$	CMe$_3$
127	SiMe$_3$	CMe$_3$

The appearance of ^{29}Si-NMR upfield signals points to the formation of these six-membered rings. In addition, this thesis is supported by X-ray analysis of **126**. Figure 10 shows the structure of compound **126**; Table XVIII gives selected bond lengths and angles.

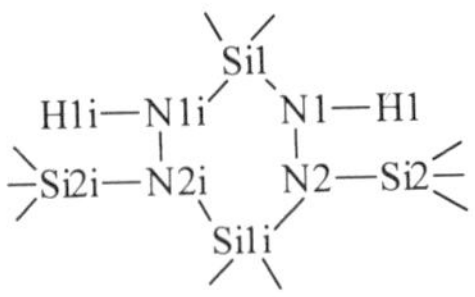

$$
\begin{array}{c}
\diagdown \diagup \\
Si1 \\
H1i-N1i \diagup \quad \diagdown N1-H1 \\
| \qquad | \\
\diagdown Si2i-N2i \quad N2-Si2 \diagdown \\
\diagdown Si1i \diagup \\
\diagup \diagdown
\end{array}
$$

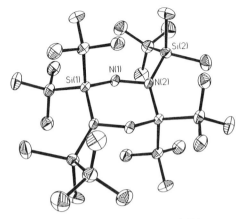

FIG. 10.   Crystal structure of **126**.

Compound **126** crystallizes in a chair conformation. Because of the inversion center, the atoms N2, Si1, N2a, and Si1a lie exactly in one plane. The atoms N1 and N1a are 47 pm above and below this plane, respectively. The sum of angles at N2 amounts to 359.8°. This nitrogen atom, in contrast to the NH-functional N1 with 349.9, is sp^2-hybridized. The angle Si2–N2–Si1 of 140.7 is comparatively increased. The average Si—N and N—N bond distances are 175 and 148 pm, close to the expected values of 174 pm for Si—N and 145 pm for N—N.[60] The same method of preparation, shown in Eq. (39), led to the formation of the first fluorosilyl functional hydrazine six-membered rings **127** and **128**.[43] The dilithiated bis(silyl)hydrazines **10** and **12** react with tetrafluorosilane to yield tetrakis-(silyl)hydrazines, which again react with the lithium derivatives.

TABLE XVIII
SELECTED BOND LENGTHS [pm] AND ANGLES [°] OF **126**

Bond lengths (pm)		Bond angles (deg)	
Si1—N1	174.4	N1–Si1–N2i	108.0
N1—N2	148.0	N2–N1–H1	106
Si2—N2	174.8	N1–N2–Si2	109.7
N1—H1	88.0	Si2–N2–Si1i	140.7
		N2–N1–Si1	128.6
		Si1–N1–H1	115

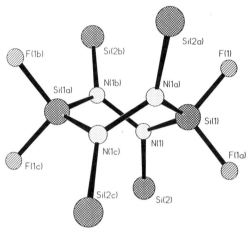

F_{IG}. 11.   Crystal structure of **127**.

$$\underset{\textbf{10, 12}}{\underset{F_3Si}{\overset{R}{\diagdown}}N-N\underset{SiF_3}{\overset{R}{\diagup}}} \xrightarrow[\text{- 2 LiF}]{+ \ (RNLi)_2} \underset{\textbf{127, 128}}{\begin{array}{c} F\diagdown \diagup F \\ Si \\ R-N \diagdown N-R \\ | \quad\quad | \\ R-N \diagup N-R \\ Si \\ F\diagup \diagdown F \end{array}} \tag{39}$$

127:   R = SiMe₂CMe₃

128:   R = SiMe₃

127 is crystalline, and its structure has been determined from X-ray diffractometer data (Fig. 11). The ring is far from planar. In contrast to **126**, this ring has a twist conformation. The exocyclic Si—N bonds are found to have lengths of 179.5 pm, significantly longer than the endocyclic bonds (170.5 pm), which are shortened by the electron-withdrawing effect of the fluorine atoms (Table XIX). The sum of angles at the nitrogen atoms is 357.6°, which means that the N atoms are sp^2-hybridized. The CMe₃SiMe₂ groups take a staggered position in the molecule.

TABLE XIX
SELECTED BOND LENGTHS [pm] AND ANGLES [°] OF **127**

Bond lengths (pm)		Bond angles (deg)	
Si1—F1	157.1	F1–Si1–F1a	101.5
Si1—N1	170.5	N1–Si1–N1a	112.4
N1—N1b	149.1	Si1–N1–N1b	108.2
N1—Si2	179.5	Si1–N1–Si2	125.8

Finally, a cyclodisilatetrazene can be formed from a disilatriazole. Silicon–nitrogen chemistry is rich in rearrangements.[7,61,62] In general, the term *silatropy* is defined as the tendency of silyl groups to be bound to the most electronegative element in the molecule. Anionic silyl group migration also occurs when the molecule consists of differently polarized atoms of the same kind. In cyclic molecules, this rearrangement leads to expansion or contraction of the ring.[61,62] Related to the five-membered ring **121**, the thermodynamically more stable ring would be the initially expected six-membered ring with equivalent N-atoms. Although S_{N_2} mechanisms dominate in reactions of silicon-containing compounds, isomerism has to be partly thermally activated. When **121** is dissolved in refluxing *n*-hexane and crystallized at room temperature, or held briefly above its melting point (184°C), the structural isomer **129** is formed [Eq. (40)].

$$ (40) $$

R= N(CHMe$_2$)$_2$

6. *Seven-Membered Silylhydrazine Ring Systems*

As shown in Scheme 3, the synthesis of a cyclo-1,3,5-trisila-2,4,6,7-tetrazane was accomplished by the coupling of a 1,3-dichlorosilane with a dilithium 1,2-hydrazine,[63,64] whereas in the analogous reaction with pure hydrazine only the equivalent six-membered ring is obtained.[64]

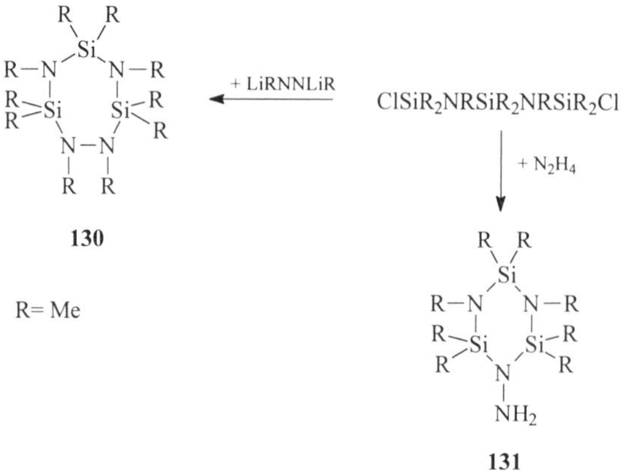

SCHEME 3

Other seven-membered silicon–nitrogen rings were prepared from the lithium derivatives of a bis(aminodimethylsilyl)-1,2-dimethylhydrazine with Me_2SiCl_2.[65]

7. Bicyclic Silylhydrazine Systems

The bicyclic six-membered ring system **132** was obtained according to Eq. (41) by the reaction of bis(N,N'-fluorodimethylsilyl-N,N'-trimethylsilyl)hydrazine **87** with dilithiated hydrazine.[39]

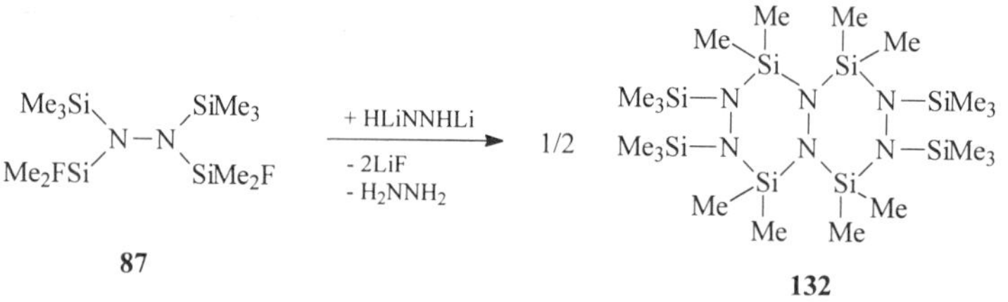

Hydrazine or its lithium derivative react with bis(halogenosilyl)amines in a 1 : 2 molar ratio to give a bicyclic compound **133**[57] [Eq. (42)].

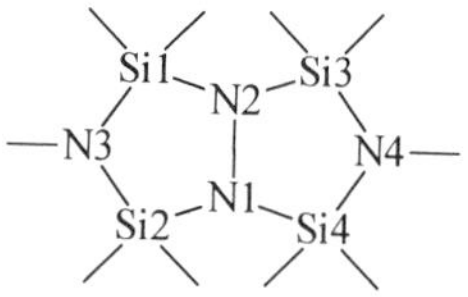

FIG. 12.   Structure of **133**.

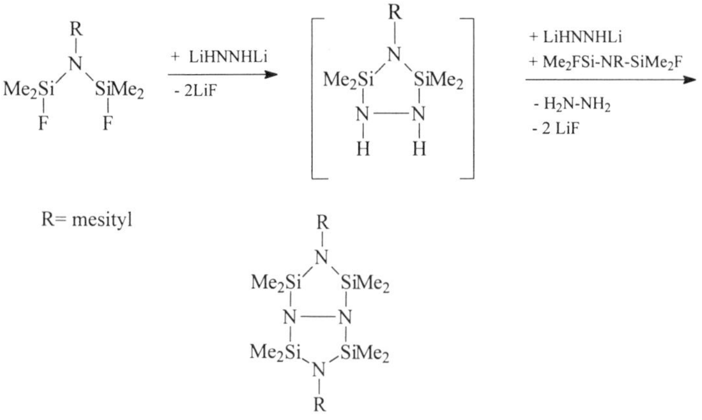

(42)

133

The synthesis of an analogous ring system in the reaction of hydrazine with *N,N*-bis(chlorodimethylsilyl)methylamine was described in 1965,[66] but here the formation of the two isomers **134a** and **134b** was discussed:

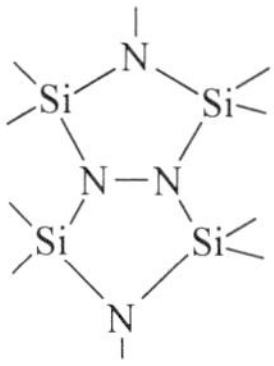

and

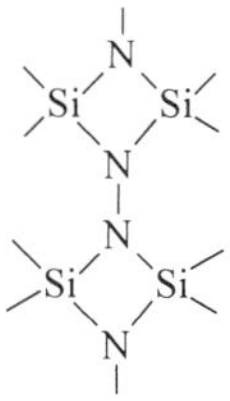

134a   **134b**

The structure of compound **133** (Fig. 12) was determined by X-ray diffraction.[57] It has an approximate D_2 symmetry with planar nitrogen atoms. An N—N unit is shared by the two five-membered rings, which are not planar. The torsion angles Si8–N11–N11′–Si8′ and Si12–N11–N11′–Si12′ are found to be −45° and −37°. (See Table XX.)

TABLE XX

Selected Bond Lengths [pm] and Angles [°] of **133**

Bond lengths (pm)		Bond angles (deg)	
N1—N2	151.9	Si3–N4–Si4	110.5
Si2—N3	174.4	N4–Si4–N1	98.3
Si2—N1	172.5	Si4–N1–N2′	107.7
		Si4–N1–Si2	142.1

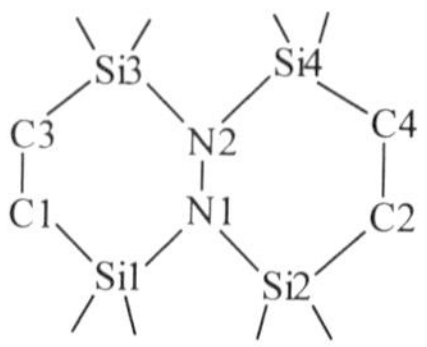

FIG. 13.   Structure of **135**.

In 1992, two different cyclization modes in the formation of silylhydrazines were reported by Schmidbaur and co-workers.[67] The reactions of 1,2-bis(bromosilyl)ethane and 1,3-bis(bromosilyl)propane with anhydrous hydrazine, shown in Eqs. (43) and (44), lead to the formation of the two bicyclic silylhydrazines 1,6-diaza-2,5,7,10-tetrasila[4.4.0]bicyclodecane **135** and bis(1-aza-2,6-disila-1-cyclohexyl) **136**.

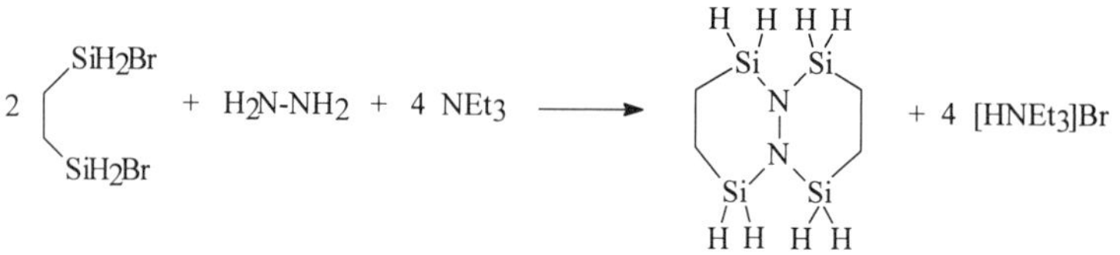

$$2 \ \text{(SiH}_2\text{Br / SiH}_2\text{Br)} + \text{H}_2\text{N-NH}_2 + 4\ \text{NEt}_3 \longrightarrow \textbf{135} + 4\ [\text{HNEt}_3]\text{Br} \tag{43}$$

and

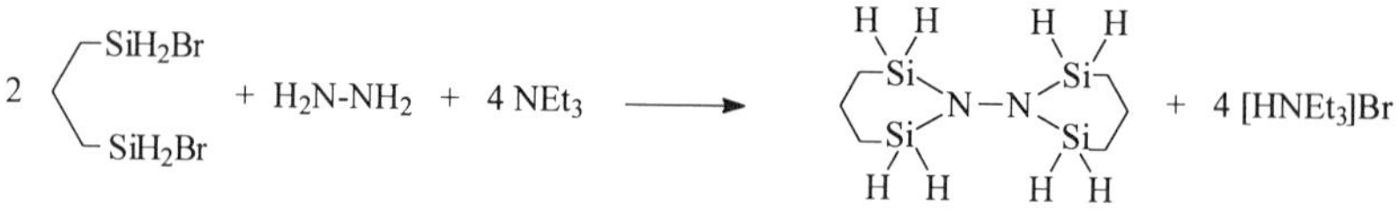

$$2 \ \text{(SiH}_2\text{Br / SiH}_2\text{Br)} + \text{H}_2\text{N-NH}_2 + 4\ \text{NEt}_3 \longrightarrow \textbf{136} + 4\ [\text{HNEt}_3]\text{Br} \tag{44}$$

For compound **135**, a distillable liquid with mp 7°C, the low-temperature crystal structure determination has shown a bicyclic configuration with

TABLE XXI

SELECTED BOND LENGTHS [pm] AND ANGLES [°] OF **135**

Bond lengths (pm)		Bond angles (deg)	
Si1—N1	172.4	Si1–N1–N2	113.5
N1—N2	149.7	Si1–N1–Si2	133.4
Si1—C1	185.1	N1–Si1–C1	108.4
		Si1–C1–C3	111.4

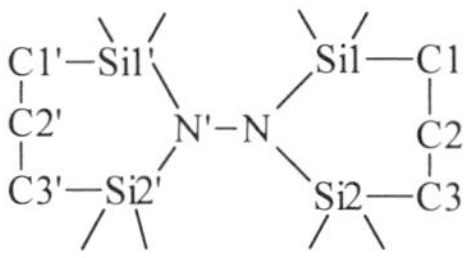

FIG. 14. Structure of **136**.

TABLE XXII

SELECTED BOND LENGTHS [pm] AND ANGLES [°] OF **136**

Bond lengths (pm)		Bond angles (deg)	
N—N'	147.4	N–Si1–C1	108.0
Si1—N	172.3	Si1–N–Si2	126.8
Si1—C1	186.4	Si1–N–N'	115.0

the N—N unit shared by two slightly twisted chair conformed heterocycles (Fig. 13). The N—NSi$_2$ units are virtually planar but not orthogonal relative to each other, owing to conformational restraints. (See Table XXI.)

In crystals of compound **136**, mp 57°C, two six-membered rings each in a half chair conformation are linked solely through the hydrazine N—N bond (Fig. 14). The two planar NNSi$_2$ units are at right angles, as expected for a nonforced tetrahydrazine geometry. The N—N bond is longer in **135** than in **136**. (See Table XXII.)

REFERENCES

(1) Eaborn, C. *Organosilicon Compounds;* Butterworths: London, 1960, p. 339.
(2) Ebsworth, E. A. V. *Volatile Silicon Compounds;* Pergamon: New York, 1963, p. 101.
(3) Fessenden, R.; Fessenden, J. S. *Chem. Rev.* **1961,** *61,* 361.
(4) MacDiarmid, A. G. *Adv. Inorg. Chem. Radiochem.* **1961,** *3,* 207.
(5) Michalowski, Z. *Wiadomosci Chem.* **1959,** *13,* 543.
(6) Wannagat, U. *Adv. of Inorg. Radiochem.* **1964,** *6,* 225.
(7) West, R. *Adv. Organomet. Chem.* **1977,** *16,* 1.
(8) Wannagat, U. *Fortschr. Chem. Forsch.* **1967,** *9,* 102.
(9) Aylett, B. J. *J. Inorg. Nucl. Chem.* **1956,** *2,* 325.
(10) Wannagat, U.; Liehr, W. *Angew. Chem.* **1957,** *69,* 783; Wannagat, U.; Liehr, W. *Z. Anorg. Allg. Chem.* **1958,** *297,* 129.
(11) Wannagat, U.; Höfler, F.; Bürger, H. *Monatsh. Chem.* **1966,** *96,* 2038; Wannagat, U.; Niederprüm, H. *Angew. Chem.* **1959,** *71,* 574.
(12) Mitzel, N. W.; Bissinger, P.; Schmidbaur, H. *Chem. Ber.* **1993,** *126,* 345.
(13) Drost, C.; Klingebiel, U. *Chem. Ber.* **1993,** *126,* 1413.

(14) Dielkus, S.; Drost, C.; Herbst-Irmer, R.; Klingebiel, U. *Angew. Chem. Int. Ed. Engl.* **1993,** *32,* 1625.
(15) Dielkus, S.; Drost, C.; Herbst-Irmer, R.; Klingebiel, U. *Organometallics* **1994,** *13,* 3985.
(16) Bode, K.; Klingebiel, U.; Witte-Abel, H.; Gluth, M. W., Noltemeyer, M.; Herbst-Irmer, R.; Schäfer, M. *Phorphorus, Sulfur and Silicon;* in press.
(17) Drost, C. Dissertation, Göttingen, 1993.
(18) Dilworth, J. R.; Rodriguez, A.; Leigh, G. J.; Murrell, J. N. *J. Chem. Soc., Dalton Trans.* **1983,** *2,* 455.
(19) Dilworth, J. R.; Latham, I. A.; Leigh, G. J.; Huttner, G.; Jibril, I. *J. Chem. Soc., Chem. Commun.* **1983,** *22,* 1368.
(20) Bailey, R. E.; West, R. *J. Am. Chem. Soc.* **1964,** *86,* 5369.
(21) West, R.; Ishikawa, M.; Bailey, R. E. *J. Am. Chem. Soc.* **1966,** *88,* 4648.
(22) West, R.; Ishikawa, M.; Bailey, R. E. *J. Am. Chem. Soc.* **1967,** 89, 4068.
(23) West, R.; Ishikawa, M.; Bailey, R. E. *J. Am. Chem. Soc.* **1967,** *89,* 4981.
(24) West, R.; Ishikawa, M.; Bailey, R. E. *J. Am. Chem. Soc.* **1967,** *89,* 4072.
(25) West, R.; Bichlmeir, B. *J. Am. Chem. Soc.* **1972,** *94,* 1649.
(26) Dewar, M. J. S.; Jennings, W. B. *J. Am. Chem. Soc.* **1973,** *95,* 1562.
(27) Fletcher, J. R.; Sutherland, J. O. *J. Chem. Soc. Commun.* **1970,** 687.
(28) Scherer, O. J.; Bultjer, U. *Angew. Chem.* **1971,** *83,* 364.
(29) Kessler, H. *Angew. Chem.* **1970,** *52,* 237.
(30) Schöllkopf, U. *Angew. Chem., Int. Ed. Engl.* **1970,** *9,* 763.
(31) Woodward, R. B.; Hoffmann, R. *The Conservation of Orbital Symmetry;* Academic Press: New York, 1970.
(32) Bode, K.; Drost, C.; Jäger, C.; Klingebiel, U.; Noltemeyer, M.; Zdirad Zak, J. *Organomet. Chem.* **1994,** *482,* 285.
(33) Nöth, H.; Sachdev, H.; Schmidt, M.; Schwenk, H. *Chem. Ber.* **1995,** *128,* 1105.
(34) Drost, C.; Jäger, C.; Freitag, S.; Klingebiel, U.; Noltemeyer, M.; Sheldrick, G. M. *Chem. Ber.* **1994,** *127,* 845.
(35) Metzler, N.; Nöth, H.; Sachdev, H. *Angew. Chem.* **1994,** 106, 1837.
(36) Wannagat, U.; Niederprüm, H. *Z. Anorg. Allg. Chem.* **1961,** *310,* 32.
(37) Klingebiel, U., unpublished results, 1994.
(38) Klingebiel, U.; Wendenburg, G.; Meller, A. *Monatsh. Chem.* **1979,** *110,* 289.
(39) Clegg, W.; Haase, M.; Hluchy, H.; Klingebiel, U.; Sheldrick, G. M. *Chem. Ber.* **1983,** *116,* 290.
(40) Bode, K.; Klingebiel, U.; Noltemeyer, M.; Witte-Abel, H. *Z. Anorg. Allg. Chem.* **1995,** *621,* 500.
(41) Veith, M.; Goffing, F.; Huch, V. *Chem. Ber.* **1988,** 121, 943.
(42) Wiberg, N.; Veith, M. *Chem. Ber.* **1971,** *104,* 3176.
(43) Drost, C.; Klingebiel, U.; Noltemeyer, M. *J. Organomet. Chem.* **1991,** *414,* 307.
(44) Seppelt, K.; Sundermeyer, W. *Chem. Ber.* **1969,** *102,* 1247.
(45) Wiberg, N.; Weinberg, E.; Joo, W. Ch. *Chem. Ber.* **1974,** *107,* 1764.
(46) Glidewell, C.; Rankin, D. W. H.; Robiette, A. G.; Sheldrick, G. M. *J. Chem. Soc.* (A), **1970,** 318.
(47) Veith, M.; Bäringhausen, H. *Acta Cryst.* **1974,** *B30,* 1806.
(48) Wiberg, N.; Joo, W. Ch.; Uhlenbrock, W. *Angew. Chem.* **1968,** *80,* 661.
(49) Niederprüm, H.; Wannagat, U. *Z. Anorg. Allg. Chem.* **1961,** *311,* 270.
(50) Sergeeva, Z. J.; Sin-Chzhan Tszyan; Tsitovich, D. D. *J. Allg. Chem.* **1960,** *30,* 694; *Chem. Abstr.* **1960,** *54,* 24349c.
(51) Klingebiel, U.; Hluchy, H.; Meller, A. *Chem. Ber.* **1978,** *111,* 906.
(52) Seppelt, K.; Sundermeyer, W. *Chem. Ber.* **1970,** *103,* 3939.

(53) Hluchy, H. Dissertation, Göttingen, 1981.

(54) Drost, C.; Jäger, C.; Klingebiel, U.; Freire-Erdbrügger, C.; Herbst-Irmer, R.; Schäfer, M. *Z. Naturforsch.* **1995,** *50b,* 76.

(55) Wannagat, U.; Niederprüm, H. *Angew. Chem.* **1958,** *70,* 745.

(56) Hluchy, J.; Klingebiel, U. *Angew. Chem.* **1982,** *21,* 301.

(57) Clegg, W.; Hluchy, H.; Klingebiel, U.; Sheldrick, G. M. *Z. Naturforsch.* **1979,** *34b,* 1260.

(58) Walter, S.; Klingebiel, U.; Schmidt-Bäse, D. *J. Organomet. Chem.* **1991,** *412,* 319.

(59) Boese, R.; Klingebiel, U. *J. Organomet. Chem.* **1986,** *315,* C17.

(60) Rademacher, P. *Strukturen anorganischer Moleküle;* VCH: Weinheim, Germany, 1987.

(61) Dippel, K.; Werner, E.; Klingebiel, U. *Phosphorus, Sulfur, Silicon* **1992,** *64,* 15.

(62) Klingebiel, U. *Inorg. React. Methods* **1990,** *17,* 116.

(63) Wannagat, U.; Bogusch, E.; Hofler, F. *J. Organomet. Chem.* **1967,** *7,* 203.

(64) Wannagat, U. *Angew. Chem.* **1966,** *12,* 648.

(65) Wannagat, U.; Schlingmann, M. *Z. Anorg. Allg. Chem.* **1974,** *406,* 7.

(66) Wannagat, U.; Bogusch, E. *Inorg. Nucl. Chem. Letters* **1965,** *1,* 13.

(67) Mitzel, N. W.; Bissinger, P.; Riede, J.; Dreihäupl, K. H.; Schmidbaur, H. *Organomet.* **1992,** *12,* 413.

ADVANCES IN ORGANOMETALLIC CHEMISTRY, VOL. 40

The Organometallic Chemistry of Halocarbonyl Complexes of Molybdenum(II) and Tungsten(II)

PAUL K. BAKER

Department of Chemistry
University of Wales
Bangor, Gwynedd LL57 2UW

I

INTRODUCTION AND SCOPE OF THE REVIEW

The organometallic chemistry of halocarbonyl complexes of molybdenum(II) and tungsten(II) has been extensively studied since the early report in 1956 by Piper and Wilkinson[1] of the cyclopentadienyl halocarbonyl complexes [MoX(CO)$_3$Cp] (X = Cl, Br, I). More than 700 references of relevance to the review were collected up to June 1995, and hence some selection had to be made. The review is concerned with molybdenum(II) and tungsten(II) complexes containing both carbon monoxide and at least one halide ligand. The review is mainly restricted to the formation of complexes with those ligands as the final products of reactions, and not reactions of this type of complex.

This article is concerned with six- and seven-coordinate halocarbonyl complexes of molybdenum(II) and tungsten(II) (Section II), of which earlier reviews by Colton et al.[2] on substituted halocarbonyls of group 6 transition metals and on steric effects in substituted halocarbonyls of molybdenum and tungsten[3] have been published. Also, reviews on the structures of seven-coordinate compounds by Drew in 1977[4] and by Melník and Sharrock in 1985[5] contain relevant material.

Section III of the review is concerned with alkylidene and alkylidyne halocarbonyl complexes, and a review on recent advances in the chemistry of metal–carbon triple bonds by Mayr and Hoffmeister in 1991 contains material of interest.[6] Section IV is concerned with halocarbonyl alkyne and alkene complexes described from 1987 to mid-1995, because an extensive and comprehensive review entitled "Four-Electron Alkyne ligands in Molybdenum(II) and Tungsten(II) Complexes" by Templeton[7] was published in 1989.

Section V describes the chemistry of π-allyl halocarbonyl complexes, mainly of the type $[MX(CO)_2(NCMe)_2(\eta^3\text{-allyl})]$ and their derivatives. In 1995, Brisdon and Walton[8] reviewed transition-metal complexes containing butadienyl ligands; their review contains relevant material. Section VI is concerned with the relatively unexplored area of η^4-diene halocarbonyl complexes of molybdenum(II) and tungsten(II).

Section VII describes Cp and related η^5-ligand halocarbonyl complexes, and Section VIII is concerned with the little-studied η^6-arene halocarbonyl complexes. Finally, Section IX briefly discusses the halocarbonyl complexes of molybdenum(II) and tungsten(II) containing η^7-cycloheptatrienyl as a ligand. As Green and Ng[9] have reviewed cycloheptatrienyl and -enyl complexes of the early transition metals, only one reference in this area is given.

II

SIX- AND SEVEN-COORDINATE HALOCARBONYL COMPLEXES OF MOLYBDENUM(II) AND TUNGSTEN(II)

A. Introduction

Since Nigam and Nyholm's report[10] in 1957 of the first dihalocarbonyl donor ligand complex, namely $[MoI_2(CO)_2(diars)]$, this area of chemistry has flourished. Although two early reviews by Colton et al.[2,3] have been published, it is appropriate in this article to review Colton's pioneering work on $[\{M(\mu\text{-}X)X(CO)_4\}_2]$ (M = Mo, W; X = Cl, Br, I) type complexes, which begins this section of the review. However, the two reviews on structural aspects of seven-coordinate complexes by Drew in 1977[4] and Melník and Sharrock[5] in 1985 discuss in detail the structural aspects of this chemistry; hence, only more recent structural results are described. Generally, the structures of seven-coordinate complexes of molybdenum(II) and tungsten(II) have (i) capped octahedral, (ii) capped trigonal prismatic, (iii) pentagonal bipyramidal, or (iv) the so-called 4 : 3 geometry. Colton and Kevekordes[11] have used ^{13}C-NMR spectroscopy to correlate solid-state structures as determined by X-ray crystallography of seven-coordinate carbonyl halide complexes, and their solution state structures. For capped octahedral complexes there is a large shift to low field for a carbonyl ligand in the capping position. Two papers have described the enumeration of isomers using Pólya's theorem for both the capped octahedral geometry[12] and pentagonal bipyramidal geometry.[13]

In 1977, a paper[14] giving a molecular orbital exploration of structure, stereochemistry, and reaction dynamics of seven-coordinate complexes was published.

B. *Synthesis and Reactions with Donor Ligands of [{M(μ-X)X(CO)₄}₂]*
   (M = Mo, W; X = Cl, Br, I)

Early attempts to prepare carbonyl halides of molybdenum failed,[15-17] until pioneering work by Colton and Tomkins in 1966[18] described the reaction of $[Mo(CO)_6]$ with liquid chlorine at $-78°C$ to yield the tetracarbonyl dimer, $[\{Mo(\mu\text{-}Cl)Cl(CO)_4\}_2]$ [Eq. (1)]. In the same year[19] the bromobridged dimer, $[\{Mo(\mu\text{-}Br)Br(CO)_4\}_2]$, was prepared in an analogous manner.

$$[Mo(CO)_6] \xrightarrow[-78°C]{Cl_2} [\{Mo(\mu\text{-}Cl)Cl(CO)_4\}_2] \tag{1}$$

The tungsten dimer was prepared in a similar manner. The tungsten derivatives $[\{W(\mu\text{-}X)X(CO)_4\}_2]$ (X = Cl, Br) were synthesized by reacting $[W(CO)_6]$ with X_2 at $-78°C$.[20] In 1969,[21] the diiodo–carbonyl complexes $[\{M(\mu\text{-}I)I(CO)_4\}_2]$ were prepared by the photochemical reaction of I_2 with $[M(CO)_6]$ at room temperature. A detailed preparation of $[\{Mo(\mu\text{-}Br)Br(CO)_4\}_2]$ has been described.[22] The X-ray crystal structure of $[\{W(\mu\text{-}Br)Br(CO)_4\}_2]$ has been determined by Cotton *et al.*[23] (see Fig. 1) and shows each *fac*-$[WBr_3(CO)_3]$ octahedron capped by a carbonyl ligand on the tricarbonyl face. Interestingly, in the same paper[23] the preparation

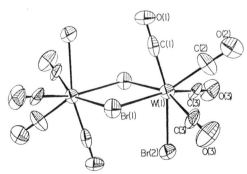

FIG. 1.   The molecular structure of $[\{W(\mu\text{-}Br)Br(CO)_4\}_2]$. [Reprinted with permission from *Inorg. Chem.* **1985**, *24*, 516. Copyright 1985 American Chemical Society.]

and molecular structure of $[(OC)_4W(\mu\text{-}Br)_3WBr(CO)_3]$ also are described, and the complex has a $4:3$ arrangement of either $(OC)_3BrWBr_3$ or $(OC)_4WBr_3$ about each tungsten centre. Colton et al.[18–21,24–39] and several other groups[40–48] have investigated the chemistry of these highly reactive species. Some selected examples are given in Scheme 1.

C. Preparation of Six- and Seven-Coordinate Complexes of Molybdenum(II) and Tungsten(II) by Halogen Oxidation of Zero-Valent Carbonyl Complexes Containing at Least One Other Donor Ligand

This is the most common method of preparing halocarbonyl complexes, and the section is divided into a number of different parts depending on the other donor ligands attached to the metal.

1. Halide Ligands

In 1964, King[49] described the preparation of the first anionic trihalocarbonyl complexes, [N-methylpyridinium][$MI_3(CO)_4$], by reaction of [N-methylpyridinium][$MI(CO)_5$] with I_2. A series of related complexes [N-methylpyridinium][$MoIX_2(CO)_4$] (X = Br, I) and [NEt_4][$MXX'X''(CO)_4$] (M = W, X = X' = X'' = Br, I; X = X' = Br, X'' = I; X = X' = I, X'' = Br; M = Mo, X = X' = X'' = Br; X = X' = I, X'' = Br) were prepared in an analogous manner by Ganorkar and Stiddard[50] in 1965.

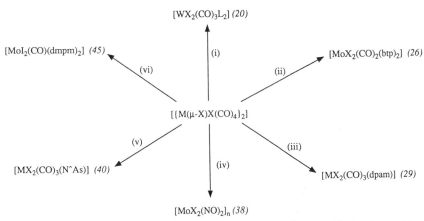

SCHEME 1.   (i) X = Cl, Br; L = PPh_3, $AsPh_3$, $SbPh_3$. (ii) X = Cl, Br; btp = N-n-butylthiopicolinamide. (iii) X = Cl, Br; dpam. (iv) X = Cl, Br; NO. (v) X = Cl, Br; N^As = 8-dimethylarsinoquinoline. (vi) dmpm.

The reactions of these trihalo-tetracarbonyl anions with monodentate,[51] bidentate,[52,53] and tridentate[53] donor ligands also have been investigated.

2. *'Butylisonitrile*

The oxidation of *fac*-[W(CO)₃(CNBuᵗ)₃] with I₂ gives the crystallographically characterized complex [WI₂(CO)₂(CNBuᵗ)₃], which has the 4 : 3 piano-stool geometry as shown in Fig. 2.[54] Similarly, the complexes *fac*-[M(CO)₃(CNR)₃] (M = Mo, W; R = Et, Buᵗ) are oxidized by Br₂ to yield the seven-coordinate complexes [MBr₂(CO)₂(CNR)₃].[55]

3. *Monodentate Nitrogen Donor Ligands*

Tripathi *et al.*[56,57] have described the reactions of [W(CO)₅(amine)] and [W(CO)₄(amine)₂] (amine = butylamine, cyclohexylamine, piperidine, morpholine) with an equimolar amount of X₂ (X = Br, I) to afford the oxidized species [WX₂(CO)₄(amine)], and [WX₂(CO)₃(amine)₂], respectively. In 1983, Moss and Smith[58] described the reactions of [M(CO)₅(py)] (M = Mo, W) with X₂ (X = Br, I) to give the anionic seven-coordinate complexes [pyH][MX₃(CO)₄], which upon reaction with bipy give the neutral complexes [MX₂(CO)₃(bipy)].

4. *Bidentate Nitrogen Donor Ligands*

In 1962, Stiddard[59] reported the reactions of [M(CO)₄(bipy)] with X₂ to give the oxidized products formulated as [MX₂(CO)₃(bipy)] (M = W,

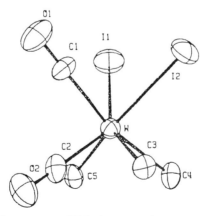

FIG. 2.   The molecular structure of [WI₂(CO)₂(CNBuᵗ)₃], oriented to show the 4 : 3 piano-stool structure. Only the coordinated carbon atoms of the *tert*-butyl isocyanide ligands are shown. [Reprinted with permission from *Inorg. Chem.* **1979**, *18*, 1907. Copyright 1979 American Chemical Society.]

X = Br, I; M = Mo, X = Br) and [{MoI$_2$(CO)$_3$(bipy)}$_2$]. Similarly, oxidation of [Mo(CO)$_2$(bipy)$_2$] and [M(CO)$_2$(phen)$_2$] (M = Mo, W) with I$_2$ gives the cationic complexes [MoI(CO)$_2$(bipy)$_2$]I and [MI(CO)$_2$(phen)$_2$]I, respectively.[60] Tripathi *et al.*[61] have described the oxidation of [W(CO)$_3$ (amine)(N^N)] (amine = cyclohexylamine, piperidine, morpholine, *n*-butylamine, hexylamine; N^N = bipy, phen) with X$_2$ (X = Br, I) to yield complexes of the type [W(CO)$_3$(amine)(N^N)] · nX$_2$. Similarly,[62] reaction of [Mo(CO)$_3$L(N^N)] (L = AsPh$_3$, SbPh$_3$; N^N = bipy, phen) with X$_2$ (X = Br, I) affords [MoX(CO)$_3$L(N^N)]X. The halogen oxidation of the Schiff-base complexes *cis*-[Mo(CO)$_4$(SB)] {SB = *N,N'*-ethylenebis (benzalideneimine), *N,N'*-ethylenebis(*p*-anisylideneimine), *N,N'*-ethyl-enebis(methylphenylketimine), *N,N'*-ethylenebis(diphenylketimine)} also have been investigated.[63] Treatment of [Mo(CO)$_4$(dab)] {dab = diacetyl-dihydrazone (ddh), glyoxalcyclohexylamine (gcy)} with X$_2$ (X = Cl, Br, I) under mild conditions generally gives the expected seven-coordinate complexes [MoX$_2$(CO)$_3$(dab)]; however, neutral or ionic binuclear complexes often are formed, such as [{Mo(CO)$_2$(ddh)$_2$(μ-Br)$_2$}$_2$Br$_2$] and [{MoX (CO)$_2$(ddh)$_2$}$_2$(μ-CO)$_2$}X$_2$] (X = Cl, Br).[64]

The six- and seven-coordinate complexes of the types [MX$_2$(CO)$_m$(N^N)] (M = Mo, W; X = Br, I; m = 2,3; N^N = H$_2$CPz$_2$, H$_2$CPz$_2'$, H$_2$CPz$_2''$, PhHCPz$_2$, PhHCPz$_2'$; Pz = pyrazol-1-yl; Pz' = 3,5-dimethylpyrazol-1-yl; Pz'' = 3,4,5-trimethylpyrazol-1-yl) have been formed by reaction of the zero-valent complexes [M(CO)$_4$(N^N)] with X$_2$. The variable coordination number of these complexes is explained by the steric and electronic effects of the ligands. The molecular structures of three of these complexes have been determined crystallographically (see Table I). The structure of [MoBr$_2$(CO)$_2$(H$_2$CPz$_2'$)] is shown in Fig. 3.[65,66] Dunn and Edwards[67] have described the preparation of the seven-coordinate complexes [MoX$_2$ (CO)$_3$(η^2-Z-paphy)] (X = Br, I; Z-paphy = Z-pyridine-2-aldehyde 2'-pyri-dylhydrazone) and [MoI$_2$(CO)$_2$(η^3-E-paphy)].

5. Tridentate Nitrogen Donor Ligands

In 1972, Dunn and Edwards[68] reported the reactions of *fac*-[M(CO)$_3$ (N^N^N)] {N^N^N = bis(2-pyridylmethyl)amine, bis(2-pyridylmethyl) methylamine, bis(2-pyridylethyl)amine} with I$_2$ to give the cationic complexes [MI(CO)$_3$(N^N^N)]I in all cases. The cationic complexes [MX(CO)$_3$ (N^N^N)]X (M = Mo, W; X = Br, I; N^N^N = 1,4,7-triazacyclononane) have been prepared; the molecular structure for the perchlorate salt [MoBr(CO)$_3$(N^N^N)][ClO$_4$] · H$_2$O has been determined crystallographi-cally and has a 4 : 3 piano-stool geometry.[69] Similarly, the 1,4,7-trimethyl-1,4,7-triazacyclononane complexes [MX(CO)$_3$(N'^N'^N')]$^+$ {M = Mo,

TABLE I

X-Ray Crystal Structure Determinations of Six- and Seven-Coordinate Halocarbonyl Complexes Reported after 1984

Complex	Structure[a]	Ref.
[{W(μ-Br)Br(CO)$_4$}$_2$]	COct	23
[(OC)$_4$W(μ-Br)$_3$WBr(CO)$_3$]	4:3	23
[MoBr$_2$(CO)$_2$(H$_2$CPz$_2'$)]	quasi-Oct	65, 66
[MoI$_2$(CO)$_2$(PhHCPz$_2'$)]	quasi-Oct	66
[WBr$_2$(CO)$_3$(H$_2$CPz$_2''$)]	COct	66
[MoBr(CO)$_3$(1,4,7-triazacyclononane)][ClO$_4$]	4:3	69
[WBr(CO)$_3$(1,4,7-triazacyclononane)][BF$_4$]	4:3	72
[MoBr(CO)$_3$Tp]	CTP or 4:3	73
[MoI$_2$(CO)$_3$(Ph$_2$POPPh$_2$)]	PBP	85
[MoBr$_2$(CO)$_3$(p-tol$_2$POPtol-p_2)]	PBP	85
[WBr$_2$(CO){P(OMe)$_3$}$_2${MeAsC(CF$_3$)=C(CF$_3$)As(Me)$_2$}]	CTP	95
[WI$_2$(CO)$_3$(NCMe)$_2$]	COct	104, 105
[WI$_2$(CO)$_3$(NCEt)$_2$]	COct	105
[WI$_2$(CO)$_3$(NCPh)$_2$]	COct	106
[WI$_2$(CO)$_3$(NCEt)(AsPh$_3$)]	CTP	107
[WI$_2$(CO)$_3$(NCPh)(AsPh$_3$)]	CTP	107
[WI$_2$(CO)$_3$⟨PhN{P(OPh)$_2$}⟩]	PBP	120
[MoI$_2$(CO)$_2${η^3-Ph$_2$PN(Pr^i)PPh(DMP)}]	CTP	121
[WI$_2$(CO)$_3${η^2-dppf(=O)}]	COct	123
[PPh$_2$CyH][WI$_3$(CO)$_3$(SbPh$_3$)]	COct	135
[WI$_2$(CO)$_3${4-MeC$_6$H$_4$S(CH$_2$)$_2$SC$_6$H$_4$Me-4}]	COct	144
[WI$_2$(CO)$_2${MeS(CH$_2$)$_2$S(CH$_2$)$_2$SMe—S,S′,S″}]	COct	144
[WBr(CO)$_2${η^4-(Me$_8$[16]aneS$_4$)}][WBr$_3$(CO)$_4$]	COct	145
[WI(CO)$_3${η^3-(Me$_8$[16]aneS$_4$)}][WI$_3$(CO)$_4$]	COct	146
[WI(CO)$_2${η^4-(Me[16]aneS$_4$)}][WI$_3$(CO)$_4$]	COct	146
[(OC)$_4$W(μ-Cl)$_3$W(SnCl$_3$)(CO)$_3$]	4:3	158
[WH(Cl)(CO)$_3$(dppe)]	COct	163
[MoCl(SnCl$_2$Bu^n)(CO)$_2${P(OMe)$_3$}$_3$]	COct or TCP	173
[Mo(CO)$_2$(PCy$_3$)(μ-Cl)(μ^3-S$_2$CPCy$_3$)SnCl$_2$Bu^n]	2	174
[MoCl(HgCl)(CO)$_3$(2,9-Me$_2$-1,10-phen)]	CTP	185
[MnMo(μ-Br)(μ-S$_2$CPPr$_3^i$)(CO)$_6$]	—	188
[(η^6-C$_6$Me$_6$)Ru(μ-Cl)(μ-S$_2$CPCy$_3$)W(CO)$_3$][PF$_6$]	—	190
[WF(CO)$_3$(C$_6$F$_4$CH=NCH$_2$CH$_2$NMe$_2$)]	—	192
[(Me$_2$PhP)$_2$(OC)$_2$W(μ-F)$_3$W(CO)$_2$(PMe$_2$Ph)$_2$][BF$_4$]	4:3	196
[(Ph$_3$P)$_2$(OC)$_2$Mo(μ-F)$_3$Mo(CO)$_2$(PPh$_3$)$_2$][BF$_4$]	4:3	197
[(Ph$_3$P)$_2$(OC)$_2$Mo(μ-F)$_2$(μ-Br)Mo(CO)$_2$(PPh$_3$)$_2$][BF$_4$]	4:3	198
[MoCl$_2$(CO)(CNBu^t)(PMe$_3$)$_3$]	—	199
[MoBr$_2$(CO)$_2$(PPh$_3$)(dppm)]	COct or CTP	201
[MoBr$_2$(CO)(bipy)(dppm)]	PBP	203
[MoI$_2$(CO)$_2$(η^1-dppm)(η^2-dppm)]	COct or CTP	209
[WCl$_2$(CO)(NCH$_2$CH=CH$_2$)(PMe$_2$Ph)$_2$]	Octahedral	210
[NEt$_4$][MoF(CO)$_2$(S$_2$CNEt$_2$)$_2$]	CTP	223
[{MoCl(CO)$_2$(dppmSe)$_2$}$_2$][Mo$_6$O$_{19}$]	COct	224
[MoBr$_2$(CO)$_2$(η^1-dppm)(η^2-dppm)]	—	225

[a] COct = capped octahedral; CTP = capped trigonal prismatic; PBP = pentagonal bipyramidal.

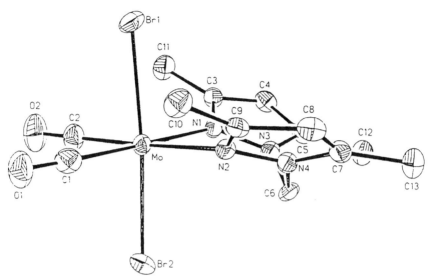

FIG. 3.   The molecular structure of $[MoBr_2(CO)_2(H_2CPz'_3)]$. [Reproduced with permission from *Organometallics* **1990**, *9*, 670. Copyright 1990 American Chemical Society.]

X = Br, I[70]; M = W, X = F, Cl, Br, I[71]} also have been reported. In 1987, Wieghardt *et al.*[72] determined the X-ray crystal structure of [WBr(CO)₃ (N^N^N)][BF₄] (see Table I).

Treatment of [Mo(CO)₃Tp]⁻ with X₂ (X = Br, I) affords the seven-coordinate complexes [MoX(CO)₃Tp], which has been characterized crystallographically for X = Br. The structure can either be viewed as a distorted capped trigonal prism with Br capping a quadrilateral face or as a 3:4 piano stool.[73]

6. *Monodentate Phosphine Ligands*

In 1967, Lewis and Whyman[74] described the reactions of [M(CO)₃ (PPh₃)₃] (M = Mo, W) with three equivalents of I₂ to yield the anionic complexes [PPh₃H][MI₃(CO)₃(PPh₃)]. Similarly, treatment of *trans*-[M(CO)₄(PPh₃)₂] or [M(CO)₃(PPh₃)₃] with three equivalents of Br₂ gives [PPh₃H][MBr₃(CO)₃(PPh₃)]. In 1970, Moss and Shaw[75] described an extensive series of seven-coordinate complexes including complexes of the type [MX₂(CO)₃L₂] (M = Mo, W; X = Cl, Br, I; L = PEt₃, PMe₂Ph, AsMe₂Ph). Oxidation of the thiocarbonyl complex [W(CO)₅(CS)] with Br₂ in the presence of PPh₃ gives [WBr₂(CO)₂(CS)(PPh₃)₂].[76] Treatment of [Mo(CO)₅L] (L = PPh₃, AsPh₃, SbPh₃) with X₂ (X = Cl, Br, I) gives the six-coordinate complexes [MoX₂(CO)₃L], whereas reaction of *cis*-

$[Mo(CO)_4L_2]$ (L = PPh_3, $AsPh_3$, $SbPh_3$) with X_2 affords the seven-coordinate complexes $[MoX_2(CO)_3L_2]$.[77] Addition of X_2 (X = Br, I) to $[W(CO)_5L]$ gives $[WBr_2(CO)_4L]$ (L = PMe_3, $AsMe_3$) and $[WI_2(CO)_4L]$ (L = PMe_3, $AsMe_3$, $SbMe_3$). The complexes $[WI_2(CO)_4L]$ (L = PBu_3^n, NMe_3, NEt_3, SMe_2) have been spectroscopically identified. The molecular structure of $[WI_2(CO)_4(PMe_3)]$ was crystallographically determined and shown to be capped octahedral with a CO ligand capping the two iodide ligands and the PMe_3 ligand.[78] The carbonyl substitution reactions of $[WX_2(CO)_4L]$ (X = Cl, Br, I; L = PMe_3, $AsMe_3$, $SbMe_3$) with L' {L' = PMe_3, $AsMe_3$, $SbMe_3$, $AsMe_2H$, $P(OMe)_3$, PPh_3, py} to give $[WX_2(CO)_3LL']$ also have been reported.[79]

7. Bidentate Phosphine Ligands

In 1965, Lewis and Whyman[80] described the oxidation of $[M(CO)_4$ (dppe)] with X_2 (X = Br, I) to yield $[MX_2(CO)_3(dppe)]$. They also investigated the reactions of $[M(CO)_2(dppe)_2]$ with X_2 (X = Br, I). Oxidation of $[Mo(CO)_4(dppm)]$ with I_2 gives $[MoI_2(CO)_3(dppm)]$, which reacts further with dppm to afford $[MoI_2(CO)_2(dppm)_2]$.[81] Connor et al.[82,83] have described the reactions of cis-$[M(CO)_4(dmpe)]$ (M = Mo, W) and fac,fac-$[M_2(CO)_6(dmpe)_3]$ with I_2 to give a range of halocarbonyl complexes. For example, oxidation of cis-$[M(CO)_4(dmpe)]$ with I_2 gives $[MI(CO)_4(dmpe)]I$, which can be transformed into $[MI_2(CO)_3(dmpe)]$. The silicon-backbone phosphine ligand complexes $[MoBr_2(CO)_n\{Ph_2P(CH_2)SiMe_2CH{=}CH_2\}]$ (n = 2,3) and $[MoBr_2(CO)_3\{Me_2Si(CH_2PPh_2)_2\}]$ have been prepared and characterized.[84] The preparation, molecular structures (X = I, R = Ph; X = Br, R = p-tol), and fluxional behavior of the complexes $[MoX_2$ $(CO)_3(R_2POPR_2)]$ (X = Br, I; R = Ph, p-tol), which were prepared by halogen oxidation of $[Mo(CO)_4(R_2POPR_2)]$, have been described.[85] The molecular structures of $[MoX_2(CO)_3(R_2POPR_2)]$ (X = I, R = Ph; X = Br, R = p-tol) both have a pentagonal bipyramidal geometry. The seven-coordinate complexes $[\{MBr_2(CO)_2\}_2(\mu\text{-mtppe})]$ and $[MoBr_2(CO)_2$ (mtppe)] (mtppe = **1**) have been prepared and characterized.[86] Reaction

1

of fac-[Mo(CO)$_3${PPh$_2$CH$_2$C(Bu^t)=N—N=C(Me)CH$_2$S(C$_6$H$_4$Me-4)}] or [Mo(CO)$_4${PPh$_2$CH$_2$C(Bu^t)=N—N=C(Me)CH$_2$S(C$_6$H$_4$Me-4)}] with Br$_2$ gives the P^N^S coordinated complex [MoBr$_2$(CO)$_2${PPh$_2$CH$_2$C(Bu^t)=N—N=C(Me)CH$_2$S(C$_6$H$_4$Me-4)}], whereas treatment of the tungsten complex fac-[W(CO)$_3${PPh$_2$CH$_2$C(Bu^t)=N—N=C(Me)CH$_2$S(C$_6$H$_4$Me-4)}] with Br$_2$ yields the cationic complex [WBr(CO)$_3${PPh$_2$CH$_2$C(Bu^t)=N—N=C(Me)CH$_2$S(C$_6$H$_4$Me-4)}]$^+$, which was isolated as its [PF$_6$]$^-$ salt.[87]

8. Arsenic Donor Ligands

Although a preliminary report in 1957 by Nigam and Nyholm[10] of the complex [MoI$_2$(CO)$_2$(diars)] was published, full details of these results were reported in 1960.[88] Nyholm $et\ al.$[88] described the reactions of [Mo(CO)$_4$(diars)] with two equivalents of X$_2$ (X = Br, I) to give [MoX$_2$(CO)$_3$(diars)], whereas reaction of [Mo(CO)$_2$(diars)$_2$] with two equivalents of X$_2$ (X = Br, I) yields [MoX(CO)$_2$(diars)$_2$]X (X = I$^-$, Br$^-$). Similar reactions of the tungsten o-phenylenebisdimethylarsine (diars) complexes [W(CO)$_4$(diars)] and [W(CO)$_2$(diars)$_2$] with X$_2$ (X = Br, I) were also reported in 1963.[89] Oxidation of [M(CO)$_3$(1-triars or v-triars)] {1-triars = methylbis(dimethylarsino-3-propyl)arsine; v-triars = tris-1,1,1-(dimethylarsinomethyl)ethane} with X$_2$ (X = Br, I) gives [MX(CO)$_3$(1-triars or v-triars)]X.[90,91] Treatment of [M(CO)$_3$(ttas)] {ttas = bis(o-dimethylarsinophenyl)methylarsine} with X$_2$ yields the cationic seven-coordinate complexes [MX(CO)$_3$(ttas)]X.[92] A variable-temperature NMR study of a series of complexes [MX$_2$(CO)$_2$(dam)$_2$] {M = Mo, W; X = Cl, Br, I; dam = bis(diphenylarsino)methane} have been described.[93] The synthesis and fluxional properties of the seven-coordinate complex [MoI$_2$(CO)$_3${o-phenylenebis(methylphenylarsine)}] have been reported.[94] Oxidation of [W(CO)$_4$(L^L)] {L^L = MeAsC(CF$_3$)=C(CF$_3$)As(Me)$_2$} with Br$_2$ gives [WBr$_2$(CO)$_3$(L^L)], which reacts further with, for example, P(OMe)$_3$ to yield the crystallographically characterized complex [WBr$_2$(CO){P(OMe)$_3$}$_2$(L^L)].[95] The structure is approximated most closely to a capped trigonal prism with a capping bromide ligand.

9. Oxygen and Sulfur Donor Ligands

In 1962, Mannerskantz and Wilkinson[96] described the reactions of [Mo(CO)$_4$(2,5-dithiahexane)] with X$_2$ (X = Br, I) to give the seven-coordinate complexes [MoX$_2$(CO)$_3$(2,5-dithiahexane)]. Treatment of the oxygen donor ligand complexes [M(CO)$_3$(RPPh)] (M = Mo, W) {RPPh = bis[(diphenylphosphinyl)methyl]phenylphosphine oxide}, [M$_2$(CO)$_6$(MDPO)] {MDPO = methylenebis[diphenylphosphine oxide]}, and [M(CO)$_4$(R

POEt)] {RPOEt = bis[(diphenylphosphinyl)methyl]ethyl phosphinate} with X_2 (X = Br, I) gives the complexes $[MX_2(CO)_3(RPPh)]$, $[MX_2(CO)_3(MDPO)]$, and $[MI_2(CO)_3(RPOEt)]$, respectively.[97] Oxidation of the macrocyclic thioether ligand complexes $[\{Mo(CO)_3\}_2(TTP)]$ (TTP = 1,4,8,11-tetrathiacyclotetradecane), $[\{Mo(CO)_3\}_2(HTO)]$ (HTO = 1,4,7, 10,13,16-hexathiacyclooctadecane) and $[W(CO)_3(TTP)]$ with I_2 gives the seven-coordinate complexes $[MoI_2(CO)_3(TTP)] \cdot CH_2Cl_2$, $[\{MoI_2(CO)_3\}_2(HTO)]$, and $[WI_2(CO)_3(TTP)] \cdot CH_2Cl_2$, respectively.[98]

D. *Synthesis and Reactions with Donor Ligands of the Highly Versatile Seven-Coordinate Complexes $[MX_2(CO)_3(NCMe)_2]$ (M = Mo, W; X = Br, I)*

This method of synthesis of seven-coordinate complexes could have been considered in Section II,C,3, i.e., halogen oxidation of carbonyl complexes containing monodentate nitrogen donor ligands. However, a separate section is given to these $[MX_2(CO)_3(NCMe)_2]$ complexes, as an extensive chemistry has been developed over the 10 years since the details of their synthesis first appeared in 1986.[99,100]

In their original paper in 1962, Tate, Knipple, and Augl[101] described the synthesis of *fac*-$[M(CO)_3(NCMe)_3]$ (M = Cr, Mo, W) by refluxing the parent hexacarbonyl in acetonitrile. They also reported that the reaction of $[W(CO)_3(NCMe)_3]$ with iodine in methanol evolves three moles of gas and gives a non-carbonyl-containing product. It should also be noted that Westland and Muriithi in 1972[102] studied the reactions of the dimeric complexes first reported by Colton et al., $[\{Mo(\mu-X)X(CO)_4\}_2]$ (X = Cl, Br), with weaker field ligands such as L (L = py, thf, NCMe, etc.) to yield the non-carbonyl-containing products $[MoX_3L_3]$ as the major products. They proposed that these reactions proceeded via molybdenum(II) carbonyl halide intermediates, and finally disproportionated to give $[Mo(CO)_6]$ and $[MoX_3L_3]$ (Scheme 2). They also describe the reaction of $[\{Mo(\mu-Cl)Cl(CO)_4\}_2]$ suspended in a mixture of Et_2O and NCMe for 12 hr to give a product that molybdenum analysis suggested to be $[MoCl_2(CO)_3(NCMe)_2]$. This complex decomposed gradually, and it was not possible to obtain satisfactory analysis of the remaining elements.

In view of these observations, Baker et al. decided to carry out the reactions of the tris(nitrile) complexes *fac*-$[M(CO)_3(NCMe)_3]$ (M = Mo, W) (prepared *in situ*) with X_2 (X = Br, I), rather than in methanol as previously attempted by Tate, Knipple, and Augl.[101] On November 15, 1984, Elaine Keys (now Elaine Armstrong) carried out the reaction of *fac*-$[Mo(CO)_3(NCMe)_3]$ (prepared *in situ*) with an equimolar amount of

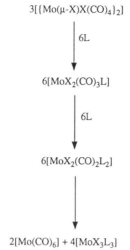

$$3[\{Mo(\mu\text{-}X)X(CO)_4\}_2]$$

$$\Big\downarrow 6L$$

$$6[MoX_2(CO)_3L]$$

$$\Big\downarrow 6L$$

$$6[MoX_2(CO)_2L_2]$$

$$\Big\downarrow$$

$$2[Mo(CO)_6] + 4[MoX_3L_3]$$

SCHEME 2. [Reproduced with permission from *J. Organomet. Chem.* **1986**, *309*, 319.]

I_2 at 0°C to give the seven-coordinate complex $[MoI_2(CO)_3(NCMe)_2]$ in quantitative yield.[99] Soon after, the tungsten analogue $[WI_2(CO)_3(NCMe)_2]$ and the less stable dibromo complexes $[MBr_2(CO)_3(NCMe)_2]$ were prepared. An improved synthesis of the dibromo complexes $[MBr_2(CO)_3$ $(NCMe)_2]$ (M = Mo, W) has been described[100] by reaction of *fac*-$[M(CO)_3$ $(NCMe)_3]$ (prepared *in situ*) with an equimolar quantity of Br_2 at −78°C. In 1991, Richmond *et al.*[103] described the preparation of $[WI_2(CO)_3$ $(NCEt)_2]$ by oxidation of *fac*-$[W(CO)_3(NCEt)_3]$ with I_2 in a suitable form as an inorganic–organometallic laboratory experiment for advanced undergraduates.

The molecular structures of three of the bis(nitrile) tungsten complexes $[WI_2(CO)_3(NCR)_2]$ {R = Me,[104,105] Et,[105] Ph[106]} have been described. The geometry of the complexes can be described best as distorted capped octahedra with a carbonyl ligand capping an octahedral face containing an iodide and the remaining two carbonyl ligands. The molecular structure of $[WI_2(CO)_3(NCMe)_2]$ is shown in Fig. 4.[104,105] As mentioned in the Introduction, in 1982 Colton and Kevekordes[11] described how ^{13}C-NMR spectroscopy can be used to correlate the solution state structure for a capped octahedral species with a carbonyl ligand in the capping position with the solid-state structure. The low-temperature ^{13}C-NMR (−70°C, CD_2Cl_2) spectrum of $[WI_2(CO)_3(NCMe)_2]$ has two carbonyl resonances at δ = 202.36 and 228.48 ppm with an approximate intensity ratio of 2 : 1. From Colton and Kevekordes' observations,[11] it is likely that the low-field reso-

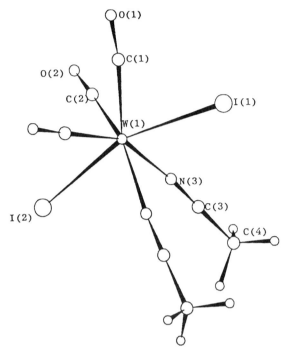

FIG. 4. The molecular structure of $[WI_2(CO)_3(NCMe)_2]$. [Reprinted from *Polyhedron* **1995**, *14*, 619. Copyright 1995, with kind permission from Elsevier Science Ltd., The Boulevard, Langford Lane, Kidlington OX5 1GB, U.K.]

nance at 228.48 ppm is due to the unique capping carbonyl and that at 202.36 ppm is in the expected octahedral range for the two equivalent octahedral carbonyl groups.[104,105]

1. Reactions of $[MI_2(CO)_3(NCMe)_2]$ (M = Mo, W) with Nitrogen Donor Ligands

Reaction of $[MI_2(CO)_3(NCMe)_2]$ with excess NCR yields the nitrile exchanged products $[MI_2(CO)_3(NCR)_2]$ (for M = Mo, R = Ph; for M = W, R = Et, Bu^t, CH$_2$Ph, Ph), which react further with one equivalent of L (L = PPh$_3$, AsPh$_3$, SbPh$_3$) to give the mixed ligand complexes $[MI_2(CO)_3(NCR)L]$. The molecular structures of $[WI_2(CO)_3(NCR)(AsPh_3)]$ (R = Et, Ph) have been determined crystallographically, and both have slightly distorted monocapped trigonal prismatic coordination geometry (Table I).[107] Treatment of $[MI_2(CO)_3(NCMe)_2]$ with one equivalent of a series of pyridines (py) (py = 2Me-py, 4Me-py, 3,5-Me$_2$-py, 3Cl-py, 3Br-py, 4Cl-py, 4Br-py) gives either [MI$_2$(CO)$_3$(NCMe)(py)] or [{M(μ-I)I(CO)$_3$ (py)}$_2$], depending on the steric and electronic properties of the substituted pyridines.[108] The preparation of the complexes [MI$_2$(CO)$_3$L$_2$] (L$_2$ = py$_2$, bipy, phen) by reaction of [MI$_2$(CO)$_3$(NCMe)$_2$] with L$_2$ has been reported.[109] The reactions of [MI$_2$(CO)$_3$(NCMe)$_2$] and [MI$_2$(CO)$_3$(NCMe)L] (L = PPh$_3$, AsPh$_3$, SbPh$_3$) with a series of substituted imidazoles[110] and pyrazole[111] have been described.

Treatment of [WI$_2$(CO)$_3$(NCMe)$_2$] with two equivalents of 4-HO$_2$C C$_5$H$_4$N in MeOH gives [WI$_2$(CO)$_3$(4-HO$_2$CC$_5$H$_4$N)$_2$], which reacts with two equivalents of NaOH to yield a completely water-soluble complex of tungsten(II), namely, [WI$_2$(CO)$_3$(4-NaO$_2$CC$_5$H$_4$N)$_2$]. This complex reacts with an equimolar amount of 3-NaO$_3$SC$_5$H$_4$N in water to afford [WI$_2$(CO)$_3$(4-NaO$_2$CC$_5$H$_4$N)(3-NaO$_3$SC$_5$H$_4$N)] and is the first room-temperature ligand substitution reaction in water of a metal carbonyl complex.[112] The 1,4-diaza-1,3-butadiene complexes [MI$_2$(CO)$_3$(RN=CHCH =NR)] (M = Mo, W; R = Bu^i, Cy, Ph, p-MeOPh) have been synthesized by acetonitrile displacement reactions of [MI$_2$(CO)$_3$(NCMe)$_2$] with RN= CHCH=NR.[113] A series of 23 neutral, cationic, and dicationic seven-coordinate complexes of molybdenum(II) and tungsten(II) that have mono- and bidentate nitrogen donor ligands attached to the metal have been prepared and characterized.[114]

2. *Reactions of [MI$_2$(CO)$_3$(NCMe)$_2$] with Phosphorus, Arsenic, Antimony, and Bismuth Donor Ligands*

The earliest reported reactions of [MI$_2$(CO)$_3$(NCMe)$_2$], described in 1986 by Baker and Fraser,[115] were with phosphites, P(OR)$_3$ (R = Me, Ph) to give [MI$_2$(CO)$_3${P(OR)$_3$}$_2$]. The complex [WI$_2$(CO)$_3${P(OMe)$_3$}$_2$] upon reflux in CHCl$_3$ loses CO to yield the "16-electron" complex [WI$_2$(CO)$_2$ {P(OMe)$_3$}$_2$]. Treatment of [MI$_2$(CO)$_3$(NCMe)$_2$] with two equivalents of L (L = PPh$_3$, AsPh$_3$, SbPh$_3$) yields [MI$_2$(CO)$_3$L$_2$]. The bis(triphenylphos- phine) complexes rearrange in CH$_2$Cl$_2$ to the anionic complexes [PPh$_3$H] [MI$_3$(CO)$_3$(PPh$_3$)].[116] The fluxionality of the complexes [MI$_2$(CO)$_3$L$_2$] (M = Mo, L = PMe$_2$Ph; M = W, L = PMe$_2$Ph, PMePh$_2$, PPh$_3$) have been investigated by ^{31}P-{^{1}H} NMR spectroscopy.[117] A series of 10 diphenylphos- phinoferrocene (PPh$_2$Fe) seven-coordinate complexes have been prepared by reactions of [MI$_2$(CO)$_3$(NCMe)$_2$] or [MI$_2$(CO)$_3$(NCMe)L] (L = PPh$_3$, AsPh$_3$, SbPh$_3$) with PPh$_2$Fe.[118]

A high-yield synthesis of the bidentate phosphine complexes [MI$_2$ (CO)$_3${Ph$_2$P(CH$_2$)$_n$PPh$_2$}] (n = 1–6) has been reported.[119] Krishnamurthy *et al.*[120] have described the reactions of [MI$_2$(CO)$_3$(NCMe)$_2$] with the di- phosphazane ligands RN{P(OPh)$_2$}$_2$ (R = Me, Ph) to give [MI$_2$(CO)$_3$(RN

{P(OPh)$_2$})], which has been crystallographically characterized for M = W, R = Ph and has unusually a slightly distorted pentagonal bipyramidal structure (Fig. 5).[120] The preparation and molecular structure of the diphosphazane complex [MoI$_2$(CO)$_2${η^3-Ph$_2$PN(Pr^i)PPh(DMP)}] (DMP = 3,5-dimethyl-1-pyrazolyl) have been reported.[121] The 1,1′-bis(diphenylphosphino)ferrocene (dppf) complexes [MoI$_2$(CO)$_2$(η^2-dppf)], [MI$_2$(CO)$_3$(η^2-dppf)] (M = Mo, W), and [MoCl$_2$(CO)$_2$(dppf)$_2$] have been prepared and characterized.[122] Yeh *et al.*[123] have described the reaction of *fac*-[W(CO)$_3$(NCMe)$_3$] with dppf to give [W(CO)$_3$(NCMe)(η^2-dppf)], which reacts with I$_2$ to give [WI$_2$(CO)$_3$(η^2-dppf)], which can be oxidized with H$_2$O$_2$ to yield the crystallographically characterized complex [WI$_2$(CO)$_3${η^2-dppf(=O)}] (Fig. 6). The geometry of this complex can be described as a distorted capped octahedron with a carbonyl ligand (C37) in the unique capping position.

Reaction of equimolar quantities of [MI$_2$(CO)$_3$(NCMe)$_2$] and Ph$_2$PC≡CPPh$_2$ gives the phosphine-bridged complexes [M$_2$I$_4$(CO)$_6$(μ-Ph$_2$PC≡CPPh$_2$)$_2$], whereas reaction of [MI$_2$(CO)$_3$(NCMe)$_2$] with two equivalents of Ph$_2$PC≡CPPh$_2$ yields the monodentately attached complexes [MI$_2$(CO)$_3$(η^1-Ph$_2$PC≡CPPh$_2$)$_2$].[124] Treatment of [MI$_2$(CO)$_3$(NCMe)$_2$] with an

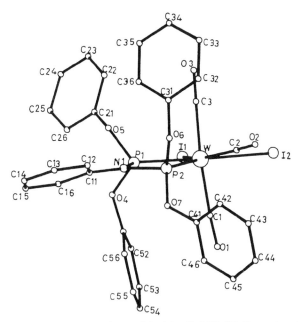

FIG. 5.   The molecular structure of [WI$_2$(CO)$_3$(PhN{P(OPh)$_2$})]. [Reproduced with permission from *Organometallics* **1991**, *10*, 2524. Copyright 1991 American Chemical Society.]

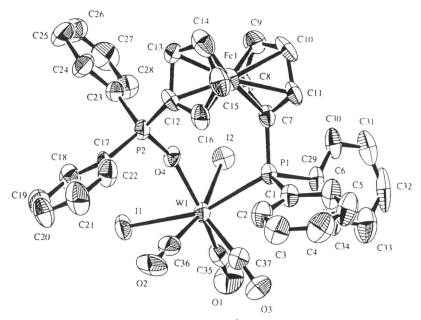

FIG. 6. The molecular structure of [WI$_2$(CO)$_3$\{η^2-dppf(=O)\}]. [Reproduced with permission from *J. Organomet. Chem.* **1995**, *492*, 122].

equimolar amount of PhP(CH$_2$CH$_2$PPh$_2$)$_2$ initially gives [MoI$_2$(CO)$_2$\{η^3-PhP(CH$_2$CH$_2$PPh$_2$)$_2$\}] and [WI$_2$(CO)$_3$\{η^2-PhP(CH$_2$CH$_2$PPh$_2$)$_2$\}], respectively. Some reactions of [WI$_2$(CO)$_3$\{η^2-PhP(CH$_2$CH$_2$PPh$_2$)$_2$\}] are shown in Scheme 3.[125] A series of triphenylbismuth complexes derived from [MI$_2$(CO)$_3$(NCMe)$_2$] have been reported[126,127] and are the first examples

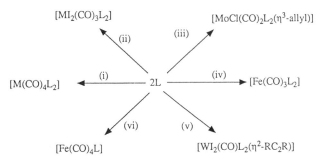

SCHEME 3. L = [WI$_2$(CO)$_3$\{η^2-PhP(CH$_2$CH$_2$PPh$_2$)$_2$\}]. (i) [M(CO)$_4$(piperidine)$_2$] (M = Mo, W); (ii) [MI$_2$(CO)$_3$(NCMe)$_2$] (M = Mo, W); (iii) [MoCl(CO)$_2$(NCMe)$_2$(η^3-allyl)]; (iv) [Fe(CO)$_3$(η^4-cot)]; (v) [WI$_2$(CO)(NCMe)(η^2-RC$_2$R)$_2$] (R = Me, Ph); (vi) [Fe$_2$(CO)$_9$].

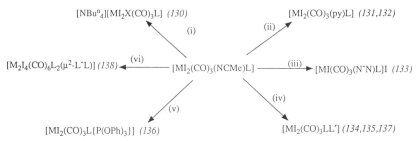

SCHEME 4. Reagents: M = Mo, W; L = PPh_3, $AsPh_3$, $SbPh_3$. (i) $[NBu^n]X$ (X = I, IBr_2, Br_3);[130] (ii) py = substituted pyridines;[131,132] (iii) N^N = bipy, phen, 1,2-phenylenediamine;[133] (iv) L′ = PPh_3, PPh_2Cy, $PPhCy_2$, $AsPh_3$, $SbPh_3$[134,135]; L′ = $SC(NH_2)_2$, $SC(NMe_2)_2$, $SC(NH_2)$ Me[137]; (v) $P(OPh)_3$;[136] (vi) $\frac{1}{2}L^L = Ph_2P(CH_2)_nPPh_2$ (n = 1, 2 or 4), $\frac{1}{2}L^L = [Fe(\eta^5 C_5H_4PPh_2)_2]$.[138]

of seven-coordinate complexes of molybdenum(II) and tungsten(II) containing $BiPh_3$ as an attached ligand.

3. *Synthesis and Reactions of $[MI_2(CO)_3(NCMe)L]$ (L = PPh_3, $AsPh_3$, $SbPh_3$)*
   and Related Complexes

The reactions of $[MI_2(CO)_3(NCMe)_2]$ with one equivalent of L (L = a series of monodentate phosphines, phosphites, $AsPh_3$, $SbPh_3$) gives either $[MI_2(CO)_3(NCMe)L]$ {L = PPh_3, $AsPh_3$, $SbPh_3$, PPh_2Cy, $P(OPh)_3$} or the iodo-bridged dimers $[\{M(\mu-I)I(CO)_3L\}_2]$, depending on the steric and electronic effects of the ligands.[128,129] The complexes $[MI_2(CO)_3(NCMe)L]$ (L = PPh_3, $AsPh_3$, $SbPh_3$) have a range of chemistry shown in Scheme 4.[130–138] The molecular structure of $[PPh_2CyH][WI_3(CO)_3(SbPh_3)]$ was determined crystallographically[135] and has a capped octahedral geometry with a capping CO ligand. The capped face has two carbonyl ligands and the triphenylantimony ligand.

4. *Reactions of $[MI_2(CO)_3(NCMe)_2]$ with Neutral Oxygen and Sulfur*
   Donor Ligands

The reactions of $[MI_2(CO)_3(NCMe)_2]$ with thiourea, N,N,N',N'-tetra-methylthiourea,[139] and thioacetamide[140] to give a number of seven-coordinate complexes have been described. A large series of 42 triphenylphosphine oxide and triphenylphosphine sulfide complexes derived from $[MI_2(CO)_3(NCMe)_2]$ have been prepared as shown in Scheme 5.[141,142] Also, a series of tricyclohexylphosphinecarbondisulfide seven-coordinate complexes derived from the reactions of $[MI_2(CO)_3(NCMe)_2]$, $[MI_2(CO)_3(NC Me)L]$ {L = PPh_3, $AsPh_3$, $SbPh_3$, $P(OPh)_3$}, and $[WI_2(CO)_3(PPh_3)_2]$ with PCy_3CS_2 have been reported.[143]

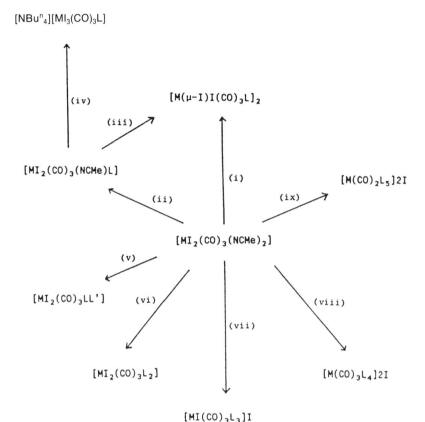

SCHEME 5.   All reactions carried out in CH_2Cl_2 at room temperature. For (i)–(vii) and (ix), M = Mo or W. Reagents: (i) L = $OPPh_3$ for 30 s. (ii) L = $SPPh_3$ for 30 s. (iii) Stirring in CH_2Cl_2 for 24 hr. (iv) L = $SPPh_3$; [NBu_4^n]I for 45 min. (v) L′ = PPh_3 (1 min), $AsPh_3$ (3 min) or $SbPh_3$ (5 min) followed by an *in situ* reaction with L = $OPPh_3$ or $SPPh_3$ for 18 hr. (vi) 2L = $OPPh_3$ or $SPPh_3$ for 18 hr. (vii) 2L = $OPPh_3$ or $SPPh_3$ for 2 hr, followed by one further equivalent of L added *in situ* for 18 hr. (viii) M = Mo, 2L = $SPPh_3$ for 2 hr followed by two further equivalents of L added *in situ* for 18 hr. M = W, 2L = $OPPh_3$ or $SPPh_3$ for 2 hr followed by two further equivalents of L added *in situ* for 18 hr. (ix) 2L = $OPPh_3$ for 2 hr followed by three further equivalents of L added *in situ* for 18 hr. [Reproduced with permission from *Inorg. Chim. Acta* **1990**, *174*, 122.]

Treatment of [$MI_2(CO)_3(NCMe)_2$] with a slight excess of $RS(CH_2)_2SR$ (R = Ph, 4-MeC_6H_4, 4-FC_6H_4) gives [$MI_2(CO)_3\{RS(CH_2)_2SR\}$], which has been characterized crystallographically for M = W, R = 4-MeC_6H_4, and has a capped octahedral geometry with a carbonyl ligand in the unique capping position.[144] Also, reaction of [$MI_2(CO)_3(NCMe)_2$] with an equimolar amount of $MeS(CH_2)_2S(CH_2)_2SMe$ gives [$MoI_2(CO)_2\{MeS(CH_2)_2$

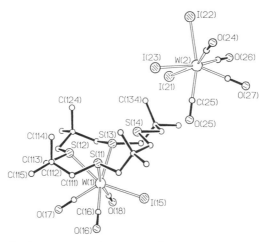

FIG. 7. The molecular structure of [WI(CO)$_3${η^3-(Me$_8$[16]aneS$_4$)}] [WI$_3$(CO)$_4$]. [Reproduced with permission from *J. Organomet. Chem.* **1994**, *469*, C23.]

S(CH$_2$)$_2$SMe-S,S',S"}]   and   [WI$_2$(CO)$_3${MeS(CH$_2$)$_2$S(CH$_2$)$_2$SMe-S,S'}], which loses CO to give the crystallographically characterized complex [WI$_2$(CO)$_2${MeS(CH$_2$)$_2$S(CH$_2$)$_2$SMe-S,S'S"}]. The structure is capped octahedral with a carbonyl ligand in the capping position.[144] Reaction of two equivalents of [MI$_2$(CO)$_3$(NCMe)$_2$] with one equivalent of Me$_8$[16]aneS$_4$ (Me$_8$[16]aneS$_4$ = 3,3,7,7,11,11,15,15-octamethyl-1,5,9,13-tetrathiacyclohexadecane) yields the unusual seven-coordinate cation/anion complexes [WI(CO)$_3${η^3-(Me$_8$[16]aneS$_4$)}][WI$_3$(CO)$_4$] and [WI(CO)$_2${η^4-(Me$_8$[16]aneS$_4$)}][WI$_3$(CO)$_4$], which both have been crystallographically characterized[145,146]; both cations have the 4:3 piano-stool geometry, as shown in Figs. 7 and 8. The anion [WI$_3$(CO)$_4$]$^-$ has capped octahedral geometry with a capping carbonyl ligand. Similarly, treatment of [{Mo(μ-Br)Br(CO)$_4$}$_2$] with Me$_8$[16]aneS$_4$ gives the crystallographically characterized complex [MoBr(CO)$_2${η^4-(Me$_8$[16]aneS$_4$)}][MoBr$_3$(CO)$_4$], which also has similar geometries for both cation and anion to the analogous iodo–tungsten complex.[145]

5. Reactions of [MI$_2$(CO)$_3$(NCMe)$_2$] with Neutral Bidentate N⌃S Donor Ligands

Reactions of [MI$_2$(CO)$_3$(NCMe)$_2$] with an equimolar amount of NH$_2$NHCSNH$_2$ or RR'CNNHCSNH$_2$ (R = R' = Me, Et; R = Me, R' = ^{i}Pr; R = Ph, R' = H) gives either the dinuclear molybdenum complexes [{Mo(μ-I)I(CO)$_2$(N⌃S)}$_2$] or the mononuclear tungsten complexes [WI$_2$(CO)$_3$(N⌃S)].[147] A series of pyridine-2-thione and pyrimidine-2-thione seven-

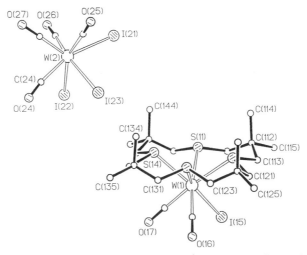

FIG. 8. The molecular structure of [WI(CO)₂{η⁴-(Me₈[16]aneS₄)}][WI₃(CO)₄]. [Reproduced with permission from *J. Organomet. Chem.* **1994,** *469,* C22.]

coordinate complexes with N^S coordination derived from reactions of [MI₂(CO)₃(NCMe)₂] and [WI₂(CO)₃L₂] {L₂ = PPh₃, Ph₂P(CH₂)ₙPPh₂ (n = 1–6)} with the N^S donor ligands have been described.[148]

6. *Reaction of [MI₂(CO)₃(NCMe)₂], [MI₂(CO)₃(NCMe)L], or [MI₂(CO)₃L₂] with Anionic Oxygen and Sulfur Donor Ligands*

Reaction of [MI₂(CO)₃(NCMe)(PPh₃)] and [MI₂(CO)₃(PPh₃)₂] with an equimolar amount of Na[O^O] (O^O = acac, hfacac, bzacac) affords the bidentately attached seven-coordinate complexes [MI(CO)₃(PPh₃)(O^O)] and [MI(CO)₂(PPh₃)₂(O^O)], respectively.[149] An extensive series of dithiocarbamate complexes has been prepared and characterized from reactions of [MI₂(CO)₃(NCMe)₂] and [MI₂(CO)₃(NCMe)L] (L = PPh₃, AsPh₃, SbPh₃) with a range of dithiocarbamate ligands.[150,151] Treatment of [MI₂(CO)₃(NCMe)L] {prepared *in situ;* L = PPh₃, AsPh₃, SbPh₃} with either [NH₄][S₂CNC₄H₈][152] or K[S₂COEt][153] yields the seven-coordinate complexes [MI(CO)₃L(S₂CNC₄H₈)] or [MI(CO)₃L(S₂COEt)], respectively. More recently, a large series of pyridine-2-thionate and pyrimidine-2-thionate complexes of molybdenum(II) and tungsten(II) have been prepared, including the crystallographically characterized complex [W(CO)₃(η²-pymS)₂] (pymS = pyrimidine-2-thionate), which has a distorted capped trigonal prismatic geometry with a carbonyl in the unique capping position.[154]

E. *Oxidation of [M(CO)₆] (M = Mo, W) and Substituted Carbonyl Complexes with Other Oxidizing Agents*

Thus far the synthesis of six- and seven-coordinate halocarbonyl complexes has been restricted to oxidation by Cl_2, Br_2 and I_2. This section deals with a range of other oxidizing agents such as ICl, IBr, $GeCl_4$, $SnCl_4$, and HgX_2, and organometallic complexes such as *fac*-[MnCl (CO)₃(S₂CPCy₃)].

1. *Oxidation of [M(CO)₆] (M = Mo, W)*

An early report in 1966 by Lange and Dehnicke[155] described the reactions of [M(CO)₆] (M = Mo, W) with ClN_3 to give the polymeric carbonyl azides [{M(μ-Cl)(μ-N₃)(CO)₂}ₓ]. In 1988, Szymańska-Buzar[156] described the photochemical oxidation of [W(CO)₆] with CCl_4 to give [{W(μ-Cl)Cl(CO)₄}₂], which is used for the preparation of [WCl₂(CO)₃(PPh₃)₂] and [WCl₂(CO)₂ (dppe)]. Similarly, Szymańska-Buzar[157] used the photochemically prepared complex [{W(μ-Cl)Cl(CO)₄}₂] to synthesize a series of nitrogen donor ligand complexes such as [WCl₂(CO)₃(N^N)] (N^N = 2py, 2NCMe, bipy, phen, 2,9-Me₂-phen). The photochemical reaction of [W(CO)₆] with $SnCl_4$ gives the crystallographically characterized complex [(OC)₄W(μ-Cl)₃ W(SnCl₃)(CO)₃].[158] Each tungsten has 4 : 3 piano-stool geometry, as shown in Fig. 9. However, the photochemical reaction of [W(CO)₆] with $SnCl_4$ in the presence of PPh₃ yields the crystallographically characterized complex [WCl₄(OPPh₃)₂].[159]

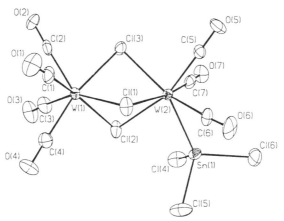

FIG. 9.   The molecular structure of [(OC)₄W(μ-Cl)₃W(SnCl₃)(CO)₃]. [Reproduced with permission from *J. Organomet. Chem.* **1995**, *489*, 210.]

2. Oxidation of Substituted Carbonyl Complexes with ICl, IBr, and ICN

Baker *et al.* have described the synthesis and reactions with donor ligands of the molybdenum complexes $[MoClI(CO)_3(NCMe)_2]^{160}$ and $[MoBrI(CO)_3(NCMe)_2]$,[161] which are prepared by reaction of *fac*-$[Mo(CO)_3(NCMe)_3]^{101}$ with ClI or BrI, respectively. Reaction of $[M(CO)_4(N^\frown N)]$ (M = Mo, W; $N^\frown N$ = bipy, phen) with ICN affords the seven-coordinate complexes $[MI(CN)(CO)_3(N^\frown N)]$.[162]

3. Oxidation of Substituted Carbonyl Complexes with Group IV Halides and Alkyl Halides

Reactions of $[M(CO)_3(NCR')(L^\frown L)]$ (M = Mo, W; $L^\frown L$ = tmeda, bipy, dppe, $2PPh_3$) with $SiHClR_2$ gave, not the expected hydrido–silyl complexes, but the hydrido–chloro complexes $[MH(Cl)(CO)_3(L^\frown L)]$, which can also be prepared by treatment of $[M(CO)_3(NCR')(L^\frown L)]$ with HCl.[163] The tungsten complex $[WH(Cl)(CO)_3(dppe)]$ has been crystallographically characterized and has a capped octahedral geometry with the hydride ligand capping the triangular face formed by a CO ligand and two phosphorus atoms of dppe.[163] A series of heterometallic complexes such as $[MX(M'X_3)(CO)_3(bipy)]$ (M = Mo, W; M' = Ge, Sn; X = Cl, Br) have been prepared by oxidation of the zero-valent complexes $[M(CO)_4(bipy)]$ with $M'X_4$.[164]

In 1989, Baker and Bury[165] described the reaction of *fac*-$[Mo(CO)_3(NCMe)_3]$ (prepared *in situ*[101]) with an equimolar amount of $SnCl_4$ to give $[MoCl(SnCl_3)(CO)_3(NCMe)_2]$ in quantitative yield. The reaction of this complex with phosphines also was reported.[165] Reactions of $[MoCl(SnCl_3)(CO)_3(NCMe)_2]$ with an equimolar amount of L {L = py, 3Br-py, 4Br-py, PPh_3, $AsPh_3$, $SbPh_3$, $P(OPh)_3$} yields the chloro-bridged dimers $[\{Mo(\mu\text{-}Cl)(SnCl_3)(CO)_3\}_2]$.[166] The reactions of $[MoCl(SnCl_3)(CO)_3(NCMe)_2]$ with bidentate nitrogen donor ligands also have been described.[167] Oxidation of *fac*-$[Mo(CO)_3(NCMe)_3]$ (prepared *in situ*) with $GeCl_4$ furnishes the seven-coordinate complex $[MoCl(GeCl_3)(CO)_3(NCMe)_2]$.[168,169] The reactions of $[MoCl(GeCl_3)(CO)_3(NCMe)_2]$ with neutral monodentate Group 5 donor ligands,[168,169] bidentate and tridentate Group 5 donor ligands,[170] and also with $OPPh_3$ and $SPPh_3$[171] have been reported.

In 1991, Riera *et al.*[172,173] reported the reactions of *fac*-$[M(CO)_3(NCR)_3]$ (M = Mo, R = Me; M = W, R = Et) with $SnCl_3R'$ (R' = Bu^n, Ph) to yield $[MCl(SnR'Cl_2)(CO)_3(NCR)_2]$. The reaction of $[MCl(SnR'Cl_2)(CO)_3(NCR)_2]$ with excess tetramethylthiourea gives $[MCl(SnR'Cl_2)(CO)_3\{SC(NMe_2)_2\}_2]$, whereas reaction of $[MCl(SnR'Cl_2)(CO)_3(NCR)_2]$ with three equivalents of $P(OR'')_3$ (R'' = Me, Et) affords $[MCl(SnR'Cl_2)(CO)_2\{P(OR'')_3\}_3]$, which has been crystallographically characterized for M =

Mo, $R' = Bu^n$, $R'' = Me$. The structure is described as intermediate between a distorted capped trigonal prism and a distorted capped octahedron. Reaction of $[MoCl(SnBu^nCl_2)(CO)_3(NCMe)_2]$ with two equivalents of S_2CPR_3 (R = Cy, Et, Bu^n) gives the complexes **2**, which are the first examples of complexes whereby S_2CPR_3 bridges between a transition and main-group metal. The molecular structure for R = Cy has been determined crystallographically.[174]

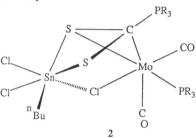

2

Cano *et al.*[175,176] have described the preparation of the heterobimetallic complexes $[MoCl(SnRCl_2)(CO)_2\{P(C_6H_4X-4)_3\}(phen)]$ (R = Cl, Me, Ph; X = F, Cl, Me, OMe) from reaction of $[Mo(CO)_3\{P(C_6H_4X-4)_3\}(phen)]$ with $SnRCl_3$. In 1993, Cano *et al.*[177] also described the synthesis and characterization of $[MoCl(SnRCl_2)(CO)_3(NCMe)_2]$ (R = Me, Ph) by oxidation of *fac*-$[Mo(CO)_3(NCMe)_3]$ with $SnRCl_3$. In 1994, they described the reactions of $[MoCl(SnRCl_2)(CO)_3(NCMe)_2]$ with $P(C_6H_4X-4)_3$ (X = H, F, Cl, Me, OMe) in several different solvents.[178] The oxidation of *fac*-$[M(CO)_3(NCMe)_3]$ (M = Mo, W) with $SnPhCl_3$ gives $[MCl(SnPhCl_2)(CO)_3(NCMe)_2]$. The reaction chemistry with phosphine and phosphite ligands also has been discussed.[179]

4. *Oxidation of Substituted Carbonyl Halides with HgX$_2$*

In 1965, Ganorkar and Stiddard[180] described the oxidation of $[W(CO)_4(bipy)]$ with $HgCl_2$ to give the seven-coordinate complex $[WCl(HgCl)(CO)_3(bipy)]$. A series of papers by Cano *et al.*[181–185] described the reactions of zero-valent substituted molybdenum and tungsten carbonyls with mercury(II) halides to give molybdenum(II) and tungsten(II) carbonyl halide derivatives. The molecular structure of the seven-coordinate complex $[MoCl(HgCl)(CO)_3(2,9-Me_2-1,10-phen)]$ has been determined crystallographically[185] and has a capped trigonal prismatic geometry (see Fig. 10).

5. *Reactions of fac-[M(CO)$_3$(NCMe)$_3$] and fac-[M(CO)$_3$(NCMe)(N$^{\wedge}$N)] with Organotransition-Metal Halides*

In 1989, Granifo and Vargas[186] described the reactions of *fac*-$[M(CO)_3(NCMe)(N^{\wedge}N)]$ (M = Mo, W; $N^{\wedge}N$ = bipy, phen, en) with [Fe

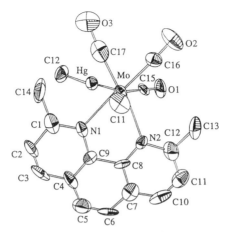

FIG. 10.   The molecular structure of [MoCl(HgCl)(CO)₃(2,9-Me₂-1,10-phen)]. [Reprinted from *Polyhedron* **1994**, *13*, 1672. Copyright 1994, with kind permission from Elsevier Science Ltd., The Boulevard, Langford Lane, Kidlington OX5 1GB, U.K.]

(HgX)(CO)₂Cp] (X = Cl, Br, I, SCN, N₃) to give the seven-coordinate molybdenum and tungsten complexes [CpFe(CO)₂HgMX(CO)₃(N⌃N)], with Fe–Hg–M–X arrays. Riera *et al.*[187,188] have described the reactions of *fac*-[M(CO)₃(NCMe)₃] with [MnBr(CO)₃(S₂CPR₃)] (R = Cy, ⁱPr) to yield the heterodinuclear complexes [MnMo(μ-Br)(μ-S₂CPR₃)(CO)₆], which have been characterized crystallographically for R = ⁱPr. These dimeric complexes react with mono- and bidentate phosphines and phosphites to give carbonyl displaced products, including [MnMo(μ-Br)(μ-S₂CPCy₃) (CO)₅(μ-tedip)] (tedip = tetraethylpyrophosphite). Similar reactions of the rhenium analogues of *fac*-[ReBr(CO)₃(S₂CPR₃)] (R = Cy, ⁱPr) with *fac*-[Mo(CO)₃(NCMe)₃] also have been published.[189]

Reaction of *fac*-[M(CO)₃(NCMe)₃] (M = Mo, W) with [RuCl (S₂CPCy₃)(η⁶-C₆Me₆)][PF₆] affords the cationic heterobimetallic complexes [(η⁶-C₆Me₆)Ru(μ-Cl)(μ-S₂CPCy₃)M(CO)₃][PF₆], which have been characterized crystallographically for M = W.[190]

6. *Intramolecular Oxidation of Zero-Valent Complexes of Molybdenum(0) and Tungsten(0)*

In 1994, Richmond *et al.*[191] published an extensive review describing the activation of C—F bonds using organotransition-metal complexes. This review includes a full discussion of C—F bond cleavage reactions of the type shown in Eq. (2).[192]

Shaw *et al.*[193] have described the reaction of the 3-diphenylphosphino-(IR)-(+)-camphor dimethylhydrazone complex [Mo(CO)₄(PPh₂C₁₀H₁₅N

$$(2)$$

NMe_2)] with HCl to yield the molybdenum(II) imine complex [$MoCl_2(CO)_3(PPh_2C_{10}H_{15}NH)$)] via a rapid redox/fission reaction. Treatment of mixed azines of the type ZE-$PPh_2CH_2C(Bu^t)$=N—N=CH $(C_6H_4X$-$2)$ (X = F, Cl, Br, I) with fac-[$W(CO)_3(NCEt)_3$] gives [$W(CO)_3(NCEt)\{PPh_2CH_2C(Bu^t)$=N—N=CH$(C_6H_4X$-$2)\}$], which rapidly undergo oxidative addition to yield the seven-coordinate complexes [$WX(CO)_3\{PPh_2CH_2(Bu^t)$=N—N=CH$(C_6H_4)\}$].[194]

F. Synthesis of Halocarbonyl Complexes of Molybdenum(II) and Tungsten(II) by Other Methods

This section describes other methods that do not easily fit into Sections II,A–E of the review.

1. Metal Carbonyl Fluorides

Fluorocarbonyl complexes of molybdenum(II) and tungsten(II) are much less common than their other halogen counterparts. However, several examples have been reported. For example, the selective fluorination of [$Mo(CO)_6$] with XeF_2 in perfluoro solvents or HF yields [$MoF_2(CO)_3$], which has been characterized by IR, mass spectrometry, and X-ray powder diffraction analysis.[195]

Reaction of [$WH_6(PMe_2Ph)_3$] with H[BF_4] · OEt_2 in a carbon monoxide atmosphere gives the crystallographically characterized complex [(Me_2 PhP)$_2(OC)_2W(\mu$-F)$_3W(CO)_2(PMe_2Ph)_2$][BF_4], whereby each tungsten atom has a basal plane containing three fluorine atoms, with the remaining four ligands in a second parallel plane, in which the two carbon monoxide ligands are mutually trans as shown in Fig. 11.[196] Beck et al.[197] described the reactions of [$MBr_2(CO)_2(PPh_3)_2$] (M = Mo, W) with two equivalents of Ag[BF_4] to yield the fluoro-bridged complexes [(Ph_3P)$_2(OC)_2M(\mu$-F)$_3M$ $(CO)_2(PPh_3)_2$][BF_4], which have been characterized crystallographically for M = Mo. Similarly, Woodward et al.[198] treated [$MoBr_2(CO)_2(PPh_3)_2$] with aqueous Na[BF_4] to afford the dimeric complexes [(Ph_3P)$_2(OC)_2$ $Mo(\mu$-X)(μ-F)$_2(CO)_2(PPh_3)_2$]$^+$ (X = F, Cl, Br, OH), which has been

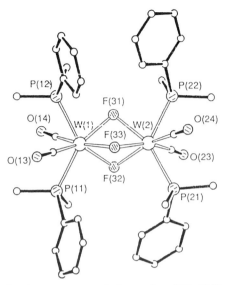

FIG. 11. The molecular structure of the cation [(Me₂PhP)₂(OC)₂W(µ-F)₃W(CO)₂(PMe₂Ph)₂]⁺ [Reproduced with permission from *J. Chem. Soc., Dalton Trans.* **1992**, 973.]

characterized crystallographically for X = Br, with fluoride as the counter-anion.

2. *Reactions of Metal Halocarbonyls with Neutral Donor Ligands*

Carmona *et al.*[199] have described the reactions of [MCl₂(CO)₂(PMe₃)₃] (M = Mo, W) with CNR in the presence of PMe₃ to give the substituted complexes [MCl₂(CO)(CNR)(PMe₃)₃] (M = Mo, R = Buᵗ, Cy, PhCH₂, 2,6-C₆H₃Me₂; M = W, R = Buᵗ), which has been characterized crystallographically for M = Mo, R = Buᵗ (see Fig. 12). In 1987, Brisdon and Hodson[200] described the reactions of [MX₂(CO)₃L₂] (for M = Mo, X = Cl, Br, I, L = PMePh₂; X = Br, L = PEtPh₂, PMe₂Ph, PEt₂Ph; for M = W, X = Br, L = PMe₂Ph, PMePh₂) in boiling acetonitrile to afford the carbonyl displaced products [MX₂(CO)₂(NCMe)L₂]. In the same year, Riera *et al.*[201] described the reactions of [MoBr₂(CO)₂(PPh₃)₂] with a series of mono- and bidentate nitrogen and phosphorus donor ligands to give a variety of seven-coordinate complexes, including [MoBr₂(CO)₂(PPh₃)(dppm)], which has been characterized crystallographically. Similarly, the reactions of [MoX₂(CO)₃(N^N)] (X = Cl, Br, I; N^N = bipy, phen) with dppm and dppe have been described.[202,203] The X-ray crystal structure of [MoBr₂(CO)(bipy)(dppm)] also has been reported[203] and has a distorted

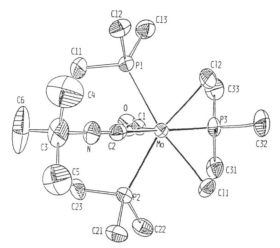

FIG. 12. The molecular structure of [MoCl$_2$(CO)(CNBu^t)(PMe$_3$)$_3$]. [Reproduced with permission from *Inorg. Chem.* **1990,** *29,* 701. Copyright 1990, American Chemical Society.]

pentagonal bipyramidal geometry with a carbon monoxide and one nitrogen atom of the 2,2'-bipyridyl ligand in the axial positions.

The reaction of [MBr$_2$(CO)$_3$(PPh$_3$)$_2$] (M = Mo, W) with RN$_3$ (R = Ph, 4MeC$_6$H$_5$) yields the phosphazide complexes [MBr$_2$(CO)$_3$(RN$_3$PPh$_3$)], N$_2$, and RN=PPh$_3$.[204,205] The tungsten complex [WBr$_2$(CO)$_3$(4MeC$_6$H$_5$N$_3$PPh$_3$)] was characterized crystallographically[205] and is seven-coordinate with the phosphazide ligand coordinated in a chelating manner through the α and γ nitrogen atoms.

3. *Reductive Carbonylation of Molybdenum and Tungsten Complexes*

In 1972, Bencze[206] described the reaction of [MoCl$_4$(PPh$_3$)$_2$] with AlEtCl$_2$ under a carbon monoxide atmosphere to give [MoCl$_2$(CO)$_3$(PPh$_3$)$_2$]. Reaction of Mg(CH$_2$SiMe$_3$)$_2$ and [MoCl$_4$(thf)$_2$] in the presence of PMe$_3$ furnishes [MoCl(CH$_2$SiMe$_3$)$_3$(PMe$_3$)], which reacts with CO to yield the crystallographically characterized chloro-bridged dimer [{Mo(μ-Cl)(η^2-COCH$_2$SiMe$_3$)(CO)$_2$(PMe$_3$)}$_2$], which has been shown to have a bidentate η^2-acyl group.[207] More recently, Carmona *et al.*[208] have described the reactions of the paramagnetic complexes [MCl$_2$(PMe$_3$)$_4$] (M = Mo, W) with CO in boiling thf to afford [MCl$_2$(CO)$_2$(PMe$_3$)$_3$], which reacts with KX to afford [MX$_2$(CO)$_2$(PMe$_3$)$_3$] (X = Br, I, NCO, NCS).

4. *Other Methods*

In 1987, Cotton and Matusz[209] described a new high-yield synthesis of [MoI$_2$(CO)$_2$(η^1-dppm)(η^2-dppm)] by reaction of [NEt$_4$][MoI$_3$(CO)$_4$] with dppm in benzene. The molecular structure of $[MoI_2(CO)_2(\eta^1\text{-dppm})(\eta^2\text{-dppm})]$ has been determined crystallographically and can be considered either as a capped octahedron with a capping carbonyl ligand, or as a capped trigonal prism with the iodo ligand capping a rectangular face. Reaction of $[WCl_2(PMePh_2)_4]$ with allyl isocyanate gives $cis,trans\text{-}[WCl_2(CO)(NCH_2CH=CH_2)(PMe_2Ph)_2]$, which has been characterized by X-ray crystallography.[210] Atwood et al.[211] have described the reactions of a mixture of $[W(CO)_6]$, HClg, H_2O, and either 18-crown-6 or bis-aza-18-crown-6 to give $[H_3O^+ \cdot 18\text{-crown-6}][WCl(CO)_5]$, $[H_3O^+ \cdot 18\text{-crown-6}][WCl_3(CO)_4]$, and $[H_2O \cdot \text{bis-}aza\text{-18-crown-6}] \cdot (H^+)_2][\{WCl_3(CO)_4\}_2]$.

G. Halocarbonyl Complexes of Molybdenum(II) and Tungsten(II) Containing Acyl and Anionic Oxygen and Sulfur Donor Ligands

1. Acyl Complexes of Molybdenum(II) and Tungsten(II)

Carmona et al.[212] have studied extensively the synthesis, molecular structures, and reactivity of a series of acyl complexes of molybdenum and tungsten. For example,[213] the five-coordinate alkyl complex $[WCl(CH_2SiMe_3)_3(PMe_3)]$ reacts with CO to afford the η^2-acyl dimeric tungsten(II) complex $[\{WCl(1\text{-}2\text{-}\eta\text{-COCH}_2SiMe_3)(CO)_2(PMe_3)\}_2]$, which has been characterized crystallographically. A number of papers in this area describe complexes containing halide, carbonyl and acyl ligands.[214–217]

2. Molybdenum(II) and Tungsten(II) Halocarbonyl Complexes Containing Anionic Oxygen and Sulfur Donor Ligands

A number of papers describing the synthesis and characterization of six- and seven-coordinate complexes of the types $[M(CO)_n(S^\wedge S)_2]$ (M = Mo, W; n = 2,3; $S^\wedge S$ = S_2CNR_2, S_2COEt, $S_2CNC_4H_4$, S_2PR_2, etc.) have been published. These are not considered here because they do not contain a halo group.

Archer et al. have described the synthesis[218] and molecular structures[219] of the seven-coordinate complexes $[WCl(CO)_3(PPh_3)(dcq)]$ (dcq = 5,7-dichloro-8-quinolato), $[W(CO)_2(PPh_3)(dcq)_2]$, and $[WCl(CO)_2(PPh_3)_2(dcq)]$. A series of complexes of the type $[MX(CO)_2L_2(O^\wedge O)]$ (M = Mo, W; X = Cl, Br, O_2CR; L = PEt_3, PPh_3; $O^\wedge O$ = acac, hfacac, tropolonate, carboxylate) have been prepared and characterized.[220] Similarly, Riera et al.[221] have described the synthesis and characterization of the complexes $[MoBr(CO)_2(PPh_3)_2(XOCR)]$ (X = O, R = H, Me, Ph, CCl_3; X = S, R = Me, Ph).

In 1977, McDonald et al.[222] described the reactions of $[MCl_2(CO)_2(PPh_3)_2]$ (M = Mo, W) with excess $HS_2P(^iPr)_2$ or an equimolar amount of $HS_2P(OEt)_2$ to afford $[MCl(CO)_2(PPh_3)_2\{\eta^2\text{-}S_2P(^iPr)_2$ or $S_2P(OEt)_2\}]$. Burgmayer and Templeton[223] have reported reactions of $[M(CO)_2(S_2CNR_2)_2]$ (R = Me, Et) with F^- or N_3^- to yield the anionic complexes $[MX(CO)_2(S_2CNR_2)_2]^-$, which has been characterized crystallographically for M = Mo, X = F, R = Et, with $[NEt_4]^+$ as the countercation. The structure is capped trigonal prismatic with fluoride occupying the capping position and is trans to both carbonyl groups.

H. Structures of Seven-Coordinate Halocarbonyl Complexes

As described in the Introduction, two extensive reviews on the structures of seven-coordinate complexes have been published, by Drew[4] in 1977 and by Melník and Sharrock[5] in 1985. Hence, this review has concentrated on the synthetic aspects of seven-coordinate halocarbonyl complexes. The structures of seven-coordinate halocarbonyl complexes described after 1984 are correlated in Table I in the order that they appear in Sections II,A–G.

The X-ray crystal structure of $[\{MoCl(CO)_2(dppmSe)\}_2][Mo_6O_{19}] \cdot 4MeNO_2$ was reported in 1988[224] by Colton et al. The molybdenum is seven-coordinate with a slightly distorted capped octahedral geometry, with a carbonyl ligand in the unique capping position. In 1991, the molecular structure of the dibromo complex $[MoBr_2(CO)_2(\eta^1\text{-}dppm)(\eta^2\text{-}dppm)]$ was reported.[225]

I. Catalytic Activity of Halocarbonyl Complexes of Molybdenum(II) and Tungsten(II)

Bencze and Markó[226,227] have observed that tungsten halocarbonyl complexes such as $[WCl_2(CO)_3(PPh_3)_2]$ in the presence of Lewis acids such as $AlEtCl_2$ are active metathesis catalysts. In 1985, Bencze and Kraut-Vass[228] also reported that the complexes $[MX_2(CO)_3L_2]$ (M = Mo, W; X = Cl, Br; L = PPh_3, $AsPh_3$) are single-component catalysts for the ring-opening polymerization of norbornene and norbornadiene. The rate-determining step in these reactions is loss of donor ligand L, followed by coordination of the alkene in an η^2 manner. The mechanism of these catalytic reactions have been studied further[229] and shown to involve a 1,2-hydrogen shift of the $M(\eta^2\text{-}alkene)$ group to give a carbene intermediate.

Szymańska-Buzar et al.[230–234] have studied how Group 6 metal halocarbonyl intermediates are involved in the catalytic activity of these complexes in the presence of CCl_4[230,231] and $ZrCl_4$[230–234] toward the transformation of both alkenes[230–233] and alkynes.[234]

III

ALKYLIDENE AND ALKYLIDYNE HALOCARBONYL COMPLEXES OF MOLYBDENUM(II) AND TUNGSTEN(II)

Several reviews have been published[235-238] concerning complexes with M≡CR fragments. An extensive review in this series by Mayr and Hoffmeister[6] describes the work carried out on the $[MX(CO)_2L_2(CR)]$ (M = Cr, Mo, W; X = Cl, Br, I; R = alkyl, aryl) type complexes, which have been considered to be in oxidation state zero. Hence, this section is restricted to carbene complexes in which the molybdenum and tungsten oxidation state is +2.

A. *Alkylidene Halocarbonyl Complexes of Molybdenum(II) and Tungsten(II)*

In 1977, Lappert and Pye[239] described the oxidation of the zero-valent dicarbene complexes *cis*-$[M(CO)_4(L^R)_2]$ (L^R = :$\overline{CN(R)(CH_2)_2N}R$) with X_2 to give the six-coordinate complexes $[MX_2(CO)_2(L^R)_2]$ (M = Mo, X = Cl, Br, I; R = Et; M = W, X = I, R = Me, Et) and the seven-coordinate $[WI_2(CO)_3(L^R)_2]$ (R = Me, Et) complexes, which are carbon monoxide carriers. However, oxidation of the mono-carbene complexes $[M(CO)_5(L^R)]$ (R = Me) with I_2 furnishes the complexes $[IL^R][MoI_3(CO)_4]$ or $[WI_2(CO)_4(L^R)]$, respectively. Churchill and Wasserman[240] described the reaction of $[W(=CH_2)Cl(PMe_3)_4][CF_3SO_3]$ with CO to afford the crystallographically characterized complex $[WCl(CH_2PMe_3)(CO)_2(PMe_3)_3]$ $[CF_3SO_3]$. The structure of this complex approximates a capped trigonal-prismatic geometry. Winter and Woodward[241] have described the reaction of $[W(SnPh_3)(=CHPh)(CO)_2Cp]$ with I_2 to yield the nonheteroatom-stabilized carbene complex $[WI(=CHPh)(CO)_2Cp]$.

Mayr *et al.*[242] have described the synthesis and molecular structure of the carbene complex $[W(=CHPh)Cl_2(CO)(PMe_3)_2]$. Mayr *et al.*[243] also have prepared some related carbene complexes as shown in Eq. (3). The molecular structure of the cation $[W(=CHPh)Cl(CO)(CNBu^t)(PMe_3)_2]^+$ is shown in Fig. 13.[243]

The reactions of $[W(CPh)Br(CO)_2(NC_5H_4Me-4)_2]$ and $[W(CBr)Br(CO)_2(NC_5H_4NMe_2-4)_2]$ with allyl or 2-methylallyl bromide give the allylidene complexes $[W(=C(Ph)CH=CHR)Br_2(CO)_2(NC_5H_4Me-4)]$ (R = H, Me) and $[W\{=C(Ph)CH=CHMe\}Br_2(CO)_2(NC_5H_4NMe_2-4)]$, which were all crystallographically characterized. All the complexes are fluxional, as shown by variable-temperature ^{13}C-NMR spectroscopy.[244]

$$\begin{array}{c} {}^{t}\text{Bu} \\ | \\ \text{N} \\ \| \\ \text{C} \quad \text{PMe}_3 \\ | \quad \, \\ \text{X}\!-\!\text{W}\!\equiv\!\text{CR} \\ | \\ \text{Me}_3\text{P} \quad \text{C} \\ \| \\ \text{O} \end{array} \xrightarrow[\text{Et}_2\text{O}]{\text{HY}} \left[\begin{array}{c} {}^{t}\text{Bu} \\ | \\ \text{N} \\ \| \\ \text{C} \quad \text{PMe}_3 \\ | \quad \,\,\,\,\,\,\, \\ \text{X}\!-\!\text{W}\!\cdots\!\text{H} \\ | \quad\backslash\!\!\backslash \\ \text{Me}_3\text{P} \quad \text{C} \quad \text{C} \\ \| \quad \backslash \\ \text{O} \qquad \text{R} \end{array} \right] \text{Y} \qquad (3)$$

(X = Cl, I; R = Me, Ph; Y = CF$_3$SO$_3$, BF$_4$)

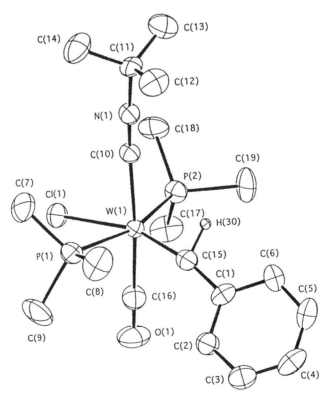

FIG. 13. The molecular structure of the cation [W($=$CHPh)Cl(CO)(CNBu^t)(PMe$_3$)$_2$]$^+$.
[Reproduced with permission from *Angew. Chem., Int. Ed. Engl.* **1993**, *32*, 744.]

IV

ALKYNE AND ALKENE HALOCARBONYL COMPLEXES OF MOLYBDENUM(II) AND TUNGSTEN(II)

As stated in the Introduction, Templeton[7] published a comprehensive review entitled "Four Electron Alkyne Ligands in Molybdenum(II) and Tungsten(II) Complexes," including halocarbonyl complexes, which dealt with the literature up to 1987. Hence, this section of the review deals with the literature on halocarbonyl alkyne and alkene complexes published from 1987 to mid-1995. The vast majority of papers that have appeared during this period are concerned with the reactions of $[MI_2(CO)_3(NCMe)_2]$ (M = Mo, W) and their derivatives with alkynes, which begin this section of the review.

A. Alkyne Complexes Derived from $[MI_2(CO)_3(NCMe)_2]$ (M = Mo, W) and Derivatives

Since the donor ligand chemistry of $[MI_2(CO)_3(NCMe)_2]$ and derivatives has been described in Sections II,D and II,E of this review, it is appropriate to describe the alkyne chemistry of these versatile complexes.

1. Reactions of $[MI_2(CO)_3(NCMe)_2]$ with One Equivalent of RC_2R'

Reaction of $[MI_2(CO)_3(NCMe)_2]$ with an equimolar amount of RC_2R' (R = R' = Me, Ph, CH_2Cl; R = Ph, R' = Me, CH_2OH; R = Me, R' =

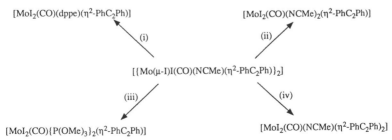

SCHEME 6. (i) 2dppe in CH_2Cl_2; (ii) 2NCMe warm to 40°C; (iii) 4P(OMe)$_3$ in CH_2Cl_2; (iv) 2PhC$_2$Ph in CH_2Cl_2.

PhS, p-tolS) initially yields the bis(nitrile) complexes [MI$_2$(CO)
(NCMe)$_2$(η^2-RC$_2$R')], which dimerize to yield the iodo-bridged complexes
[{M(μ-I)I(CO)(NCMe)(η^2-RC$_2$R')}$_2$].[245] The reactions of the diphenylacet-
ylene molybdenum complex with donor ligands and alkynes are shown
in Scheme 6.

2. Reactions of [MI$_2$(CO)$_3$(NCR)$_2$] with Excess RC$_2$R'

In 1988, Baker et al.[246] described the reactions of [MI$_2$(CO)$_3$(NCMe)$_2$]
with an excess of RC$_2$R' (R = R' = Ph; R = Me, R' = Ph; for M = W
only, R = R' = Me, CH$_2$Cl, p-tol; R = Ph, R' = CH$_2$OH) to give the
monomeric alkyne complexes [MI$_2$(CO)(NCMe)(η^2-RC$_2$R')$_2$] and the dimeric molybdenum complex [{Mo(μ-I)I(CO)(η^2-MeC$_2$Me)$_2$}$_2$]. The molecular structures of [WI$_2$(CO)(NCMe)(η^2-RC$_2$R)$_2$] (R = Me, Ph) have been
determined crystallographically and the structure for the bis(2-butyne)
derivative is shown in Fig. 14.[246]

More recently, in 1994[247] they described the reactions [WI$_2$(CO)$_3$(NCR)$_2$]
(R = Et, Bu^t, Ph, CH$_2$Ph) with excess R'C$_2$R' (R' = Me, Ph) to yield the
related complexes [WI$_2$(CO)(NCR)(η^2-R'C$_2$R')$_2$], which has been characterized crystallographically for R = Bu^t, R' = Me (Table II).

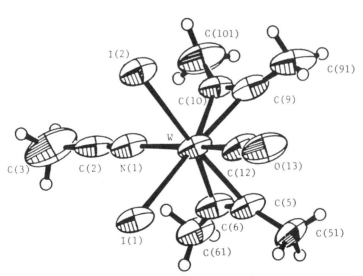

FIG. 14.   The molecular structure of [WI$_2$(CO)(NCMe)(η^2-MeC$_2$Me)$_2$]. [Reprinted with
permission from *Organometallics* **1988**, 7, 322. Copyright 1988 American Chemical Society.]

3. *Reactions of [WI$_2$(CO)(NCMe)(η^2-RC$_2$R)$_2$] (R = Me, Ph) with Neutral Donor Ligands*

An extensive chemistry of [WI$_2$(CO)(NCMe)(η^2-RC$_2$R)$_2$] (R = Me, Ph) has been developed in recent years. This section is divided according to the type of neutral donor ligands.

a. *Carbon donor ligands.* Reaction of [WI$_2$(CO)(NCMe)(η^2-RC$_2$R)$_2$] (R = Me, Ph) with CO gives the crystallographically characterized 2-butyne complex [WI$_2$(CO)$_2$(η^2-MeC$_2$Me)$_2$] and eventually the iodo-bridged dimer [{W(μ-I)I(CO)(η^2-PhC$_2$Ph)$_2$}$_2$], respectively.[248] Treatment of [WI$_2$(CO)(NCMe)(η^2-RC$_2$R)$_2$] with one to five equivalents of CNBu^t affords the

TABLE II

Complex	Ref.
[WI$_2$(CO)(NCMe)(η^2-MeC$_2$Me)$_2$]	246
[WI$_2$(CO)(NCMe)(η^2-PhC$_2$Ph)$_2$]	246
[WI$_2$(CO)(NCBu^t)(η^2-MeC$_2$Me)$_2$]	247
[WI$_2$(CO)$_2$(η^2-MeC$_2$Me)$_2$]	248
[WI$_2$(CO){NCCH$_2$(3-C$_4$H$_3$S)}(η^2-MeC$_2$Me)$_2$]	251
[WI(CO)(bipy)(η^2-MeC$_2$Me)$_2$][BPh$_4$]	254
[WI$_2$(CO)(dppm)(η^2-MeC$_2$Me)]	257
[WI$_2$(CO){P(OMe)$_3$}$_2$(η^2-MeC$_2$Me)]	259
[WI$_2$(CO)(η^2-triphos)(η^2-MeC$_2$Me)]	261
[WI(CO)(C$_4$H$_3$N$_2$S)(η^2-MeC$_2$Me)$_2$]	267
[WI(CO)(S$_2$CNC$_4$H$_8$)(η^2-MeC$_2$Me)$_2$]	268
[WI(CO){P(OPr^i)$_3$}(dppm)(η^2-MeC$_2$Me)][BF$_4$]$_{0.5}$[OH]$_{0.5}$ · 0.5H$_2$O	271
[WI(CO){SC(NH$_2$)$_2$}(dppm)(η^2-MeC$_2$Me)][ClO$_4$]	275
[Cl(OC)(Ph$_3$P)(η^2-MeC$_2$Me)Mo(μ-SCN)(μ-NCS)MoCl(CO)(PPh$_3$)(η^2-MeC$_2$Me)]	288
[MoBr$_2$(CO)(PMePh$_2$)$_2$(η^2-MeC$_2$Me)]	289
[{W[C(CF$_3$)=C(CF$_3$)P(OEt)$_2$O](μ-Br)(CO)(η^2-CF$_3$C$_2$CF$_3$)}$_2$]	290
[WCl$_2$(CO)(PMe$_3$)$_2$(η^2-PhC$_2$Ph)]	292
[WCl$_2$(CO)(PMe$_2$Ph)$_2$(η^2-PhC$_2$Ph)]	292
[WI(CO)(η^2-PhC$_2$Me)Tp′]	293
[WCl$_3$(CO)(PMe$_3$)(η^2-PhCH$_2$CNBu^t)]	295
[WCl$_2$(CO)(PMe$_3$)$_2$(η^2-PhC$_2$NHBu^t)]	295
[WCl$_2$(CO)(PMe$_3$)$_2$(η^2-PhC$_2$OH)]	296
[WCl$_2$(CO)(PMe$_3$)$_2${η^2-PhC$_2$OC(O)C$_6$H$_4$OMe-4}]	296
[WI$_2$(CO)(PMe$_3$)$_2$(η^2-PhC$_2$SH)]	299
[WI(CO)(PMe$_3$)$_2$(CNC$_6$H$_3$Me$_2$-2,6)(η^2-PhC$_2$S)]	299
[Mo$_2$(μ-Cl){μ-Ph$_2$PC(H)=CH}(CO)$_2$Cp$_2$]	302
[Mo$_2$Cl$_2$(μ-PPh$_2$)(μ-Ph$_2$PC=CHMe)Cp$_2$]	302

complexes $[WI_2(CO)(CNBu^t)(\eta^2\text{-}RC_2R)_2]$, $[WI_2(CO)(CNBu^t)_2(\eta^2\text{-}RC_2R)]$, $[WI_2(CNBu^t)_3(\eta^2\text{-}RC_2R)]$, $[WI(CNBu^t)_4(\eta^2\text{-}RC_2R)]I$, and $[W(CNBu^t)_5$ $(\eta^2\text{-}RC_2R)]2I$, respectively.[249]

b. *Nitrogen donor ligands.* The nitrile exchange reactions of $[WI_2$ $(CO)(NCMe)(\eta^2\text{-}RC_2R)_2]$ with $NCPh^{[250]}$ or thiophene-3-acetonitrile[251] to give $[WI_2(CO)(NCR)(\eta^2\text{-}RC_2R)_2]$ have been described. The molecular structure of the thiophene-3-acetonitrile complex $[WI_2(CO)\{NCCH_2$ $(3\text{-}C_4H_3S)\}(\eta^2\text{-}MeC_2Me)_2]$ has been determined crystallographically.[251] Treatment of $[WI_2(CO)(NCMe)(\eta^2\text{-}PhC_2Ph)_2]$ in refluxing acetonitrile gives $[WI_2(NCMe)_2(\eta^2\text{-}PhC_2Ph)_2]$.[252] The preparation and characterization of the imidazole or pyrazole complexes $[WI_2(CO)L(\eta^2\text{-}RC_2R)_2]$ (L = imidazole, pyrazole) have been reported.[253] The reactions of $[WI_2(CO)$ $(NCMe)(\eta^2\text{-}RC_2R)_2]$ with a series of pyridine and bidentate nitrogen donor

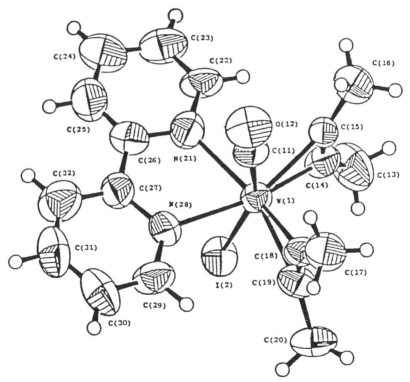

FIG. 15. The molecular structure of the cation $[WI(CO)(bipy)(\eta^2\text{-}MeC_2Me)_2]^+$. [Reprinted with permission from *Inorg. Chem.* **1988**, *27*, 2291. Copyright 1988 American Chemical Society.]

ligands have been described; they give a range of products including the cationic complex [WI(CO)(bipy)(η^2-MeC$_2$Me)$_2$][BPh$_4$], which has been characterized crystallographically and is shown in Fig. 15.[254] Similarly, reaction of [WI$_2$(CO)(NCMe)(η^2-RC$_2$R)$_2$] with 1,4-diaza-1,3-butadienes, R'N=CHCH=N'R (R' = Cy, ^{i}Pr, Ph, p-MePh), gives the cationic complexes [WI(CO)(R'N=CHCH=NR')(η^2-RC$_2$R)$_2$]I.[255] Finally, reaction of 2,3-bis(2-pyridyl)pyrazine (bpp) with two equivalents of [WI$_2$(CO)(NCMe) (η^2-RC$_2$R)$_2$] furnishes the dicationic bpp-bridged complexes [W$_2$(μ-bpp)I$_2$ (CO)$_2$(η^2-RC$_2$R)$_4$]2I.[256]

   c. *Phosphorus donor ligands.* Treatment of [WI$_2$(CO)(NCMe) (η^2-RC$_2$R)$_2$] with one equivalent of either L$_2$ {L$_2$ = Ph$_2$P(CH$_2$)$_n$PPh$_2$ (n = 1, 2, 3, 4, or 6) or [Fe(η^5-C$_5$H$_4$PPh$_2$)$_2$]} (for R = Me only), or with two equivalents of L {L = PMe$_3$, PEt$_3$, PBu$_3^n$, PMe$_2$Ph, PMePh$_2$, PEt$_2$Ph, PEtPh$_2$, PPh$_2$(CH$_2$CH=CH$_2$), ⟨PPh$_3$, PPh$_2$Cy, for R = Me only⟩} gives the highly colored mono(alkyne) complexes [WI$_2$(CO)L$_2$(η^2-RC$_2$R)].[257] The molecular structure of [WI$_2$(CO)(dppm)(η^2-MeC$_2$Me)] has been determined crystallographically and is shown in Fig. 16. The barrier to 2-butyne rotation of a series of these complexes has been determined and the results discussed in terms of the steric and electronic effects of the various ligands,

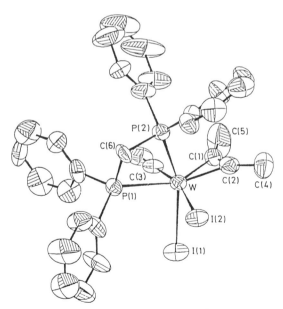

FIG. 16.   The molecular structure of [WI$_2$(CO)(dppm)(η^2-MeC$_2$Me)]. [Reproduced with permission from *J. Chem. Soc., Dalton Trans.* **1989,** 297.]

and a graph of the barrier to 2-butyne rotation ($\Delta G\ddagger$) and the cone angle[258] of a series of phosphines is shown in Graph 1.

A series of bis(phosphite) complexes of the type [WI$_2$(CO){P(OR')$_3$}$_2$ (η^2-RC$_2$R)] (R = Me, Ph; R' = Me, Et, ^{i}Pr, ^{n}Bu) have been prepared and characterized crystallographically for R = R' = Me.[259] Another series of bis(phosphine) complexes, [WI$_2$(CO)L$_2$(η^2-MeC$_2$Me)] {L = PPh$_2$Np (Np = napthyl), trans-Ph$_2$PCH=CHPPh$_2$; L$_2$ = cis-Ph$_2$PCH=CHPPh$_2$, $R(+)$-Ph$_2$P{CHMeCH$_2$}PPh$_2$, Me$_2$P(CH$_2$)$_2$PMe$_2$}, also has been prepared and characterized.[260]

Treatment of [WI$_2$(CO)(NCMe)(η^2-MeC$_2$Me)$_2$] with one equivalent of triphos {triphos = PhP(CH$_2$CH$_2$PPh$_2$)$_2$} gives the new organometallic phosphine ligand, [WI$_2$(CO)(η^2-triphos)(η^2-MeC$_2$Me)], which has been characterized crystallographically; two crystallographically independent molecules are present, which are diastereoisomers as shown in Fig. 17a and b. The chemistry of this and related complexes as a phosphine ligand is being explored by Baker et al.[261]

d. *Oxygen and sulfur donor ligands.* The reactions of [WI$_2$(CO) (NCMe)(η^2-RC$_2$R)$_2$] with urea, thiourea, N,N,N',N'-tetramethylthiourea,

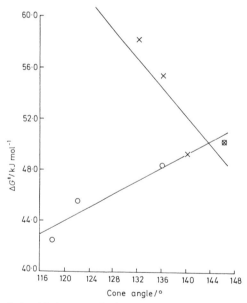

GRAPH 1. The relationship between the barrier to 2-butyne rotation ($\Delta G\ddagger$) and the cone angle of the phosphine ligand in the series of complexes [WI$_2$(CO)L$_2$(η^2-MeC$_2$Me)] {L = PMe$_3$, PMe$_2$Ph, PMePh$_2$, PPh$_3$(O); PEt$_3$, PEt$_2$Ph, PEtPh$_2$, PPh$_3$(X)}. [Reproduced with permission from *J. Chem. Soc., Dalton Trans.* **1989**, 299.]

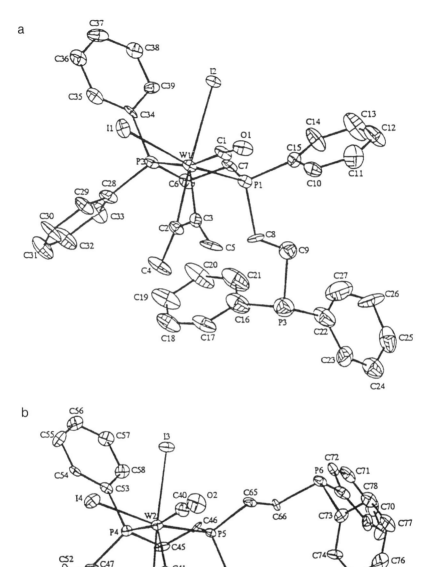

FIG. 17. The molecular structure of [WI$_2$(CO)(η^2-triphos)(η^2-MeC$_2$Me)] showing the two independent molecules that are diastereoisomers. [Reproduced with permission from *J. Organomet. Chem.* **1995,** *503,* C9.]

thioacetamide,[262] and $OPPh_3$ and $SPPh_3$[263] donor ligands have been reported.

Treatment of $[WI_2(CO)(NCMe)(\eta^2\text{-}MeC_2Me)_2]$ with Ph(H)CNN HCONH$_2$ or R(R')CNNHCSNH$_2$ (R = R' = Me, Et; R = H, R' = Ph; R = Me, R' = Pr^n, Bu^t, Ph) gives the cationic complexes $[WI(CO)\{Ph(H)CNNHCONH_2\}(\eta^2\text{-}MeC_2Me)_2]I$ or $[WI(CO)\{R(R')CNNHCSNH_2\}(\eta^2\text{-}MeC_2Me)_2]I$, respectively.[264]

4. Reactions of $[WI_2(CO)(NCMe)(\eta^2\text{-}RC_2R)_2]$ (R = Me, Ph) with Anionic Donor Ligands

In 1988, Armstrong and Baker[265] described the reactions of $[WI_2(CO)(NCMe)(\eta^2\text{-}MeC_2Me)_2]$ with one and two equivalents of Na[acac] to give the bidentately attached acac complexes $[WI(CO)(acac)(\eta^2\text{-}MeC_2Me)_2]$ and $[W(CO)(acac)_2(\eta^2\text{-}MeC_2Me)]$, respectively. Treatment of $[WI_2(CO)(NCMe)(\eta^2\text{-}MeC_2Me)_2]$ with $K[SC_5H_4N]$ (SC$_5$H$_4$N = pyridine-2-thionate) affords the N^S coordinated complexes $[WI(CO)(\eta^2\text{-}SC_5H_4N)(\eta^2\text{-}MeC_2Me)_2]$, which reacts further with Na[BPh$_4$] in acetonitrile to give the crystallographically characterized complex $[W(CO)(NCMe)(SC_5H_4N)(\eta^2\text{-}MeC_2Me)_2][BPh_4]$.[266] Similarly, the preparation, fluxional properties, and X-ray crystal structure of $[WI(CO)(SC_4H_3N_2)(\eta^2\text{-}MeC_2Me)_2]$ (SC$_4$H$_3$N$_2$ = pyrimidine-2-thionate) have been reported.[267]

The reactions of $[WI_2(CO)(NCMe)(\eta^2\text{-}RC_2R)_2]$ with a series of dithiocarbamates and related ligands have been described.[268] The molecular structure of one of the complexes, $[WI(CO)(S_2CNC_4H_8)(\eta^2\text{-}MeC_2Me)_2]$, has been determined crystallographically (Fig. 18), and variable-temperature ^1H NMR studies also show the complex to be fluxional in solution.[268]

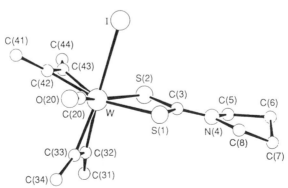

FIG. 18.   The molecular structure of $[WI(CO)(S_2CNC_4H_8)(\eta^2\text{-}MeC_2Me)_2]$. [Reproduced with permission from *J. Chem. Soc., Dalton Trans.* **1990**, 2538.]

5. *Reactions of [WI$_2$(CO)L$_2$(η^2-RC$_2$R)] (L = phosphines; R = Me, Ph) with Ag[BF$_4$]*

Reaction of [WI$_2$(CO)(dppm)(η^2-MeC$_2$Me)] with an equimolar amount of Ag[BF$_4$] in CH$_2$Cl$_2$ at room temperature gives the iodo-bridged dimer [{W(μ-I)(CO)(dppm)(η^2-MeC$_2$Me)}$_2$][BF$_4$]$_2$, which reacts with two equivalents of L {L = CNBu^t, P(O^{i}Pr)$_3$, SC(NH$_2$)$_2$} to afford the iodo-bridge cleaved products [WI(CO)L(dppm)(η^2-MeC$_2$Me)][BF$_4$].[269] In contrast, reaction of [WI$_2$(CO)L$_2$(η^2-RC$_2$R)] (L = PEt$_3$, PMe$_2$Ph; R = Me, Ph) with an equimolar amount of Ag[BF$_4$] in acetonitrile gives the cationic complexes [WI(CO)(NCMe)L$_2$(η^2-RC$_2$R)][BF$_4$].[270] Similarly, the reaction of [WI$_2$(CO){Ph$_2$P(CH$_2$)$_n$PPh$_2$}(η^2-RC$_2$R)] (R = Me, Ph; n = 1–6) with an equimolar amount of Ag[BF$_4$] in NCMe yields [WI(CO)(NCMe){Ph$_2$P(CH$_2$)$_n$PPh$_2$}(η^2-RC$_2$R)][BF$_4$].[271] The dppm complex [WI(CO)(NCMe)(dppm)(η^2-MeC$_2$Me)][BF$_4$] reacts with one equivalent of P(OR)$_3$ (R = Me, Et, Pr^i, Bu^n) to give [WI(CO){P(OR)$_3$}(dppm)(η^2-MeC$_2$Me)][BF$_4$], which has been characterized crystallographically for R = Pr^i. The cation is shown in Fig. 19.[271] An extensive chemistry of [WI(CO)(NCMe)(dppm)(η^2-MeC$_2$Me)][BF$_4$] has been developed and is shown in Scheme 7.[272–281]

6. *Reactions of [MI$_2$(CO)$_3$(NCMe)$_2$] with 1,8-Cyclotetradecadiyne and 1,4-Diphenylbutadiyne*

Reaction of [MI$_2$(CO)$_3$(NCMe)$_2$] with one equivalent of 1,8-cyclotetradecadiyne (1,8-CTDiyne) gives [MI$_2$(CO)(NCMe)(η^2,$\eta^{2'}$-1,8-CTDiyne)]. The reaction chemistry of these complexes with both neutral and anionic

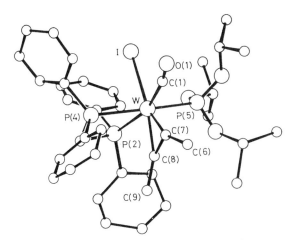

FIG. 19. The molecular structure of the cation [WI(CO){P(OPr^i)$_3$}(dppm)(η^2-MeC$_2$Me)]$^+$. [Reproduced with permission from *J. Chem. Soc., Dalton Trans.* **1989**, 1906.]

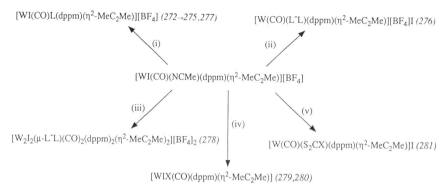

[WI(CO)L(dppm)(η^2-MeC$_2$Me)][BF$_4$] *(272→275,277)*   [W(CO)(L^L)(dppm)(η^2-MeC$_2$Me)][BF$_4$]I *(276)*

(i)   (ii)

[WI(CO)(NCMe)(dppm)(η^2-MeC$_2$Me)][BF$_4$]

(iii)   (iv)   (v)

[W$_2$I$_2$(μ-L^L)(CO)$_2$(dppm)$_2$(η^2-MeC$_2$Me)$_2$][BF$_4$]$_2$ *(278)*   [W(CO)(S$_2$CX)(dppm)(η^2-MeC$_2$Me)]I *(281)*

[WIX(CO)(dppm)(η^2-MeC$_2$Me)] *(279,280)*

SCHEME 7.   (i) L = CO, CNBu^t.[272] L = imidazole, 2Me-imidazole, 4Me-imidazole.[273] L = series of monodentate phosphines.[274] L = SC(NH$_2$)$_2$, OC(NH$_2$)$_2$, etc.[275] L^L' = semicarbazones and thiosemicarbazones.[277] (ii) L^L = bipy, phen, 5,6-Me$_2$-phen, RN=CH—CH =NR (R = Cy, Bu^t, C$_6$H$_4$OMe-p).[276] (iii) L^L = Ph$_2$P(CH$_2$)$_n$PPh$_2$ (n = 1–6) or [Fe(η^5-C$_5$H$_4$PPh$_2$)$_2$].[278] (iv) X$^-$; X = Cl, Br, I, NO$_2$, NO$_3$, NCS, OH.[279] SR$^-$; X = SR (R = Et, Bu^t, Ph, CH$_2$Ph).[280] (v) S$_2$CS$^-$; X = NMe$_2$, NEt$_2$, NC$_4$H$_8$, N(CH$_2$Ph)$_2$, OEt.[281]

ligands also is described.[282] The 1,4-diphenylbutadiene complexes [{W(μ-I)I(CO)(NCMe)(η^2-PhC$_2$C$_2$Ph)}$_2$] and [WI$_2$(CO)(NCMe)(η^2-PhC$_2$C$_2$Ph)$_2$] have been prepared by reaction of [WI$_2$(CO)$_3$(NCMe)$_2$] with one and two equivalents of 1,4-diphenylbutadiyne, respectively.[283] The reactions of these complexes with donor ligands also is discussed. Equimolar quantities of [WI$_2$(CO)$_3$(NCMe)$_2$] and [WI$_2$(CO)(NCMe)(η^2-PhC$_2$C$_2$Ph)$_2$] react to give the bimetallic 1,4-diphenylbutadiyne-bridged complex **3**.[283]

3

7. *Reactions of [MXX'(CO)$_3$(NCMe)$_2$] with Alkynes*

Thus far, the reactions of the diiodo-complexes [MI$_2$(CO)$_3$(NCMe)$_2$] with alkynes have been described. In this section the reactions of other seven-coordinate molybdenum(II) and tungsten(II) halocarbonyls with alkynes are described.

In 1994, Baker *et al.*[284] described the reactions of [WBr$_2$(CO)$_3$(NCMe)$_2$] with excess RC$_2$R (R = Me, Ph, CH$_2$Cl) to give the bis(alkyne) complexes [WBr$_2$(CO)(NCMe)(η^2-RC$_2$R)$_2$]. The reaction of the bis(2-butyne) complex [WBr$_2$(CO)(NCMe)(η^2-MeC$_2$Me)$_2$] with neutral donor ligands also was discussed.[284] The mixed halide complex [WBrI(CO)$_3$(NCMe)$_2$], synthesized by reacting *fac*-[W(CO)$_3$(NCMe)$_3$] (prepared *in situ*) with one equivalent of IBr, reacts with one or two equivalents of RC$_2$R (R = Me, Ph, CH$_2$Cl) to give [{W(μ-I)Br(CO)(NCMe)(η^2-RC$_2$R)}$_2$] or [WBrI(CO)(NCMe)(η^2-RC$_2$R)$_2$], respectively.[285]

Reaction of [MoCl(GeCl$_3$)(CO)$_2$(NCMe)$_2$(PPh$_3$)] with RC$_2$R (R = Me, Ph) gives the alkyne complexes [MoCl(GeCl$_3$)(CO)(NCMe)(PPh$_3$)(η^2-RC$_2$R)], which, unusually, have six different monodentate ligands attached to the metal.[286,287] The reaction of the 2-butyne complex with both neutral monodentate and bidentate donor ligand also has been discussed.[287] Treatment of [MoCl(GeCl$_3$)(CO)(NCMe)(PPh$_3$)(η^2-MeC$_2$Me)] with [NBu$_4^n$]X (X = Br, I, I$_3$, SCN) affords the anionic complexes [NBu$_4^n$][MoCl(GeCl$_3$)X(CO)(PPh$_3$)(η^2-MeC$_2$Me)], which for the case of X = SCN rearranges in acetone with a trace of water to the highly novel M$_{ABCDEF}$ dimeric complex [Cl(OC)(Ph$_3$P)(η^2-MeC$_2$Me)Mo(μ-SCN)(μ-NCS)MoCl(CO)(PPh$_3$)(η^2-MeC$_2$Me)], which has unsymmetrically bridged-thiocyanate ligands.[288] The structure of this complex is shown in Fig. 20.

B. *Other Halocarbonyl Alkyne Complexes of Molybdenum(II) and Tungsten(II) Described Since 1987*

In 1990, Brisdon *et al.*[289] described the synthesis of a series of haloalkyne complexes of the type [MX$_2$(CO)L$_2$(η^2-RC$_2$R)] (M = Mo, X = Br, R = Me, L = PEt$_2$Ph, PEtPh$_2$, PMe$_2$Ph, PMePh$_2$; M = Mo, X = Cl, I, L = PMePh$_2$; M = Mo or W, X = Br, R = Ph, L = PMePh$_2$; M = W, R = Me, X = Br, L = PMePh$_2$) by reaction of [MX$_2$(CO)$_2$(NCMe)L$_2$] with the appropriate alkyne. The molecular structure of [MoBr$_2$(CO)(PMePh$_2$)$_2$(η^2-MeC$_2$Me)] has been determined crystallographically.[289]

Davidson *et al.*[290] have described the reaction of [{W(μ-Br)Br(CO)$_4$}$_2$] with CF$_3$C$_2$CF$_3$ to yield [WBr$_2$(CO)(η^2-CF$_3$C$_3$CF$_3$)$_2$], which reacts further with P(OR)$_3$ (R = Me, Et) to afford the dimeric complexes **4** via a Michaelis–Arbuzov elimination of alkyl bromide. The complex for R =

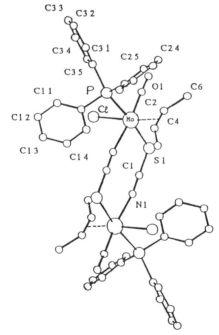

4

Et has been characterized crystallographically. Treatment of $[\{W(\mu\text{-Br})Br(CO)(\eta^2\text{-MeC}_2Me)_2\}_2]$ with an equimolar amount of $Na[S_2CNMe_2]$, $Na[S_2PMe_2]$, or $Tl[acac]$ gives the monobromo complexes $[WBr(CO)(\eta^2\text{-}S_2CNMe_2, S_2PMe_2, \text{ or } acac)(\eta^2\text{-MeC}_2Me)_2]$.[291]

FIG. 20. The molecular structure of $[Cl(OC)(Ph_3P)(\eta^2\text{-MeC}_2Me)Mo(\mu\text{-SCN})(\mu\text{-NCS})\text{-}MoCl(CO)(PPh_3)(\eta^2\text{-MeC}_2Me)]$. [Reprinted with permission from *Inorg. Chem.* **1993**, *32*, 3395. Copyright 1993 American Chemical Society.]

Nielson et al.[292] have described the reduction of $[WCl_3L_2(\eta^2\text{-PhC}_2Ph)]$ (L = PMe_3, PMe_2Ph) under CO to afford the dichloro-diphenylacetylene complexes $[WCl_2(CO)L_2(\eta^2\text{-PhC}_2Ph)]$, which have both been characterized crystallographically.

A series of {hydrotris(3,5-dimethylpyrazolyl)borato=Tp'} complexes of the type $[WI(CO)(\eta^2\text{-RC}_2R')Tp']$ (R = R' = Me; R = Ph, R' = H, Me) have been prepared and characterized, including the complex in which R = Ph, R' = Me, by X-ray crystallography.[293] Templeton et al.[294] have also described the deprotonation of $[WI(CO)(\eta^2\text{-PhC}_2Me)Tp']$ to give a nucleophilic propargl synthon, which reacts with MeI or $PhCH_2Br$ to afford an alkyne complex. Deprotonation of $[WI(CO)(\eta^2\text{-PhC}_2CH_2Bz)Tp']$ followed by methylation gives a single diastereoisomer, whereas deprotonation of $[WI(CO)(\eta^2\text{-PhC}_2CH_2Me)Tp']$ followed by benzylation yields the opposite diastereoisomer. Benzaldehyde or pivaldehyde reacts with $Li[WI(CO)(\eta^2\text{-PhC}=C=CHMe)Tp']$ to give alcohols.

Mayr and Bastos[295] have described the reaction of $[W(\equiv CPh)Cl(CO)(CNBu^t)(PMe_3)_2]$ with three equivalents of dry HCl to yield two crystallographically characterized products, $[WCl_3(CO)(PMe_3)(\eta^2\text{-PhCNBu}^t)]$ (major product) and the amino alkyne complex $[WCl_2(CO)(PMe_3)_2(\eta^2\text{-PhC}_2CNHBu^t)]$, which is shown in Fig. 21. Similarly, the photoinduced alkylidyne–carbonyl coupling by electrophiles of tungsten alkylidyne complexes is shown in Eq. (4).[296]

$$ (4) $$

R = Me, E = H (crystallographically characterized),
$C(O)Bu^t$, $C(O)C_6H_4OMe$-4, $Si(Bu^t)_2Ph$; R = Ph, E = $C(O)Bu^t$

Treatment of $[W(\equiv CR)Br(CO)_3L]$ (L = CO, R = C_6H_4OMe-4, $C_6H_4NMe_2$-4; L = PPh_3, R = Ph, $C_6H_4NMe_2$-4) with methyllithium at −78°C yields the alkylidyne–acyl complexes $Li[W(\equiv CR)Br(COMe)(CO)_3L]$, which react further with $C_2O_2Br_2$ followed by PPh_3 to afford the alkyne complexes $[WBr_2(CO)(PPh_3)_2(\eta^2\text{-MeC}_2R)]$ via the coupling of two alkylidyne groups.[297] The synthesis of alkynyltrithiocarbonato ligands from reaction of alkylidyne complexes and CS_2 as shown in Eq. (5) has been described.[298] Reaction of $[W(=CHPh)I_2(CO)(PMe_3)_2]$ with CS_2 and PPh_3 gives the crystallographically characterized η^2-alkynethiol complex $[WI_2(CO)(PMe_3)_2(\eta^2\text{-PhC}_2SH)]$, which reacts further with $Li[Bu^n]$

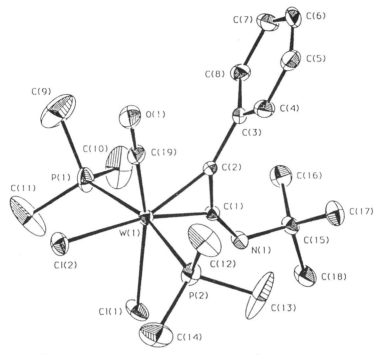

FIG. 21.   The molecular structure of [WCl$_2$(CO)(PMe$_3$)$_2$(η^2-PhC$_2$CNHBu^t)]. [Reproduced with permission from *J. Am. Chem. Soc.* **1990**, *112*, 7798. Copyright 1990 American Chemical Society.]

and isocyanides to yield the thioketenyl complexes [WI(CO)(CNR) (PMe$_3$)$_2$(η^2-PhC$_2$S)] (R = Bu^t, 2,6-C$_6$H$_3$Me$_2$).[299] Filippou[300] has described the reaction of [W($\equiv$CNEt$_2$)I(CO)$_2$(CNBu^t)$_2$] with an equimolar amount of HI to give the alkyne complexes [WI$_2$(CO)$_2$(CNBu^t){η^2-(Bu^t)HNC$_2$NEt$_2$}] (major product) and [WI$_2$(CO)(CNBu^t)$_2${η^2-(Bu^t)HNC$_2$NEt$_2$}] (minor product).

$$\underset{\text{Me}_3\text{P}}{\overset{\overset{\displaystyle\text{O}}{\overset{\displaystyle\text{C}}{|}}}{\text{X}-\text{W}\equiv\text{CPh}}} \quad \xrightarrow[\text{thf}]{2\text{CS}_2/-2\text{SPMe}_3} \quad \text{(5)}$$

X = Cl, Br, I, SBu^t, SCy

In 1993, Green *et al.*[301] described the reactions of [Mo(CO)(η^2-R'C$_2$R'')$_2$(η^5-Cp or C$_9$H$_7$)][BF$_4$] (R' = R'' = Me; R' = Me, R'' = Ph; R' = Bu^t, R'' = H) with either MgRX or MgX$_2$ to give [MoX(CO)(η^2-R'C$_2$R'')(η^5-Cp or C$_9$H$_7$)] {X = Cl, Br, I; R = R' = Me; (Cp). X = Br, R' = R'' = Me; (C$_9$H$_7$). X = Br, R' = Me, R'' = Ph; R' = Bu^t, R'' = H; (Cp)}. The complex [MoBr(CO)(η^2-MeC$_2$Me)Cp] reacts with isoprene and Ag[BF$_4$] to yield the crystallographically characterized complex [Mo{η^3,η^3-CH(Me)C(Me)CHCHC(Me)CH$_2$}(CO)Cp][BF$_4$] via a novel carbon–carbon coupling reaction. Mays *et al.*[302] have described the reaction of [Mo$_2$(CO)$_4$(μ-R'C$_2$R'')Cp$_2$] (R',R'' = H, alkyl, aryl) with PPh$_2$Cl to give several complexes including [Mo$_2$(μ-Cl){μ-Ph$_2$PC(H)=CH}(CO)$_2$Cp$_2$] and [Mo$_2$Cl$_2$(μ-PPh$_2$)(μ-Ph$_2$PC=CHMe)Cp$_2$], which have been characterized crystallographically.

V

π-ALLYL HALOCARBONYL COMPLEXES OF MOLYBDENUM(II) AND TUNGSTEN(II)

Most of the allyl halocarbonyl complexes described in this section are derived from complexes of the type [MX(CO)$_2$L$_2$(η^3-allyl)], and the synthesis, reactions, structures, and applications of these complexes are discussed.

A. Synthesis of π-Allyl Halocarbonyl Complexes of Molybdenum(II) and Tungsten(II)

1. Reactions of the Anions [MX(CO)$_3$L$_2$]$^-$ with Allyl Halides and Haloalkynes

In 1965, Murdoch[303] described the synthesis of a series of halo-bridged allyl carbonyl complexes of the type [NEt$_4$][M$_2$(μ-X)$_3$(CO)$_4$(η^3-C$_3$H$_5$)$_2$] (5), which are prepared by reaction of [NEt$_4$][MX(CO)$_5$] with C$_3$H$_5$X. A year later, Murdoch and Henzi[304] described the bridge-splitting reactions of (5) (M = Mo, X = Cl) with nitrogen donors such as pyridine or bipy to yield complexes of the type 6.

More recently, in 1978 Brisdon *et al.*[305] described the reactions of *fac*-[MX(CO)$_3$(L$^\smallfrown$L)]$^-$ (M = Mo, W; X = Cl, Br, I; L$^\smallfrown$L = phen, bipy) with various allyl halides to give the neutral complexes [MX(CO)$_2$(L$^\smallfrown$L)(η^3-allyl)]. Similarly, a series of complexes of the type [MX(CO)$_2$L$_2$(η^3-allyl)] {M = Mo, W; X = Cl, Br, I, NCS, MeCO$_2$, CF$_3$CO$_2$, PhSO$_2$, *p*-MeC$_6$H$_4$SO$_2$; L$_2$ = bipy, di(2-pyridyl)amine} have been prepared by four

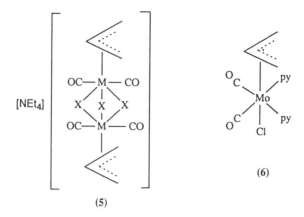

(5)

(6)

different routes, including the reaction of the anions $[MX(CO)_3L_2]^-$ with allyl halides. The reaction of the complexes $[MX(CO)_2L_2(\eta^3\text{-allyl})]$ with chelating anions also has been described.[306] In 1977, Doyle[307] described the reactions of $[M(CO)_4(\text{diket})]^-$ (M = Mo, W; diket = acac, 3Me-acac, etc.) with allyl chloride to give the anionic π-allyl complexes $[MCl(CO)_2(\text{diket})(\eta^3\text{-}C_3H_5)]^-$.

In 1986, Brisdon et al.[308] reported the reaction of $[PPh_4][MoX(CO)_3(\text{bipy})]$ with prop-2-ynyl chloride or bromide to afford the neutral complexes $[MoX(CO)_2(\text{bipy})(\eta^3\text{-}C_5H_7O_2)]$, which probably have an η^3-oxacyclobutenyl moiety attached to the molybdenum. Brisdon et al.[309] also have described the reaction of $[PPh_4][MoCl(CO)_3(\text{bipy})]$ with $ClCH_2C_2CH_2Cl$ in wet methanol to give the crystallographically characterized complex $[MoCl(CO)_2(\text{bipy})(\eta^3\text{-}CH_2\text{-}C\text{-}(CO_2Me)=C=CH_2)]$, which has an η^3-bonded trans-butadienyl ligand. Reaction of $[PPh_4][MoCl(CO)_3(\text{bipy})]$ with $ClCH_2C_2CH_2Cl$ under anhydrous conditions yields the crystallographically characterized complex $[MoCl(CO)_2(\text{bipy})\{\eta^3\text{-}CH_2\text{-}C(CO_2Me)\text{-}C(Me)(OMe)\}]$ (Table III). This work and other related papers[310–312] on $\eta^3(3e)$-butadienyl complexes of molybdenum(II) and tungsten(II) are included in a review article by Brisdon and Walton.[8]

2. Reactions of Zero-Valent Substituted Metal Carbonyls of Molybdenum and Tungsten with Allyl Halides

Treatment of $[W(CO)_6]$ with allyl halides under UV radiation gives the π-allyl complexes $[W_2Cl_3(CO)_6(\eta^3\text{-allyl})]$, $[WBr(CO)_4(\eta^3\text{-allyl})]$, and $[WI(CO)_4(\eta^3\text{-allyl})]$, respectively.[313] A large series of π-allyl complexes of the type $[MX(CO)_2L_2(\eta^3\text{-allyl})]$ have been prepared by reaction of the zerovalent complexes $[M(CO)_4L_2]$ with allyl halides.[314]

In 1968, Hayter[315] described the oxidation of fac-$[M(CO)_3(NCMe)_3]$ with

TABLE III

X-Ray Crystal Structure Determinations of π-Allyl Halocarbonyl
Complexes of Molybdenum(II) and Tungsten(II)

Complex	Ref.
$[\text{MoCl(CO)}_2(\text{bipy})(\eta^3\text{-CH}_2\text{---C---(CO}_2\text{Me)---C=CH}_2)]$	309
$[\text{MoCl(CO)}_2(\text{bipy})\{(\eta^3\text{-CH}_2\text{---C(CO}_2\text{Me)---C(Me)(OMe)}\}]$	309
$[\text{MoBr(CO)}_2(\text{PhHCPz}'_2)(\eta^3\text{-C}_3\text{H}_5)]$	318
$[\text{MoBr(CO)}_2(\text{bipy})(\eta^3\text{-C}_3\text{Ph}_3)]$	320
$[\text{MoBr(CO)}_2(\text{bipy})(\eta^3\text{-C}_4\text{Ph}_3\text{O})]$	320
$[\text{MoBr(CO)}_2(\text{NCMe})_2(\eta^3\text{-C}_5\text{H}_7)]$	322
$[\text{MoCl(CO)}_2(\text{dppe})(\eta^3\text{-C}_5\text{H}_7)]$	322
$[\text{MoBr(CO)}_2(\text{dppe})(\eta^3\text{-C}_5\text{H}_7)]$	322
$[\text{WBr(CO)}_2(\text{dmpe})(syn\text{-}\eta^3\text{-pentadienyl})]$	323
$[\text{MoBr(CO)}_2(\text{NCMe})_2(\eta^3\text{-C}_5\text{H}_6\text{Me})]$	324
$[\text{MoBr(CO)}_2\{\eta^2\text{-HN=CMe(pz)}\}(\eta^3\text{-C}_3\text{H}_5)]$	329
$[(\eta^3\text{-C}_3\text{H}_5)(\text{CO})_2\text{Mo}(\mu\text{-Br})(\mu\text{-S}_2\text{CPCy}_3)\text{Mo(CO)}_3]$	343
$[\{\text{Mo(CO)}_2(\text{bipy})(\eta^3\text{-C}_3\text{H}_5)\}_2(\mu\text{-Cl})]^+$	344
$[\text{Mo(NCS)(CO)}_2(\text{bipy})(\eta^3\text{-C}_3\text{H}_5)]$	345
$[\text{Mo(NCS)(CO)}_2(\text{phen})(\eta^3\text{-C}_3\text{H}_4\text{Me-2})]$	346
$[\text{MoCl(CO)}_2(\text{CyN=CHCH=NCy})(\eta^3\text{-C}_3\text{H}_4\text{Me-2})]$	347
$[\text{MoCl(CO)}_2(\text{CyN=CHCH=NCy})(\eta^3\text{-C}_3\text{H}_5)]$	348
$[\text{MoCl(CO)}_2\{\text{P(OMe)}_3\}_2(\eta^3\text{-C}_3\text{H}_5)]$	349
$[\text{Mo(CO)}_2(\text{NCMe})_3(\eta^3\text{-C}_3\text{H}_5)][\text{Mo}_2\text{Cl}_3(\text{CO})_4(\eta^3\text{-C}_3\text{H}_5)_2]$	350
$[\text{MoBr(CO)}_2(\text{Bu}^t\text{N=CHCH=NBu}^t)(\eta^3\text{-C}_3\text{H}_5)]$	351
$[\text{MoBr(CO)}_2(\text{CyN=CHCH=NCy})(\eta^3\text{-C}_3\text{H}_5)]$	352
$[\text{MoBr(CO)}_2(\text{pyrazole})_2(\eta^3\text{-C}_3\text{H}_4\text{Ph-1})]$	353
$[\text{MoBr(CO)}_2(\text{NCMe})_2(\eta^3\text{-C}_3\text{H}_5)]$	354

allyl halides to give the labile complexes $[\text{MX(CO)}_2(\text{NCMe})_2(\eta^3\text{-allyl})]$ (M = Mo, W; X = Cl, Br; allyl = C_3H_5, $2\text{MeC}_3\text{H}_4$, $\text{C}_3\text{H}_4\text{Cl}$). In the same year, Tom Dieck and Friedel[316] described an extensive series of molybdenum(II) complexes of the type $[\text{MoX(CO)}_2(\text{NCMe})_2(\eta^3\text{-allyl})]$. More recently, Baker[317] described the reaction of fac-$[\text{Mo(CO)}_3(\text{NCMe})_3]$ with $\text{CH}_2=\text{C(CH}_2\text{Cl)}_2$ to give $[\text{MoCl(CO)}_2(\text{NCMe})_2\{\eta^3\text{-C}_3\text{H}_4(2\text{-CH}_2\text{Cl})\}]$.

In 1991, Shiu et al.[318] and Sarkar et al.[319] reported the reactions of zero-valent bis(3,5-dimethylpyrazolyl)methane type complexes $[\text{Mo}(\text{CO})_4(\text{N}^\wedge\text{N})]$ with allyl halides to give a series of π-allyl complexes $[\text{MoX}(\text{CO})_2(\text{N}^\wedge\text{N})(\eta^3\text{-allyl})]$ ($\text{N}^\wedge\text{N}$ = H_2CPz_2, $\text{H}_2\text{CPz}'_2$, PhHCPz_2, PhHCPz'_2). These complexes also can be made by reaction of $[\text{MoBr(CO)}_2(\text{NCMe})_2(\eta^3\text{-allyl})]$ with $\text{N}^\wedge\text{N}$.[318]

Treatment of $[\text{Mo(CO)}_4\text{L}_2]$ (L_2 = bipy, phen, 2,2'-dipyridylamine) with triphenylcyclopropenyl bromide yields two types of complexes, namely $[\text{MoBr(CO)}_2\text{L}_2(\eta^3\text{-C}_3\text{Ph}_3)]$ and $[\text{MoBr(CO)}_2\text{L}_2(\eta^3\text{-C}_4\text{Ph}_3\text{O})]$. The molecular structures of the η^3-cyclopropenyl complex [MoBr(CO)$_2$(bipy)(η^3-C$_3$Ph$_3$)]·NCMe and the η^3-oxocyclobutenyl complex [MoBr(CO)$_2$(bipy)(η^3-C$_4$Ph$_3$O)]·thf have been determined crystallographically (Table III).[320] The reaction of [Mo(CO)$_4$(phen)] with allyl bromide to give [MoBr(CO)$_2$(phen)(η^3-C$_3$H$_5$)] has been shown to obey the rate law $-d$[Mo(CO)$_4$(phen)]/dt = κ_1[Mo(CO)$_4$(phen)]. The mechanism of the reaction probably proceeds via slow solvent-assisted fission of a Mo—CO bond, followed by reaction with allyl bromide.[321]

Oxidative addition of fac-[Mo(CO)$_3$(NCMe)$_3$] with 1-halopenta-2,4-diene affords [MoX(CO)$_2$(NCMe)$_2$(η^3-C$_5$H$_7$)] {X = Cl, Br (crystallographically characterized)}, which reacts with both dmpe and dppe to yield [MoCl(CO)$_2$(dmpe)(η^3-C$_5$H$_7$)] and [MoBr(CO)$_2$(dppe)(η^3-C$_5$H$_7$)] (both crystallographically characterized; see Table III), respectively.[322] A similar series of tungsten complexes [WBr(CO)$_2$(NCMe)$_2$(η^3-C$_5$H$_7$)] and their dmpe and dppe derivatives also have been reported, together with the molecular structure of [WBr(CO)$_2$(dmpe)(syn-η^3-pentadienyl)]. The $\eta^3 \rightleftharpoons \eta^5$ equilibrium for the pentadienyl ligand also has been investigated by ^{31}P-NMR spectroscopy.[323] Related work[324] on the oxidative addition of fac-[Mo(CO)$_3$(NCMe)$_3$] with 1-halohexa-2,4-dienes gives [MoX(CO)$_2$(NCMe)$_2$(η^3-C$_5$H$_6$Me)] {X = Cl, Br (crystallographically characterized)}, which react with L⌃L (L⌃L = bipy, phen, dppe) to afford the acetonitrile displaced products [MoX(CO)$_2$(L⌃L)(η^3-C$_5$H$_6$Me)].

B. *Reactions of [MX(CO)$_2$(NCMe)$_2$(η^3-allyl)] Type Complexes*

As has been described in the previous section,[316,318,322–324] the complexes [MX(CO)$_2$(NCMe)$_2$(η^3-allyl)] are labile, and they have been reacted with a wide range of donor ligands as described next.

1. *Reactions with Carbon Donor Ligands*

In 1974, King and Saran[325] described the reactions of [MoCl(CO)$_2$(NCMe)$_2$(η^3-C$_3$H$_5$)] with two equivalents of CNR (R = Me, Et, Pr^i, Bu^t, Neopentyl, Cy) to give the acetonitrile displaced products [MoCl(CO)$_2$(CNR)$_2$(η^3-C$_3$H$_5$)]. Further reaction of [MoCl(CO)$_2$(CNBu^t)$_2$(η^3-C$_3$H$_5$)] with one and two equivalents of CNBu^t gives [MoCl(CO)(CNBu^t)$_3$(η^3-C$_3$H$_5$)] and [{Mo(μ-Cl)(CNBu^t)$_4$}$_2$], respectively.[325] Deaton and Walton[326] have shown that the complexes [MoCl(CO)$_2$(CNR)$_2$(η^3-C$_3$H$_5$)] (R = Me, Bu^t, Cy) react with two further equivalents of L to give the zero-valent complexes [Mo(CO)$_2$(CNR)$_2$L$_2$] (R = Me, Bu^t, Cy; L = PEt$_3$, PPr$_3^n$, PMePh$_2$, PEtPh$_2$).

2. Reactions with Nitrogen Donor Ligands

In 1976, Hsieh and West[327] described the reactions of $[MX(CO)_2$ $(NCMe)_2(\eta^3\text{-}C_3H_4R)]$ with $R'N{=}CHCH{=}NR'$ to give the acetonitrile displaced products $[MX(CO)_2(R'N{=}CHCH{=}NR')(\eta^3\text{-}C_3H_4R)]$ (M = Mo, W; X = Cl, Br, I, NCS; R = H, Me; R' = Me, Et, Pr^i, Bu^t, Cy, Ph, p-MeC$_6$H$_4$, p-MeOC$_6$H$_4$). Treatment of $[MoCl(CO)_2(NCMe)_2(\eta^3\text{-}C_3H_4R)]$ (R = H, Me) in deoxygenated water with bi- and tridentate nitrogen donor ligands gives $[MoCl(CO)_2(N^\frown N)(\eta^3\text{-}C_3H_4R)]$ (N$^\frown$N = bipy, 2,2'-bipyridyl-amine) or the cationic complexes $[Mo(CO)_2(N^\frown N^\frown N)(\eta^3\text{-}C_3H_4R)]^+$ {N$^\frown$N$^\frown$N = dien, bis(2-pyridylmethyl)amine}, respectively.[328] The synthesis and molecular structure of the amidine complex $[MoBr(CO)_2\{\eta^2\text{-}HN{=}CMe(pz)\}(\eta^3\text{-}C_3H_5)]$ have been reported.[329] Reaction of $[MoBr(CO)_2(NCMe)_2(\eta^3\text{-}C_5H_7)]$ with N$^\frown$N (N$^\frown$N = bipy, phen) yields the complexes $[MoBr(CO)_2(N^\frown N)(\eta^3\text{-}C_5H_7)]$; their reaction chemistry also has been discussed.[330]

3. Reactions with Phosphorus and Arsenic Donor Ligands

Reaction of $[MoCl(CO)_2(NCMe)_2(\eta^3\text{-}C_3H_4R)]$ (R = H, Me) with L (L = PMe$_2$Ph, PMePh$_2$) initially gives the acetonitrile displaced product $[MoCl(CO)_2L_2(\eta^3\text{-}C_3H_4R)]$, which reacts further by reduction to yield cis-$[Mo(CO)_2L_4]$ or $[Mo(CO)_2(NCMe)(PMePh_2)_3]$. The reduction step, which is first order in the concentrations of both molybdenum complex and L, probably involves initial nucleophilic attack on the allyl ligand.[331] Similarly, reaction of $[MoX(CO)_2(NCMe)_2(\eta^3\text{-allyl})]$ with PPh$_3$ gives the thermally unstable zero-valent complexes $[Mo(CO)_2(NCMe)_2(PPh_3)_2]$.[332] Treatment of $[MCl(CO)_2(NCMe)_2(\eta^3\text{-}C_3H_4R)]$ (M = Mo, R = H, Me; M = W; R = H) with L (L = PHPh$_2$, PMePh$_2$) initially gives $[MCl(CO)_2L_2(\eta^3\text{-}C_3H_4R)]$, which reacts in NCMe with excess L to give mer-$[M(CO)_2(NCMe)L_3]$ via reductive elimination of allyl halide.[333]

Brisdon[334] has described the reactions of $[MX(CO)_2(NCMe)_2(\eta^3\text{-}C_3H_5)]$ with L$^\frown$L (L$^\frown$L = dppm, dppe) to give $[MX(CO)_2(L^\frown L)(\eta^3\text{-}C_3H_5)]$ (M = Mo, W; X = Cl, Br). Similarly, Brisdon and Paddick[335] have described the synthesis of $[MoX(CO)_2\{Ph_2As(CH_2)_2AsPh_2\}(\eta^3\text{-}C_3H_5)]$ (X = Cl, Br, I) and the dimeric complexes $[\{MoX(CO)_2(\eta^3\text{-}C_3H_5)\}_2\{\mu\text{-}Ph_2As(CH_2)AsPh_2\}]$ by an analogous method. Kapoor et al. have described the reactions of $[MoX(CO)_2(NCMe)_2(\eta^3\text{-}C_3H_5)]$ (X = Cl, Br) with L$^\frown$L' {L$^\frown$L' = R$_2$P(CH$_2$)$_2$PPh$_2$, R = m-C$_6$H$_4$F, p-C$_6$H$_4$F, m-C$_6$H$_4$CF$_3$}[336] or (p-MeC$_6$H$_4$)$_2$P(CH$_2$)$_2$P(C$_6$H$_4$Me-p)$_2$[337] to give the acetonitrile displaced complexes $[MoX(CO)_2(L^\frown L')(\eta^3\text{-}C_3H_5)]$[336] or $[MoX(CO)_2\{(p\text{-MeC}_6H_4)_2$ $P(CH_2)_2P(C_6H_4Me\text{-}p)_2\}(\eta^3\text{-}C_3H_5)]$,[337] respectively.

4. *Reactions with Bidentate O^N, S^N, and S^S Donor Ligands*

Campbell *et al.*[338,339] have described the synthesis of the semicarbazone complexes $[MoX(CO)_2\{(R'R'')CNNHCONH_2\}(\eta^3\text{-}C_3H_4R)]$ (X = Cl, Br, I; R = H, 2Me; R',R'' = H or Me and Me, Et, Pr^n, Ph) by reaction of $[MX(CO)_2(NCMe)_2(\eta^3\text{-}C_3H_4R)]$ with an equimolar amount of the appropriate semicarbazone. Similarly, thiosemicarbazone complexes of the type $[MoX(CO)_2\{(R'R'')CNNHCSNH_2\}(\eta^3\text{-}C_3H_4R)]$ (X = Cl, Br, I; R = H, 2Me; R' = H, Me, R'' = Me, Et, Pr^n, Ph)[340,341] have been prepared in an analogous manner. Reaction of $[MoX(CO)_2(NCMe)_2(\eta^3\text{-}C_3H_4R)]$ (X = Cl, Br; R = H, 2Me) with thiosemicarbazide gives $[MoX(CO)_2(H_2NNHCSNH_2)(\eta^3\text{-}C_3H_4R)]$.[342]

Treatment of $[MBr(CO)_2(NCMe)_2(\eta^3\text{-}C_3H_5)]$ (M = Mo, W) with S_2CPR_3 gives $[MBr(CO)_2(S_2CPR_3)(\eta^3\text{-}C_3H_5)]$ (R = Cy, Pr^i), which reacts further with *fac*-$[M'(CO)_3(NCMe)_3]$ (M' = Mo, W) to yield the dinuclear complexes $[(\eta^3\text{-}C_3H_5)(CO)_2M(\mu\text{-}Br)(\mu\text{-}S_2CPR_3)M'(CO)_3]$, which has been characterized crystallographically for M = M' = Mo, R = Cy.[343]

C. *X-Ray Crystal Structures of π-Allyl Halocarbonyl Complexes of Molybdenum(II) and Tungsten(II)*

The reaction of $[MoCl(CO)_2(bipy)(\eta^3\text{-}C_3H_5)]$ with Ag[BF$_4$] in CH$_2$Cl$_2$ gives the crystallographically characterized dimeric cation $[\{Mo(CO)_2(bipy)(\eta^3\text{-}C_3H_5)\}_2(\mu\text{-}Cl)]^+$ shown in Fig. 22.[344] A chronological survey of papers that are mainly concerned with the determination of the molecular structures of π-allyl halocarbonyl complexes is given in Table III.[345–354] Most of them have the same basic structure as the complex shown earlier as **6**; however, the bis(trimethylphosphite) complex $[MoCl(CO)_2\{P(OMe)_3\}_2(\eta^3\text{-}C_3H_5)]$[349] has an unusual pentagonal bipyramidal geometry, as shown in Fig. 23.

D. *Other Aspects of π-Allyl Halocarbonyl Complexes of Molybdenum(II) and Tungsten(II)*

Curtis and Eisenstein[355] have made a molecular orbital analysis of the regioselectivity of the addition of nucleophiles to π-allyl complexes and on the conformation of the η^3-allyl ligand in $[MoX(CO)_2L_2(\eta^3\text{-allyl})]$ type complexes. A detailed study of the chirality retention in rearrangements of complexes of the type $[MX(CO)_2(dppe)(\eta^3\text{-}C_3H_5)]$ has been made.[356] Studies of the photoelectron spectra,[357] electrochemical properties,[358] infrared spectroelectrochemistry,[359] and fast atom bombardment mass spec-

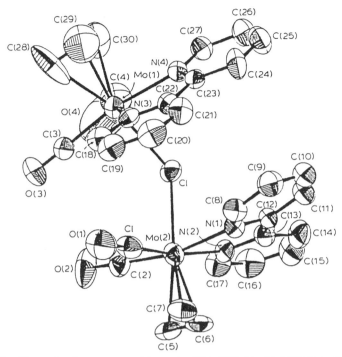

FIG. 22.   The molecular structure of the cation [{Mo(CO)$_2$(bipy)(η^3-C$_3$H$_5$)}$_2$(μ-Cl)]$^+$. [Reproduced with permission from *J. Organomet. Chem.* **1984**, *272*, 47.]

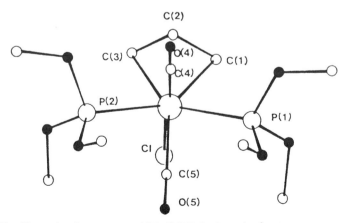

FIG. 23.   The molecular structure of [MoCl(CO)$_2$\{P(OMe)$_3$\}$_2$(η^3-C$_3$H$_5$)]. [Reproduced with permission from *Inorg. Chim. Acta* **1979**, *35*, L381.]

trometry[360] of a series of complexes of the type $[MX(CO)_2L_2(\eta^3\text{-allyl})]$ have been reported.

Trost *et al.*[361–363] have used complexes of the type $[MX(CO)_2L_2(\eta^3\text{-allyl})]$ as catalysts for allylic alkylations.

VI

DIENE HALOCARBONYL COMPLEXES OF MOLYBDENUM(II) AND TUNGSTEN(II)

Very few η^4-diene halocarbonyl complexes of molybdenum(II) and tungsten(II) had been reported up to mid-1995. This section describes conjugated diene complexes, in Section VI,A, and nonconjugated diene complexes, in Section VI,B.

A. *Conjugated Diene Halocarbonyl Complexes of Molybdenum(II) and Tungsten(II)*

In 1963, Maitlis and Games[364] reported the first π-cyclobutadiene tungsten complex, namely $[WBr(CO)(\eta^4\text{-C}_4Ph_4)Cp]$, which was prepared in 0.7% yield by reaction of $[\{W(CO)_3Cp\}_2]$ with $[\{Pd(\mu\text{-Br})Br(\eta^4\text{-C}_4Ph_4)\}_2]$ in benzene. This work was extended by Maitlis and Efraty in 1965[365] to include the improved-yield synthesis of $[WBr(CO)(\eta^4\text{-C}_4Ph_4)Cp]$ and the molybdenum complexes $[MoX(CO)(\eta^4\text{-C}_4Ph_4)Cp]$ (X = Cl, Br). Reaction of $[Mo(CO)_6]$ with $[\{Pd(\mu\text{-Br})Br(\eta^4\text{-C}_4Ph_4)\}_2]$ in refluxing xylene gives the dimeric complex **7**,[366] the structure of which was confirmed by X-ray crystallography.[367]

(7)

In 1975, Davidson and Sharp[368] described the reactions of [MoX
(CO)$_3$Cp] (X = Cl, Br, I) with PhC$_2$Ph to afford the alkyne complexes
[MoX(CO)(η^2-PhC$_2$Ph)Cp] and the π-cyclobutadiene complexes [MoX
(CO)(η^4-C$_4$Ph$_4$)Cp]. In 1976, Stone *et al.*[369] described the reaction of
[MoCl(CO)$_3$Cp] with CF$_3$C$_2$CF$_3$ to give the cyclopentadienone complex
[MoCl(CO)(η^4-C$_8$F$_{12}$CO)Cp]. In contrast, treatment of [MoCl(CO)$_3$Cp]
with MeC$_2$Me yields, depending on reaction conditions, either [MoCl(η^2-
MeC$_2$Me)$_2$Cp] or the duroquinone complex [MoCl(CO){η^4-C$_4$Me$_4$
(CO)$_2$}Cp]. Treatment of [MoCl(CO)(η^2-PhC$_2$Ph)Cp] with 1,3-butadiene
gives [MoCl(CO)(η^4-C$_4$H$_6$)Cp] or [MoCl$_2$(η^4-C$_4$H$_6$)Cp], depending on the
solvent.[370] Reaction of [MoI(CO)$_3$Cp] with CF$_3$C$_2$CF$_3$ in hexane at 100°C
gives [MoI(η^2-CF$_3$C$_2$CF$_3$)$_2$Cp], [MoI(CO){η^4-C$_4$(CF$_3$)$_4$}Cp], and the cyclopentadienone complex [MoI(CO){η^4-C$_4$(CF$_3$)$_4$CO}Cp]. Variable-temperature ^{19}F-NMR studies indicate a high barrier to rotation of the C$_4$ ring in
[MoI(CO){η^4-C$_4$(CF$_3$)$_4$}Cp].[371,372] In 1984, Hughes *et al.*[373] described an
unusual synthetic pathway to π-cyclobutadiene halocarbonyl complexes
of the type [M(CO)$_2$(η^4-C$_4$R$_3$Me)Cp][BF$_4$], as shown in Scheme 8. Further
reaction of [Mo(CO)$_2$(η^4-C$_4$MePh$_3$)Cp][BF$_4$] with chloride gives the crystallographically characterized complex [MoCl(CO)(η^4-C$_4$MePh$_3$)Cp].

B. *Noncongugated Diene Halocarbonyl Complexes of Molybdenum(II) and Tungsten(II)*

Davidson and Vasapollo[374,375] have described the reactions of [{W(μ-
Br)Br(CO)$_4$}$_2$] with three dienes (diene = cod, nbd and cot) to yield the
coordinately unsaturated diene complexes [WBr$_2$(CO)$_2$(η^4-diene)]. These
react with one equivalent of L {L = CNBu^t, PMe$_2$Ph, P(OMe)$_3$} to afford
[WBr$_2$(CO)L(η^4-diene)] via the coordinatively saturated complexes

SCHEME 8.   M = Mo, W; R = Ph; L = Cp; M = Mo, R = Me, L = Cp*.

[WBr$_2$(CO)$_2$L(η^4-diene)], which have been isolated for the norbornadiene complexes. Also, reaction of [WBr$_2$(CO)$_2$(η^4-nbd)] with two equivalents of L$_2$ {L = P(OMe)$_3$, PMePh$_2$; L$_2$ = bipy} gives the monocarbonyl complexes [WBr$_2$(CO)L$_2$(η^4-nbd)]. In 1984, Cotton and Meadows[376] reported the molecular and electronic structures of the 16-electron complexes [WBr$_2$(CO)$_2$(η^4-nbd)] and [WBr$_2$(CO)$_2$(PPh$_3$)$_2$]. The X-ray crystal structure of the norbornadiene complex [WBr$_2$(CO)$_2$(η^4-nbd)] is shown in Fig. 24.

Davidson *et al.*[377] also have described the reaction of [WBr$_2$(CO)$_2$(η^4-nbd)] with TlCp to give [WBr(CO)(η^4-nbd)Cp]. Similarly, reaction of [WBr$_2$(CO)$_2$(η^4-nbd)] with Tl[O$_2$CMe] and Tl[O(S)CMe] affords [WBr(CO)(O$_2$CMe)(η^4-nbd)] and [WBr(CO){O(S)CMe}(η^4-nbd)], respectively. However, reaction of equimolar amounts of [WBr$_2$(CO)$_2$(η^4-nbd)] with Tl[SC$_6$F$_5$] gives the crystallographically characterized complex [WBr(SC$_6$F$_5$)(CO)$_2$(η^4-nbd)], which has a distorted octahedral geometry. Davidson *et al.*[378] have carried out reactions of [WBr$_2$(CO)$_2$(η^4-diene)] (diene = nbd, cot) with one equivalent of bidentate anionic ligands L^L to give seven-coordinate complexes [WBr(CO)$_2$(L^L)(η^4-nbd)] {L^L = S$_2$CNMe$_2$, SC$_5$H$_4$N, S$_2$PMe$_2$, S$_2$P(OMe)$_2$} and [WBr(CO)$_2$(S$_2$CNMe$_2$)(η^4-cot)] or six-coordinate complexes [WBr(CO)(L^L)(η^4-cot)](L^L = S$_2$PMe$_2$, SC$_5$H$_4$N). The molecular structures of [WBr(CO)$_2${S$_2$P(OMe)$_2$}(η^4-nbd)], [WBr(CO)(PMe$_2$Ph)(S$_2$CNMe$_2$)(η^4-nbd)], and [W(CO)(S$_2$CPMe$_2$)$_2$(η^4-nbd)] have all been determined crystallographically[378] and have seven-coordination with pentagonal bipyramidal structures.

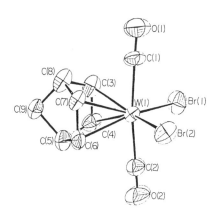

FIG. 24. The molecular structure of [WBr$_2$(CO)$_2$(η^4-nbd)]. [Reprinted with permission from *Inorg. Chem.* **1984**, *23*, 4690. Copyright 1984 American Chemical Society.]

Davidson *et al.*[379] have also found that mixtures of $[WBr_2(CO)_2(\eta^4\text{-}diene)]$ (diene = nbd, cod, cot)/$AlCl_2Et$ initiate the metathesis of *cis*-2-pentene to yield 2-butene and 3-hexene. For the norbornadiene complex, the reactions have been found to be catalytic.

VII

Cp AND RELATED η^5-LIGAND HALOCARBONYL COMPLEXES OF MOLYBDENUM(II) AND TUNGSTEN(II)

Since the early report by Piper and Wilkinson[1] of the molybdenum(II) complexes $[MoX(CO)_3Cp]$ (X = Cl, Br, I), the reaction chemistry of these complexes has been studied extensively. This review is restricted to the syntheses and reactions in which the complexes generally retain a carbonyl and halide in the final product.

A. *Synthesis and Structures of Cp and Related η^5-ligand Halocarbonyl Complexes of Molybdenum(II) and Tungsten(II)*

Treatment of $[MMe(CO)_3Cp^*]$ with iodine yields $[MI(CO)_3Cp^*]$ (M = Mo, W) and the tungsten(IV) complex $[WI_3(CO)_2Cp^*]$.[380] Photolysis of $[\{Mo(CO)_3Cp\}_2]$ with visible light in CHX_3 or CX_4 (X = Cl, Br) gives the halocarbonyl complexes $[MoX(CO)_3Cp]$.[381] Similarly, 520-nm photolysis of $[\{W(CO)_3Cp\}_2]$ generates the radical $[W(CO)_3Cp]^\bullet$, which reacts with various chlorocarbon trapping agents in solution to afford [WCl $(CO)_3Cp]$.[382] Laser photolysis of an acetonitrile solution of $[\{W(CO)_3Cp\}_2]$ that contains $[RhX(dmgH)_2(PPh_3)]$ (X = Cl, Br; dmgH = dimethylglyoximato) gives $[WX(CO)_3Cp]$ and $[Rh(dmgH)(PPh_3)_2]$.[383] Photolysis of [MCl $(CO)_3Cp]$ (M = Mo, W) in argon, methane, nitrogen, carbon monoxide, and 5% C_2H_4/CH_4 matrices at 12K gives, via ejection of CO, the 16-electron complex $[MCl(CO)_2Cp]$ as the primary process.[384]

The complexes $[MMe(CO)_3(\eta^5\text{-}C_5H_4I)]$ (M = Mo, W), in the presence of palladium, undergo methyl-iodide exchange to yield the complexes $[MI(CO)_3(\eta^5\text{-}C_5H_4Me)]$.[385] More recently, the reactions of $Li[M(CO)_3\{\eta^5\text{-}C_5(CH_2Ph)_5\}]$ (M = Mo, W) with PCl_3, PBr_3, or I_2 to give $[MX(CO)_3\{\eta^5\text{-}C_5(CH_2Ph)_5\}]$ have been described, and the molecular structure of $[MoI (CO)_3\{\eta^5\text{-}C_5(CH_2Ph)_5\}]$ has been determined crystallographically.[386]

Treatment of the Mo≡Mo bonded complex $[Mo_2(CO)_4Cp_2]$ with I_2 or HCl yields the complexes $[Mo_2I_2(CO)_4Cp_2]$ or $[Mo_2(\mu\text{-}H)(\mu\text{-}Cl)(CO)_4Cp_2]$, respectively.[387] The preparation, electrochemical properties, and molecular structure of the dimeric complex $[\{Mo(\mu\text{-}Br)(\mu\text{-}CO)Cp^\#\}_2]$ $\{Cp^\# =$

1-(2,5-dimethoxyphenyl)-2,3,4,5-tetraphenylcyclopentadienyl} have been reported.[388] Treatment of [{Mo(CO)$_3$(η^5-C$_9$H$_7$)}$_2$] (C$_9$H$_7$ = indenyl) with I$_2$ gives [MoI(CO)$_3$(η^5-C$_9$H$_7$)].[389] Reaction of [W(CO)$_3$(η^6-C$_7$H$_8$)] with HCl gives the π-cycloheptadienyl complex [WCl(CO)$_3$(η^5-C$_7$H$_9$)].[390]

The molecular structures of the complexes [MoCl(CO)$_3$Cp][391] and [MoI (CO)$_3$(η^5-C$_9$H$_7$)][392] have been determined crystallographically.

B. Reactions of [MX(CO)₃Cp] and Related Complexes with Neutral Donor Ligands

The synthesis of [MoX(CO)$_3$(η^5-C$_9$H$_7$)] (X = Cl, Br) by photolysis of [{Mo(CO)$_3$(η^5-C$_9$H$_7$)}$_2$] in CHX$_3$ has been described. The reactions of [MoX(CO)$_3$(η^5-C$_9$H$_7$)] (X = Cl, Br, I) with a series of phosphorus donor ligands (L) to give the carbonyl displaced products [MoX(CO)$_2$L(η^5-C$_9$H$_7$)] have been reported.[393] The mechanism and kinetics of the reactions of the π-tetrahydroindenyl complex [MoCl(CO)$_3$(η^5-C$_9$H$_{11}$)] with phosphorus donor ligands have been described and compared with those of the analogous cyclopentadienyl and indenyl complexes.[394] The reaction of [MCl (CO)$_3$Cp] (M = Mo, W) with methylaminobis(difluorophosphine) in various solvents to give carbonyl displaced products has been reported.[395] Treatment of [MX(CO)$_3$Cp] with phosphines in the presence of ONMe$_3$ gives high yields of [MX(CO)$_2$(phosphine)Cp]. The selectivity, yields, and rates of reaction are greatly enhanced by using ONMe$_3$.[396]

Photolysis of [MX(CO)$_3$Cp] (M = Mo, X = Br, I; M = W, X = Cl, Br, I) in benzene solutions of PPh$_3$ affords cis-[MoX(CO)$_2$(PPh$_3$)Cp] in a stereospecific manner.[397] Treatment of [MoI(CO)$_3$Cp] with CNR in the presence of [{Mo(CO)$_3$Cp}$_2$] as catalyst gives [MoI(CO)$_{3-n}$(CNR)$_n$Cp] (n = 1–3) via a radical mechanism.[398] Reaction of [MCl(CO)$_3$Cp] (M = Mo, W) with optically active bis(diphenylphosphino)-α-(S)-amino acid methyl esters yields the asymmetric metal complexes [MCl(CO){(Ph$_3$P)$_2$ NCHRCO$_2$Me}Cp].[399]

C. Other Aspects of [MX(CO)₃Cp] and Related Complexes

Faller et al.[400] have investigated the conformational equilibria of a series of phosphine complexes of the type [MX(CO)$_2$L(η^5-Cp or Indenyl)] by NMR spectroscopy. Photolysis of [MX(CO)$_2$(PPh$_3$)Cp] (M = Mo, W) at 366 nm in benzene undergo cis–trans isomerization and disproportionation to the complexes [MX(CO)$_3$Cp] and [MX(CO)(PPh$_3$)$_2$Cp].[401] The low-temperature (12 K) photochemistry of phosphorus donor ligand complexes of the type [MX(CO)$_2$LCp] has been investigated.[402]

Chromatographic methods have been used to separate the complexes cis- and trans-$[MX(CO)_2(PPh_3)Cp]$.[403] ^{13}C-NMR spectral data on a series of complexes $[MX(CO)_3Cp]$ and their phosphine-substituted products have been reported.[404] The electrooxidation of complexes of the type $[MoX(CO)_{3-n}(PR_3)_nCp]$ has been studied by Gipson et al.[405,406]

VIII

ARENE HALOCARBONYL COMPLEXES OF MOLYBDENUM(II) AND TUNGSTEN(II)

Stiddard et al.[407,408] have described the reactions of $[M(CO)_3(\eta^6\text{-arene})]$ (M = Cr, Mo, W; arene = C_6Me_6; M = Mo, W; arene = mesitylene, durene, p-cymene) with $SbCl_5$ to afford the cationic complexes $[MCl(CO)_3(\eta^6\text{-arene})][SbCl_6]$, which for M = Mo, W and arene = C_6Me_6 have been converted to the tetraphenylborate salts $[MCl(CO)_3(\eta^6\text{-arene})][BPh_4]$. Reaction of $[M(CO)_3(\eta^6\text{-arene})]$ (M = Mo, W; arene = C_6Me_6, mesitylene) with I_2 gives cationic complexes of the type $[MI(CO)_3(\eta^6\text{-arene})]^+$.[409] The complex $[WI(CO)_3(\eta^6\text{-}C_6Me_6)]I_3$ has been characterized crystallographically and has approximately C_s symmetry.[409]

Calderazzo et al.[410] have reinvestigated the reactions of $[M(CO)_3(\eta^6\text{-arene})]$ complexes with I_2 under a range of different reaction conditions and stoichiometry of reagents and shows these reactions are not as simple as previously reported. It has been observed that the cation $[MI(CO)_3(\eta^6\text{-arene})]^+$ is always obtained; however, the counteranion varies according to the reaction time and the molar ratio of the reagents. The molecular structures of $[MI(CO)_3(\eta^6\text{-}C_6H_3Me_3\text{-}1,3,5)][M_2I_5(CO)_6]$ (M = Mo, W) and $[MoI(CO)_3(\eta^6\text{-}C_6Me_6)][MoI_3(CO)_4]$ have been determined crystallographically.[410] The molecular structure of $[MoI(CO)_3(\eta^6\text{-}C_6H_3Me_3\text{-}1,3,5)][Mo_2I_5(CO)_6]$ is shown in Fig. 25a and b.[410]

Treatment of the ionic complexes $[MoI(CO)_3(\eta^6\text{-arene})][Mo_2I_5(CO)_6]$ (arene = C_6H_5Me, $1,4\text{-}C_6H_4Me_2$, $1,3,5\text{-}C_6H_3Me_3$) with CO yields the crystallographically characterized complex $[\{Mo(\mu\text{-}I)I(CO)_4\}_2]$.[411] It has been observed that the carbonylation of $[Mo(CO)_3(\eta^6\text{-arene})]$ is catalyzed by I_2, or by $[\{Mo(\mu\text{-}I)I(CO)_4\}_2]$ or $[MoI(CO)_3(\eta^6\text{-arene})][Mo_2I_5(CO)_6]$.[412] Reaction of $[MoI(CO)_3(\eta^6\text{-}C_6Me_6)]I_3$ with $[\{Mo(\mu\text{-}I)I(CO)_4\}_2]$ gives the triiodide ligand complex $[MoI(CO)_3(\eta^6\text{-}C_6Me_6)][MoI_2(I_3)(CO)_4]$ and a small quantity of the pentaiodide complex $[MoI(CO)_3(\eta^6\text{-}C_6Me_6)]I_5$, both of which have been characterized crystallographically.[413] The pentaiodide complex $[MoI(CO)_3(\eta^6\text{-}C_6Me_6)]I_5$ also can be prepared by reaction of $[MoI(CO)_3(\eta^6\text{-}C_6Me_6)]I_3$ with an equimolar quantity of I_2 or by reaction of $[Mo(CO)_3$

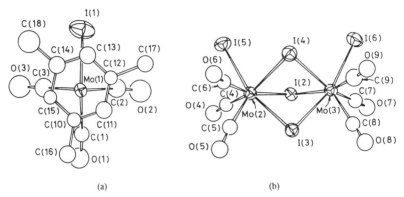

FIG. 25.   The molecular structure of [MoI(CO)$_3$(η^6-C$_6$H$_3$Me$_3$-1,3,5)][Mo$_2$I$_5$(CO)$_6$]. [Reproduced with permission from *J. Chem. Soc., Dalton Trans.* **1986**, (a) [MoI(CO)$_3$(η^6-C$_6$H$_3$Me$_3$-1,3,5)]$^+$, page 2573; (b) [Mo$_2$I$_5$(CO)$_6$]$^-$, page 2574.]

(η^6-C$_6$Me$_6$)] with the three equivalents of iodine.[413] Treatment of either [{Mo(μ-I)I(CO)$_4$}$_2$] or [Mo(CO)$_3$(η^6-C$_6$H$_5$Me)] with iodine in thf affords the crystallographically characterized octahedral complex [MoI$_3$(thf)$_3$].[414]

IX

η^7-CYCLOHEPTATRIENYL HALOCARBONYL COMPLEXES OF MOLYBDENUM(II) AND TUNGSTEN(II)

In view of Green and Ng's extensive review[9] entitled "Cycloheptatriene and Enyl Complexes of the Early Transition Metals," which includes halocarbonyl complexes of molybdenum(II) and tungsten(II) such as [MoI(CO)$_2$(η^7-C$_7$H$_7$)], it was not considered useful to review this area. One paper by Whiteley *et al.*[415] which was not included in Green and Ng's review,[9] describes the synthesis of the bicycloheptatrienyl complex [MoI(CO)$_2$(η^7-C$_{14}$H$_{12}$)][PF$_6$], which has an unattached, pendent cycloheptatrienyl ring and reacts with *fac*-[Mo(CO)$_3$(NCMe)$_3$] to give the crystallographically characterized complex [Mo$_2$(μ-I)(CO)$_4$(μ-η^7:η^7-C$_{14}$H$_{12}$)][PF$_6$]. The structure shows two crystallographically distinct cations that have a twisted conformation of the bicycloheptatrienyl group.[415]

ACKNOWLEDGMENTS

I thank Alec I. Clark, Margaret M. Meehan, and John Szewczyk for helping with the collection of references for this review. I also thank Mrs. Helen Hughes for both cheerful and efficient preparation of this article.

REFERENCES

(1) Piper, T. S.; Wilkinson, G. *J. Inorg. Nucl. Chem.* **1956,** *3,* 104.
(2) Anker, M. W.; Colton, R.; Tomkins, I. B. *Pure Appl. Chem.* **1968,** *18,* 23.
(3) Colton, R. *Coord. Chem. Rev.* **1971,** *6,* 269.
(4) Drew, M. G. B. *Prog. Inorg. Chem.* **1977,** *23,* 67.
(5) Melník, M.; Sharrock, P. *Coord. Chem. Rev.* **1985,** *65,* 49.
(6) Mayr, A.; Hoffmeister, H. *Adv. Organomet. Chem.* **1991,** *32,* 227.
(7) Templeton, J. L. *Adv. Organomet. Chem.* **1989,** *29,* 1.
(8) Brisdon, B. J.; Walton, R. A. *Polyhedron* **1995,** *14,* 1259.
(9) Green, M. L. H.; Ng, D. K. P. *Chem. Rev.* **1995,** *95,* 439.
(10) Nigam, H. L.; Nyholm, R. S. *Proc. Chem. Soc.* **1957,** 321.
(11) Colton, R.; Kevekordes, J. *Aust. J. Chem.* **1982,** *35,* 895.
(12) Haigh, C. W.; Baker, P. K. *Polyhedron* **1994,** *13,* 417.
(13) Haigh, C. W. *Polyhedron* **1994,** *13,* 2703.
(14) Hoffmann, R.; Beier, B. F.; Muetterties, E. L.; Rossi, A. R. *Inorg. Chem.* **1977,** *16,* 511.
(15) Hieber, H.; Romberg, E. *Z. Anorg. Allg. Chem.* **1935,** *221,* 321.
(16) Klemm, W.; Steinberg, H. *Z. Anorg. Allg. Chem.* **1936,** *227,* 193.
(17) Cook, C. D.; Nyholm, R. S.; Tobe, M. L. *J. Chem. Soc.* **1965,** 4194 and references cited therein.
(18) Colton, R.; Tomkins, I. B. *Aust. J. Chem.* **1966,** *19,* 1143.
(19) Colton, R.; Tomkins, I. B. *Aust. J. Chem.* **1966,** *19,* 1519.
(20) Anker, M. W.; Colton, R.; Tomkins, I. B. *Aust. J. Chem.* **1967,** *20,* 9.
(21) Colton, R.; Rix, C. J. *Aust. J. Chem.* **1969,** *22,* 305.
(22) Broomhead, J. A.; Budge, J.; Grumley, W. *Inorg. Synth.* **1990,** *28,* 145.
(23) Cotton, F. A.; Falvello, L. R.; Meadows, J. H. *Inorg. Chem.* **1985,** *24,* 514.
(24) Colton, R.; Commons, C. J. *Aust. J. Chem.* **1973,** *26,* 1493.
(25) Colton, R.; Tomkins, I. B. *Aust. J. Chem.* **1967,** *20,* 13.
(26) Colton, R.; Scollary, G. R.; Tomkins, I. B. *Aust. J. Chem.* **1968,** *21,* 15.
(27) Anker, M. W.; Colton, R.; Tomkins, I. B. *Aust. J. Chem.* **1968,** *21,* 1143.
(28) Colton, R.; Rix, C. J. *Aust. J. Chem.* **1968,** *21,* 1155.
(29) Anker, M. W.; Colton, R.; Tomkins, I. B. *Aust. J. Chem.* **1968,** *21,* 1159.
(30) Colton, R.; Scollary, G. R. *Aust. J. Chem.* **1968,** *21,* 1427.
(31) Colton, R.; Scollary, G. R. *Aust. J. Chem.* **1968,** *21,* 1435.
(32) Bowden, J. A.; Colton, R. *Aust. J. Chem.* **1968,** *21,* 2657.
(33) Anker, M. W.; Colton, R.; Rix, C. J.; (in part) Tomkins, I. B. *Aust. J. Chem.* **1969,** *22,* 1341.
(34) Colton, R.; Rix, C. J. *Aust. J. Chem.* **1969,** *22,* 2535.
(35) Colton, R.; Rix, C. J. *Aust. J. Chem.* **1970,** *23,* 441.
(36) Anker, M. W.; Colton, R. *Aust. J. Chem.* **1971,** *24,* 2223.
(37) Bowden, J. A.; Colton, R. *Aust. J. Chem.* **1972,** *25,* 17.
(38) Bowden, J. A.; Colton, R.; Commons, C. J. *Aust. J. Chem.* **1972,** *25,* 1393.
(39) Colton, R.; Panagiotidou, P. *Aust. J. Chem.* **1987,** *40,* 13.
(40) Broadbent, R. F.; Kingston, J. V. *J. Inorg. Nucl. Chem.* **1970,** *32,* 2919.
(41) Westland, A. D.; Muriithi, N. *Inorg. Chem.* **1973,** *12,* 2356.
(42) Beall, T. W.; Houk, L. W. *Inorg. Chem.* **1974,** *13,* 2549.
(43) Fong, L. K.; Fox, J. R.; Foxman, B. M.; Cooper, N. J. *Inorg. Chem.* **1986,** *25,* 1880.
(44) Cotton, F. A.; Dunbar, K. R.; Poli, R. *Inorg. Chem.* **1986,** *25,* 3700.
(45) Cotton, F. A.; Poli, R. *Inorg. Chem.* **1986,** *25,* 3703.

(46) Cotton, F. A.; Falvello, L. R.; Poli, R. *Polyhedron* **1987**, *6*, 1135.
(47) Lee, S. W.; Trogler, W. C. *Organometallics* **1990**, *9*, 1470.
(48) Backhouse, J. R.; Lowe, H. M.; Sinn, E.; Suzuki, S.; Woodward, S. *J. Chem. Soc., Dalton, Trans.* **1995**, 1489.
(49) King, R. B. *Inorg. Chem.* **1964**, *3*, 1039.
(50) Ganorkar, M. C.; Stiddard, M. H. B. *J. Chem. Soc.* **1965**, 3494.
(51) Bowden, J. A.; Colton, R. *Aust. J. Chem.* **1969**, *22*, 905.
(52) Colton, R.; Howard, J. J. *Aust. J. Chem.* **1969**, *22*, 2543.
(53) Tsang, W. S.; Meek, D. W.; Wojcicki, A. *Inorg. Chem.* **1968**, *7*, 1263.
(54) Dreyer, E. B.; Lam, C. T.; Lippard, S. J. *Inorg. Chem.* **1979**, *18*, 1904.
(55) Filippou, A. C.; Grünleitner, W. *J. Organomet. Chem.* **1990**, *398*, 99.
(56) Tripathi, S. C.; Srivastava, S. C.; Mani, R. P. *J. Organomet. Chem.* **1976**, *105*, 239.
(57) Tripathi, S. C.; Srivastava, S. C.; Shrimal, A. K. *Inorg. Chim. Acta* **1976**, *18*, 231.
(58) Moss, J. R.; Smith, B. J. *S. Afr. J. Chem.* **1983**, *36*, 32.
(59) Stiddard, M. H. B. *J. Chem. Soc.* **1962**, 4712.
(60) Behrens, V. H.; Rosenfelder, J. *Z. Anorg. Allg. Chem.* **1967**, *352*, 61.
(61) Tripathi, S. C.; Srivastava, S. C.; Singh, C. P. *Transition-Met. Chem.* **1977**, *2*, 198.
(62) Tripathi, S. C.; Srivastava, S. C.; Pandey, D. P.; Shrivastava, P. K. *J. Ind. Chem. Soc.* **1982**, *LIX*, 238.
(63) Tripathi, S. C.; Srivastava, S. C.; Shrimal, A. K.; Singh, O. P. *Transition-Met. Chem.* **1984**, *9*, 478.
(64) Iglesias, M.; Llorente, A.; Del Pino, C.; Santos, A. *J. Organomet. Chem.* **1984**, *263*, 193.
(65) Shiu, K.-B.; Liou, K.-S.; Wang, S.-L.; Cheng, C. P.; Wu, F.-J. *J. Organomet. Chem.* **1989**, *359*, C1.
(66) Shiu, K.-B.; Liou, K.-S.; Wang, S.-L.; Wei, S.-C. *Organometallics* **1990**, *9*, 669.
(67) Dunn, J. G.; Edwards, D. A. *J. Chem. Soc. (A)* **1971**, 988.
(68) Dunn, J. G.; Edwards, D. A. *J. Organomet. Chem.* **1972**, *36*, 153.
(69) Chaudhuri, P.; Wieghardt, K.; Tsai, Y.-H.; Krüger, C. *Inorg. Chem.* **1984**, *23*, 427.
(70) Dahmann, G. B.-; Herrmann, W.; Wieghardt, K.; Weiss, J. *Inorg. Chem.* **1985**, *24*, 485.
(71) Dahmann, G. B.-; Wieghardt, K. *Inorg. Chem.* **1985**, *24*, 4049.
(72) Schreiber, P.; Wieghardt, K.; Flörke, U.; Haupt, H.-J. *Z. Naturforsch., Sect. B* **1987**, *42*, 1391.
(73) Curtis, D.; Shiu, K.-B. *Inorg. Chem.* **1985**, *24*, 1213.
(74) Lewis, J.; Whyman, R. *J. Chem. Soc.* **1967**, 77.
(75) Moss, J. R.; Shaw, B. L. *J. Chem. Soc. (A)* **1970**, 595.
(76) Dombek, B. D.; Angelici, R. J. *Inorg. Chem.* **1976**, *15*, 2397.
(77) Tripathi, S. C.; Srivastava, S. C.; Pandey, D. P. *Transition-Met. Chem.* **1977**, *2*, 52.
(78) Umland, P.; Vahrenkamp, H. *Chem. Ber.* **1982**, *115*, 3555.
(79) Umland, P.; Vahrenkamp, H. *Chem. Ber.* **1982**, *115*, 3565.
(80) Lewis, J.; Whyman, R. *J. Chem. Soc.* **1965**, 5486.
(81) Colton, R.; Howard, J. J. *Aust. J. Chem.* **1970**, *23*, 223.
(82) Connor, J. A.; McEwen, G. K.; Rix, C. J. *J. Less-Comm. Met.* **1974**, *36*, 207.
(83) Connor, J. A.; McEwan, G. K.; Rix, C. J. *J. Chem. Soc., Dalton Trans.* **1974**, 589.
(84) Alyea, E. C.; Shakya, R. P.; Vougioukas, A. E. *Transition-Met. Chem.* **1985**, *10*, 435.
(85) Bradley, F. C.; Wong, E. H.; Gabe, E. J.; Lee, F. L.; Lepage, Y. *Polyhedron* **1987**, *6*, 1103.
(86) Al-Dulaymmi, M. F. M.; Hitchcock, P. B.; Richards, R. L. *Polyhedron* **1989**, *8*, 1876.
(87) Perera, S. D.; Shamsuddin, M.; Shaw, B. L. *Can. J. Chem.* **1995**, *73*, 1010.
(88) Nigam, H. L.; Nyholm, R. S.; Stiddard, M. H. B. *J. Chem. Soc.* **1960**, 1806.

(89) Lewis, J.; Nyholm, R. S.; Pande, C. S.; Stiddard, M. H. B. *J. Chem. Soc.* **1963**, 3600.
(90) Mašek, J.; Nyholm, R. S.; Stiddard, M. H. B. *Coll. Czech. Chem. Comm.* **1964,** *29,* 1714.
(91) Nyholm, R. S.; Snow, M. R.; Stiddard, M. H. B. *J. Chem. Soc.* **1965,** 6570.
(92) Cook, C. D.; Nyholm, R. S.; Tobe, M. L. *J. Chem. Soc.* **1965,** 4194.
(93) Anker, M. W.; Colton, R.; Rix, C. J. *Aust. J. Chem.* **1971,** *24,* 1157.
(94) Henrick, K.; Wild, S. B. *J. Chem. Soc., Dalton Trans.* **1974,** 2500.
(95) Mihichuk, L.; Pizzey, M.; Robertson, B.; Barton, R. *Can. J. Chem.* **1986,** *64,* 991.
(96) Mannerskantz, H. C. E.; Wilkinson, G. *J. Chem. Soc.* **1962,** 4454.
(97) Planinić, P.; Meider, H. *Polyhedron* **1990,** *9,* 1099.
(98) Sevdić, D.; Ćurić, M.; Božić, Lj. T.- *Polyhedron* **1989,** *8,* 505.
(99) Baker, P. K.; Fraser, S. G.; Keys, E. M. *J. Organomet. Chem.* **1986,** *309,* 319.
(100) Baker, P. K.; Hursthouse, M. B.; Karaulov, A. I.; Lavery, A. J.; Malik, K. M. A.; Muldoon, D. J.; Shawcross, A. *J. Chem. Soc., Dalton Trans.* **1994,** 3493.
(101) Tate, D. P.; Knipple, W. R.; Augl, J. M. *Inorg. Chem.* **1962,** *1,* 433.
(102) Westland, A. D.; Muriithi, N. *Inorg. Chem.* **1972,** *11,* 2971.
(103) Lucht, B.; Poss, M. J.; Richmond, T. G. *J. Chem. Ed.* **1991,** *68,* 786.
(104) Drew, M. G. B.; Baker, P. K.; Armstrong, E. M.; Fraser, S. G. *Polyhedron* **1988,** *7,* 245.
(105) Drew, M. G. B.; Baker, P. K.; Armstrong, E. M.; Fraser, S. G.; Muldoon, D. J.; Lavery, A. J.; Shawcross, A. *Polyhedron* **1995,** *14,* 617.
(106) Baker, P. K.; Muldoon, D. J.; Hursthouse, M. B.; Coles, S. J.; Lavery, A. J.; Shawcross, A., *Z. Naturforsch. Sect. B* **1996,** *51,* 263.
(107) Baker, P. K.; Harman, M. E.; Hursthouse, M. B.; Karaulov, A. I.; Lavery, A. J.; Malik, K. M. A.; Muldoon, D. J.; Shawcross, A. *J. Organomet. Chem.* **1995,** *494,* 205.
(108) Baker, P. K.; Fraser, S. G. *J. Organomet. Chem.* **1987,** *329,* 209.
(109) Baker, P. K.; Fraser, S. G. *Inorg. Chim. Acta* **1986,** *116,* L3.
(110) Baker, P. K.; van Kampen, M. *Inorg. Chim. Acta* **1993,** *204,* 247.
(111) Baker, P. K.; van Kampen, M.; Roos, C.; Spaeth, J. *Transition-Met. Chem.* **1994,** *19,* 165.
(112) Baker, P. K.; Jenkins, A. E.; Lavery, A. J.; Muldoon, D. J.; Shawcross, A. *J. Chem. Soc., Dalton Trans.* **1995,** 1525.
(113) Baker, P. K.; Barfield, J.; van Kampen, M. *Inorg. Chim. Acta* **1989,** *156,* 179.
(114) Baker, P. K.; Howells, L. L.; Fraser, S. G.; Rogers, G. W.; Snowden, M. J. *Transition-Met. Chem.* **1990,** *15,* 71.
(115) Baker, P. K.; Fraser, S. G. *J. Organomet. Chem.* **1986,** *299,* C23.
(116) Baker, P. K.; Fraser, S. G. *Inorg. Chim. Acta* **1986,** *116,* L1.
(117) Baker, P. K.; Beckett, M. A.; Severs, L. M. *J. Organomet. Chem.* **1991,** *409,* 213.
(118) Baker, P. K.; Sharp, D. J. T. *J. Coord. Chem.* **1988,** *16,* 389.
(119) Baker, P. K.; Fraser, S. G. *Inorg. Chim. Acta* **1987,** *130,* 61.
(120) Balakrishna, M. S.; Krishnamurthy, S. S.; Manohar, H. *Organometallics* **1991,** *10,* 2522.
(121) Babu, R. P. K.; Krishnamurthy, S. S.; Nethaji, M. *Organometallics* **1995,** *14,* 2047.
(122) Baker, P. K.; Fraser, S. G.; Harding, P. *Inorg. Chim. Acta* **1986,** *116,* L5.
(123) Hsu, S. C. N.; Yeh, W.-Y.; Chiang, M. Y. *J. Organomet. Chem.* **1995,** *492,* 121.
(124) Baker, P. K.; Armstrong, E. M. *Polyhedron* **1990,** *9,* 801.
(125) Baker, P. K.; Sherlock, D. J. *Polyhedron* **1994,** *13,* 525.
(126) Baker, P. K.; Fraser, S. G. *J. Coord. Chem.* **1987,** *16,* 97.
(127) Baker, P. K.; Fraser, S. G.; Matthews, T. M. *Inorg. Chim. Acta* **1988,** *150,* 217.
(128) Baker, P. K.; Fraser, S. G. *Polyhedron* **1986,** *5,* 1381.
(129) Baker, P. K.; Fraser, S. G. *Transition-Met. Chem.* **1987,** *12,* 560.
(130) Baker, P. K.; Fraser, S. G. *Transition-Met. Chem.* **1988,** *13,* 284.

(131) Baker, P. K.; Fraser, S. G. *J. Coord. Chem.* **1987**, *15*, 405.
(132) Baker, P. K.; Fraser, S. G.; Snowden, M. J. *Inorg. Chim. Acta* **1988**, *148*, 247.
(133) Baker, P. K.; Bamber, M.; Rogers, G. W. *J. Organomet. Chem.* **1989**, *367*, 101.
(134) Baker, P. K.; Fraser, S. G. *Transition-Met. Chem.* **1986**, *11*, 34.
(135) Baker, P. K.; Fraser, S. G.; Drew, M. G. B. *J. Chem. Soc., Dalton Trans.* **1988**, 2729.
(136) Baker, P. K.; Fraser, S. G. *J. Coord. Chem.* **1986**, *15*, 185.
(137) Baker, P. K.; Flower, K. R.; Thompson, S. M. L. *Transition-Met. Chem.* **1987**, *12*, 349.
(138) Baker, P. K.; van Kampen, M.; ap Kendrick, D. *J. Organomet. Chem.* **1991**, *421*, 241.
(139) Baker, P. K.; Fraser, S. G. *Polyhedron* **1987**, *6*, 2081.
(140) Baker, P. K.; Fraser, S. G. *Inorg. Chim. Acta* **1988**, *142*, 219.
(141) Baker, P. K.; ap Kendrick, D. *J. Coord. Chem.* **1988**, *17*, 355.
(142) Baker, P. K.; ap Kendrick, D. *Inorg. Chim. Acta* **1990**, *174*, 119.
(143) Baker, P. K.; ap Kendrick, D. *J. Organomet. Chem.* **1991**, *411*, 215.
(144) Baker, P. K.; Harris, S. D.; Durrant, M. C.; Hughes, D. L.; Richards, R. L. *J. Chem. Soc., Dalton Trans.* **1994**, 1401.
(145) Durrant, M. C.; Hughes, D. L.; Richards, R. L.; Baker, P. K.; Harris, S. D. *J. Chem. Soc., Dalton Trans.* **1992**, 3399.
(146) Baker, P. K.; Durrant, M. C.; Goerdt, B.; Harris, S. D.; Hughes, D. L.; Richards, R. L. *J. Organomet. Chem.* **1994**, *469*, C22.
(147) Baker, P. K.; Fraser, S. G. *J. Coord. Chem.* **1989**, *20*, 267.
(148) Baker, P. K.; Hughes, S. *J. Coord. Chem.* **1995**, *35*, 1.
(149) Baker, P. K.; ap Kendrick, D. *Polyhedron* **1991**, *10*, 433.
(150) Baker, P. K.; Fraser, S. G. *Transition-Met. Chem.* **1986**, *11*, 273.
(151) Baker, P. K.; Fraser, S. G.; ap Kendrick, D. *J. Chem. Soc., Dalton Trans.* **1991**, 131.
(152) Baker, P. K.; Fraser, S. G. *Synth. React. Inorg. Met.-Org. Chem.* **1987**, *17*, 371.
(153) Baker, P. K.; Flower, K. R. *J. Coord. Chem.* **1987**, *15*, 333.
(154) Baker, P. K.; Harman, M. E.; Hughes, S.; Hursthouse, M. B.; Malik, K. M. A. *J. Organomet. Chem.* **1995**, *498*, 257.
(155) Lange, V. G.; Dehnicke, K. *Z. Anorg. Allg. Chem.* **1966**, *344*, 167.
(156) Szymańska-Buzar, T. *Inorg. Chim. Acta* **1988**, *145*, 231.
(157) Szymańska-Buzar, T. *J. Organomet. Chem.* **1989**, *375*, 85.
(158) Szymańska-Buzar, T.; Głowiak, T. *J. Organomet. Chem.* **1995**, *489*, 207.
(159) Szymańska-Buzar, T.; Glowiak, T. *J. Organomet. Chem.* **1995**, *490*, 203.
(160) Baker, P. K.; Birkbeck, T.; Bräse, S.; Bury, A.; Naylor, H. M. *Transition-Met. Chem.* **1992**, *17*, 401.
(161) Baker, P. K.; Flower, K. R.; Naylor, H. M.; Voigt, K. *Polyhedron* **1993**, *12*, 357.
(162) Memering, M. N.; Dobson, G. R. *J. Inorg. Nucl. Chem.* **1973**, *35*, 665.
(163) Piana, H.; Schubert, U. *J. Organomet. Chem.* **1991**, *411*, 303.
(164) Kummer, R.; Graham, W. A. G. *Inorg. Chem.* **1968**, *7*, 310.
(165) Baker, P. K.; Bury, A. *J. Organomet. Chem.* **1989**, *359*, 189.
(166) Baker, P. K.; Bury, A. *Polyhedron* **1989**, *8*, 917.
(167) Baker, P. K.; Quinlan, A. J. *Inorg. Chim. Acta* **1989**, *162*, 179.
(168) Baker, P. K.; ap Kendrick, D. *Inorg. Chim. Acta* **1991**, *188*, 5.
(169) Baker, P. K.; ap Kendrick, D. *J. Organomet. Chem.* **1994**, *466*, 139.
(170) Baker, P. K.; ap Kendrick, D. *J. Coord. Chem.* **1993**, *30*, 305.
(171) Baker, P. K.; ap Kendrick, D. *Polyhedron* **1994**, *13*, 39.
(172) Miguel, D.; Pérez-Martínez, J. A.; Riera, V.; García-Granda, S. *Polyhedron* **1991**, *10*, 1717.
(173) Miguel, D.; Pérez-Martínez, J. A. Riera, V.; García-Granda, S. *J. Organomet. Chem.* **1993**, *455*, 121.

(174) Miguel, D.; Pérez-Martínez, J. A.; Riera, V.; García-Granda, S. *Angew. Chem., Int. Ed. Engl.* **1992**, *31*, 76.

(175) Cano, M.; Panizo, M. *Polyhedron* **1990**, *9*, 2863.

(176) Cano, M.; Panizo, M.; Campo, J. A.; Tornero, J.; Menéndez, N. *Polyhedron* **1994**, *13*, 1835.

(177) Cano, M.; Panizo, M.; Campo, J. A.; Tornero, J.; Menéndez, N. *J. Organomet. Chem.* **1993**, *463*, 121.

(178) Cano, M.; Campo, J. A.; Panizo, M.; Tornero, J.; Menéndez, N. *Polyhedron* **1994**, *13*, 3309.

(179) Wang, J. T.; Zhang, Y. W.; Xu, Y. M.; Cui, C. M. *Chin. Chem. Lett.* **1993**, *4*, 641.

(180) Ganorkar, M. C.; Stiddard, M. H. B. *Chem. Commun.* **1965**, 22.

(181) Lopez, A.; Panizo, M.; Cano, M. *J. Organomet. Chem.* **1986**, *311*, 145.

(182) Campo, J. A.; Cano, M.; Perpiñán, M. F.; Sánchez-Peláez, A. E. *J. Organomet. Chem.* **1988**, *345*, 299.

(183) Cano, M., Campo, J. A., Ovejero, P.; Heras, J. V. *Inorg. Chim. Acta* **1990**, *170*, 139.

(184) Cano, M.; Campo, J. A. *Polyhedron* **1991**, *10*, 2611.

(185) Cano, M.; Panizo, M.; Campo, J. A.; Gutiérrez-Puebla, E.; Monge, M. A.; Ruiz-Valero, C. *Polyhedron* **1994**, *13*, 1669.

(186) Granifo, J.; Vargas, M. E. *Polyhedron* **1989**, *8*, 1471.

(187) Miguel, D.; Pérez-Martínez, J. A.; Riera, V.; García-Granda, S. *J. Organomet. Chem.* **1991**, *420*, C12.

(188) Miguel, D.; Pérez-Martínez, J. A.; Riera, V.; García-Granda, S. *Organometallics* **1993**, *12*, 1394.

(189) López, E. M.; Miguel, D.; Pérez-Martínez, J. A.; Riera, V. *J. Organomet. Chem.* **1994**, *467*, 231.

(190) Cuyas, J.; Miguel, D.; Pérez-Martínez, J. A.; Riera, V.; García-Grada, S. *Polyhedron* **1992**, *11*, 2713.

(191) Kiplinger, J. L.; Richmond, T. G.; Osterberg, C. E. *Chem. Rev.* **1994**, *94*, 373.

(192) Osterberg, C. E.; King, M. A.; Arif, A. M.; Richmond, T. G. *Angew. Chem., Int. Ed. Engl.* **1990**, *29*, 888.

(193) Perera, S. D.; Shaw, B. L.; Thornton-Pett, M. *J. Organomet. Chem.* **1992**, *428*, 59.

(194) Perera, S. D.; Shaw, B. L. *J. Organomet. Chem.* **1994**, *479*, 117.

(195) Misra, S. N.; Venkatasubramanian, K. *Bull. Chem. Soc. Jpn.* **1988**, *61*, 4482.

(196) Dawson, D. M.; Henderson, R. A.; Hills, A.; Hughes, D. L. *J. Chem. Soc., Dalton Trans.* **1992**, 973.

(197) Rosendorfer, P.; Robl, C.; Beck, W. *Gazz. Chim. Ital.* **1993**, *123*, 145.

(198) Ford, A.; Sinn, E.; Woodward, S. *Polyhedron* **1994**, *13*, 635.

(199) Carmona, E.; Contreras, L.; Gutiérrez-Puebla, E.; Monge, A., Sánchez, L. *Inorg. Chem.* **1990**, *29*, 700.

(200) Brisdon, B. J.; Hodson, A. G. W. *Inorg. Chim. Acta* **1987**, *128*, 51.

(201) Arnáiz, F. J.; García, G.; Riera, V.; Foces Foces, C.; Cano, F. H.; Martinez-Ripoll, M. *J. Organomet. Chem.* **1987**, *332*, 299.

(202) Arnáiz, F. J.; Bartolomé, C. S.; García, G.; Pérez, M. *Inorg. Chim. Acta* **1988**, *141*, 57.

(203) Arnáiz, F. J.; García, G.; Riera, V.; Dromzée, Y.; Jeannin, Y. *J. Chem. Soc., Dalton Trans.* **1987**, 819.

(204) Hillhouse, G. L.; Haymore, B. L. *J. Organomet. Chem.* **1978**, *162*, C23.

(205) Hillhouse, G. L.; Goeden, G. V.; Haymore, B. L. *Inorg. Chem.* **1982**, *21*, 2064.

(206) Bencze, L. *J. Organomet. Chem.* **1972**, *37*, C37.

(207) Guzman, E. C.; Wilkinson, G.; Atwood, J. L.; Rogers, R. D.; Hunter, W. E.; Zaworotko, M. J. *J. Chem. Soc., Chem. Commun.* **1978**, 465.

(208) Carmona, E.; Doppert, K.; Marín, J. M.; Poveda, M. L.; Sánchez, L.; Sánchez-Delgado, R. *Inorg. Chem.* **1984**, *23*, 530.
(209) Cotton, F. A.; Matusz, M. *Polyhedron* **1987**, *6*, 261.
(210) Forster, G. D.; Hogarth, G. *J. Organomet. Chem.* **1994**, *471*, 161.
(211) Atwood, J. L.; Bott, S. G.; Junk, P. C.; May, M. T. *J. Organomet. Chem.* **1995**, *487*, 7.
(212) Contreras, L.; Pizzano, A.; Sánchez, L.; Carmona, E. *J. Organomet. Chem.* **1995**, *500*, 61.
(213) Carmona, E.; Marín, J. M.; Poveda, M. L.; Sánchez, L.; Rogers, R. D.; Atwood, J. L. *J. Chem. Soc., Dalton Trans.* **1983**, 1003.
(214) Carmona, E.; Sánchez, L.; Marín, J. M.; Poveda, M. L.; Atwood, J. L.; Priester, R. D.; Rogers, R. D. *J. Am. Chem. Soc.* **1984**, *106*, 3214.
(215) Carmona, E.; Sánchez, L. *Polyhedron* **1988**, *7*, 163.
(216) Carmona, E.; Muñoz, M. A.; Rogers, R. D. *Inorg. Chem.* **1988**, *27*, 1598.
(217) Carmona, E.; Contreras, L.; Poveda, M. L.; Sánchez, L. J.; Atwood, J. L.; Rogers, R. D. *Organometallics* **1991**, *10*, 61.
(218) Batschelet, W. H.; Archer, R. D.; Whitcomb, D. R. *Inorg. Chem.* **1979**, *18*, 48.
(219) Day, R. O.; Batschelet, W. H.; Archer, R. D. *Inorg. Chem.* **1980**, *19*, 2113.
(220) Brower, D. C.; Winston, P. B.; Tonker, T. L.; Templeton, J. L. *Inorg. Chem.* **1986**, *25*, 2883.
(221) Riera, V.; Arnaiz, F. J.; Herbosa, G. G. *J. Organomet. Chem.* **1986**, *315*, 51.
(222) Chen, G. J.-J.; Yelton, R. O.; McDonald, J. W. *Inorg. Chim. Acta* **1977**, *22*, 249.
(223) Burgmayer, S. J. N.; Templeton, J. L. *Inorg. Chem.* **1985**, *24*, 2224.
(224) Colton, R.; Hoskins, B. F.; Panagiotidou, P. *Aust. J. Chem.* **1988**, *41*, 1295.
(225) Shiu, K.-B.; Yih, K.-H., Wang, S.-L.; Liao, F.-L. *J. Organomet. Chem.* **1991**, *414*, 165.
(226) Bencze, L.; Markó, L. *J. Organomet. Chem.* **1971**, *28*, 271.
(227) Bencze, L.; Markó, L. *J. Organomet. Chem.* **1974**, *69*, C19.
(228) Bencze, L.; Kraut-Vass, A. *J. Mol. Catal.* **1985**, *28*, 369.
(229) Bencze, L.; Kraut-Vass, A.; Prókai, L. *J. Chem. Soc., Chem. Commun.* **1985**, 911.
(230) Borowczak, D.; Szymańska-Buzar, T.; Ziółkowski, J. J. *J. Mol. Catal.* **1984**, *27*, 355.
(231) Szymańska-Buzar, T.; Ziółkowski, J. J. *J. Mol. Catal.* **1987**, *43*, 161.
(232) Szymańska-Buzar, T. *J. Mol. Catal.* **1988**, *48*, 43.
(233) Szymańska-Buzar, T. *J. Mol. Catal.* **1991**, *68*, 177.
(234) Szymańska-Buzar, T. *J. Mol. Catal.* **1994**, *93*, 137.
(235) Kim, H. P.; Angelici, R. J. *Adv. Organomet. Chem.* **1987**, *27*, 51.
(236) Fischer, H.; Hofmann, P.; Kreissl, F. R.; Schrock, R. R.; Schubert, U.; Weiss, K. *Carbyne Complexes;* VCH Publishers: Weinheim, Germany, 1988.
(237) Stone, F. G. A. *Adv. Organomet. Chem.* **1990**, *31*, 53.
(238) Mayr, A. *Comments Inorg. Chem.* **1990**, *10*, 227.
(239) Lappert, M. F.; Pye, P. L. *J. Chem. Soc., Dalton Trans.* **1977**, 1283.
(240) Churchill, M. R.; Wasserman, H. J. *Inorg. Chem.* **1982**, *21*, 3913.
(241) Winter, M. J.; Woodward, S. *J. Organomet. Chem.* **1989**, *361*, C18.
(242) Mayr, A.; Asaro, M. F.; Kjelsberg, M. A.; Lee, K. S.; Van Engen, D. *Organometallics* **1987**, *6*, 432.
(243) Bastos, C. M.; Daubenspeck, N.; Mayr, A. *Angew. Chem., Int. Ed. Engl.* **1993**, *32*, 743.
(244) Mayr, A.; Asaro, M. F.; Glines, T. J.; Van Engen, D.; Tripp, G. M. *J. Am. Chem. Soc.* **1993**, *115*, 8187.
(245) Armstrong, E. M.; Baker, P. K.; Fraser, S. G. *J. Chem. Res. (S)* **1988**, 52; *J. Chem. Res. (M)* **1988**, 0410.
(246) Armstrong, E. M.; Baker, P. K.; Drew, M. G. B. *Organometallics* **1988**, *7*, 319.

(247) Baker, P. K.; Harman, M. E.; Hursthouse, M. B.; Lavery, A. J.; Malik, K. M. A.; Muldoon, D. J.; Shawcross, A. *J. Organomet. Chem.* **1994**, *484*, 169.
(248) Armstrong, E. M.; Baker, P. K.; Drew, M. G. B. *J. Organomet. Chem.* **1987**, *336*, 377.
(249) Armstrong, E. M.; Baker, P. K.; Callow, T.; Flower, K. R.; Jackson, P. D.; Severs, L. M. *J. Organomet. Chem.* **1992**, *434*, 321.
(250) Armstrong, E. M.; Baker, P. K. *Inorg. Chim. Acta* **1988**, *141*, 17.
(251) Baker, P. K.; Drew, M. G. B.; Edge, S.; Ridyard, S. D. *J. Organomet. Chem.* **1991**, *409*, 207.
(252) Baker, P. K.; Cartwright, G. A.; Jackson, P. D.; Flower, K. R.; Galeotti, N.; Severs, L. M. *Polyhedron* **1992**, *11*, 1043.
(253) Baker, P. K.; Ridyard, S. D. *Synth. React. Inorg. Met.-Org. Chem.* **1994**, *24*, 345.
(254) Baker, P. K.; Armstrong, E. M.; Drew, M. G. B. *Inorg. Chem.* **1988**, *27*, 2287.
(255) Baker, P. K.; Armstrong, E. M. *Polyhedron* **1988**, *7*, 63.
(256) Armstrong, E. M.; Baker, P. K. *Synth. React. Inorg. Met.-Org. Chem.* **1988**, *18*, 1.
(257) Armstrong, E. M.; Baker, P. K.; Harman, M. E.; Hursthouse, M. B. *J. Chem. Soc., Dalton Trans.* **1989**, 295.
(258) Tolman, C. A. *Chem. Rev.* **1977**, *77*, 313.
(259) Baker, P. K., Armstrong, E. M.; Drew, M. G. B. *Inorg. Chem.* **1989**, *28*, 2406.
(260) Baker, P. K.; Jackson, P. D. *Inorg. Chim. Acta* **1994**, *219*, 99.
(261) Baker, P. K.; Coles, S. J.; Hursthouse, M. B.; Meehan, M. M.; Ridyard, S. D. *J. Organomet. Chem.* **1995**, *503*, C8.
(262) Baker, P. K.; Armstrong, E. M. *Polyhedron* **1989**, *8*, 351.
(263) Baker, P. K.; Ridyard, S. D. *Polyhedron* **1993**, *12*, 915.
(264) Baker, P. K.; Ridyard, S. D. *Polyhedron* **1993**, *12*, 2105.
(265) Armstrong, E. M.; Baker, P. K. *Inorg. Chim. Acta* **1988**, *143*, 13.
(266) Baker, P. K.; Jackson, P. D.; Drew, M. G. B. *J. Chem. Soc., Dalton Trans.* **1994**, 37.
(267) Baker, P. K.; Jackson, P. D.; Harman, M. E.; Hursthouse, M. B., *J. Organomet. Chem.* **1994**, *468*, 171.
(268) Armstrong, E. M.; Baker, P. K.; Flower, K. R.; Drew, M. G. B. *J. Chem. Soc., Dalton Trans.* **1990**, 2535.
(269) Baker, P. K.; Flower, K. R. *Polyhedron* **1990**, *9*, 2135.
(270) Baker, P. K.; Armstrong, E. M.; Flower, K. R. *Polyhedron* **1988**, *7*, 2769.
(271) Baker, P. K.; Flower, K. R.; Drew, M. G. B.; Forsyth, G. *J. Chem. Soc., Dalton Trans.* **1989**, 1903.
(272) Baker, P. K.; Flower, K. R. *Inorg. Chim. Acta* **1989**, *165*, 241.
(273) Baker, P. K.; Flower, K. R. *Synth. React. Inorg. Met.-Org. Chem.* **1990**, *20*, 451.
(274) Baker, P. K.; Flower, K. R. *Polyhedron* **1990**, *9*, 2233.
(275) Baker, P. K.; Flower, K. R.; Bates, P. A.; Hursthouse, M. B. *J. Organomet. Chem.* **1989**, *372*, 263.
(276) Baker, P. K.; Flower, K. R. *J. Organomet. Chem.* **1994**, *465*, 221.
(277) Baker, P. K.; Flower, K. R. *Z. Naturforsch., Sect. B* **1993**, *48*, 1715.
(278) Baker, P. K.; Flower, K. R. *Polyhedron* **1990**, *9*, 2507.
(279) Baker, P. K.; Flower, K. R. *Polyhedron* **1994**, *13*, 3265.
(280) Baker, P. K.; Flower, K. R. *Z. Naturforsch., Sect. B* **1994**, *49*, 1544.
(281) Baker, P. K.; Flower, K. R. *J. Organomet. Chem.* **1990**, *397*, 59.
(282) Armstrong, E. M.; Baker, P. K. *J. Organomet. Chem.* **1988**, *352*, 133.
(283) Ajayi-Obe, T.; Armstrong, E. M.; Baker, P. K.; Prakash, S. *J. Organomet. Chem.* **1994**, *468*, 165.
(284) Baker, P. K.; Muldoon, D. J.; Lavery, A. J.; Shawcross, A. *Polyhedron* **1994**, *13*, 2915.

(285) Baker, P. K.; Bury, A.; Flower, K. R. *Polyhedron* **1989**, *8*, 2587.
(286) Baker, P. K.; ap Kendrick, D. *Polyhedron* **1991**, *10*, 2519.
(287) Baker, P. K.; ap Kendrick, D. *J. Chem. Soc., Dalton Trans.* **1993**, 1039.
(288) Baker, P. K.; Harman, M. E.; ap Kendrick, D.; Hursthouse, M. B. *Inorg. Chem.* **1993**, *32*, 3395.
(289) Brisdon, B. J.; Hodson, A. G .W.; Mahon, M. F.; Molloy, K. C.; Walton, R. A. *Inorg. Chem.* **1990**, *29*, 2701.
(290) Davidson, J. L.; Vasapollo, G.; Millar, J. C.; Muir, K. W. *J. Chem. Soc., Dalton Trans.* **1987**, 2165.
(291) Davidson, J. L.; Vasapollo, G. *J. Chem. Soc., Dalton Trans.* **1988**, 2855.
(292) Clark, G. R.; Nielson, A. J.; Rae, A. D.; Rickard, C. E. F. *J. Chem. Soc., Dalton Trans.* **1994**, 1783.
(293) Feng, S. G.; Philipp, C. C.; Gamble, A. S.; White, P. S.; Templeton, J. L. *Organometallics* **1991**, *10*, 3504.
(294) Collins, M. A.; Feng, S. G.; White, P. A.; Templeton, J. L. *J. Am. Chem. Soc.* **1992**, *114*, 3771.
(295) Mayr, A.; Bastos, C. M. *J. Am. Chem. Soc.* **1990**, *112*, 7797.
(296) Mayr, A.; Bastos, C. M.; Chang, R. T.; Haberman, J. X.; Robinson, K. S.; Belle-Oudry, D. A. *Angew. Chem., Int. Ed. Engl.* **1992**, *31*, 747.
(297) Mayr, A.; Bastos, C. M.; Daubenspeck, N.; McDermott, G. A. *Chem. Ber.* **1992**, *125*, 1583.
(298) Mayr, A.; Lee, T.-Y. *Angew. Chem., Int. Ed. Engl.* **1993**, *32*, 1726.
(299) Lee, T.-Y., Mayr, A. *J. Am. Chem. Soc.* **1994**, *116*, 10300.
(300) Filippou, A. C. *Polyhedron* **1990**, *9*, 727.
(301) Carfagna, C.; Green, M.; Nagle, K. R.; Williams, D. J.; Woolhouse, C. M. *J. Chem. Soc., Dalton Trans.* **1993**, 1761.
(302) Conole, G.; Hill, K. A.; McPartlin, M.; Mays, M. J.; Morris, M. J. *J. Chem. Soc., Chem. Commun.* **1989**, 688.
(303) Murdoch, H. D. *J. Organomet. Chem.* **1965**, *4*, 119.
(304) Murdoch, H. D.; Henzi, R. *J. Organomet. Chem.* **1966**, *5*, 552.
(305) Brisdon, B. J.; Edwards, D. A.; White, J. W. *J. Organomet. Chem.* **1978**, *156*, 427.
(306) Brisdon, B. J.; Griffin, G. F. *J. Chem. Soc., Dalton Trans.* **1975**, 1999.
(307) Doyle, G. *J. Organomet. Chem.* **1977**, *132*, 243.
(308) Brisdon, B. J.; Brown, D. W.; Willis, C. R.; Drew, M. G. B. *J. Chem. Soc., Dalton Trans.* **1986**, 2405.
(309) Drew, M. G. B.; Brisdon, B. J.; Brown, D. W.; Willis, C. R. *J. Chem. Soc., Chem. Commun.* **1986**, 1510.
(310) Brisdon, B. J.; Hodson, A. G. W.; Mahon, M. F.; Molloy, K. C. *J. Organomet. Chem.* **1988**, *344*, C8.
(311) Brisdon, B. J.; Deeth, R. J.; Hodson, A. G. W.; Kemp, C. M.; Mahon, M. F.; Molloy, K. C. *Organometallics* **1991**, *10*, 1107.
(312) Brisdon, B. J.; Hodson, A. G. W.; Mahon, M. F. *Organometallics* **1994**, *13*, 2566.
(313) Holloway, C. E.; Kelly, J. D.; Stiddard, M. H. B. *J. Chem. Soc.* (A) **1969**, 931.
(314) Hull, C. G.; Stiddard, M. H. B. *J. Organomet. Chem.* **1967**, *9*, 519.
(315) Hayter, R. G. *J. Organomet. Chem.* **1968**, *13*, P1.
(316) tom Dieck, H.; Friedel, H. *J. Organomet. Chem.* **1968**, *14*, 375.
(317) Baker, P. K. *Inorg. Chim. Acta* **1986**, *118*, L3.
(318) Shiu, K.-B.; Chang, C.-J.; Wang, S.-L.; Liao, F.-L. *J. Organomet. Chem.* **1991**, *407*, 225.
(319) Joshi, V. S.; Sarkar, A.; Rajamohanan, P. R. *J. Organomet. Chem.* **1991**, *409*, 341.

(320) Drew, M. G. B.; Brisdon, B. J.; Day, A. J. Chem. Soc., Dalton Trans. **1981**, 1310.
(321) Rousche, J.-C.; Dobson, G. R. J. Organomet. Chem. **1978**, *150*, 239.
(322) Lee, G.-H.; Peng, S.-M.; Liu, F.-C.; Mu, D.; Liu, R.-S. Organometallics **1989**, *8*, 402.
(323) Wu, S. L.; Cheng, C.-Y.; Wang, S.-L.; Liu, R.-S. Inorg. Chem. **1991**, *30*, 311.
(324) Paz-Sandoval, M. A.; Saavedra, P. J.; Pomposo, G. D.; Joseph-Nathan, P.; Powell, P. J. Organomet. Chem. **1990**, *387*, 265.
(325) King, R. B.; Saran, M. S. Inorg. Chem. **1974**, *13*, 2453.
(326) Deaton, J. C.; Walton, R. A. J. Organomet. Chem. **1981**, *219*, 187.
(327) Hsieh, A. T. T.; West, B. O. J. Organomet. Chem. **1976**, *112*, 285.
(328) Brisdon, B. J.; Cartwright, M.; Hodson, A. G. J. Organomet. Chem. **1984**, *277*, 85.
(329) Gracey, G. D.; Rettig, S. J.; Storr, A.; Trotter, J. Can. J. Chem. **1987**, *65*, 2469.
(330) Lush, S.-F.; Wang, S.-H.; Lee, G.-H.; Peng, S.-M.; Wang, S.-L.; Liu, R.-S. Organometallics **1990**, *9*, 1862.
(331) Clark, D. A.; Jones, D. L.; Mawby, R. J. J. Chem. Soc., Dalton Trans. **1980**, 565.
(332) tom Dieck, H.; Friedel, H. J. Chem. Soc., Chem. Commun. **1969**, 411.
(333) Bridson, B. J.; Edwards, D. A.; Paddick, K. E. Transition-Met. Chem. **1981**, *6*, 83.
(334) Brisdon, B. J. J. Organomet. Chem. **1977**, *125*, 225.
(335) Brisdon, B. J.; Paddick, K. E. J. Organomet. Chem. **1978**, *149*, 113.
(336) Kapoor, P. N.; Pathak, D. D.; Gaur, G.; Mercykutty, P. C. J. Organomet. Chem. **1987**, *322*, 71.
(337) Kapoor, P. N.; Mercykutty, P. C. J. Organomet. Chem. **1988**, *339*, 97.
(338) Campbell, M. J. M.; Morrison, E.; Rogers, V.; Baker, P. K. Inorg. Chim. Acta **1987**, *127*, L17.
(339) Campbell, M. J. M.; Morrison, E. C.; Rogers, V.; Baker, P. K. Polyhedron **1988**, *7*, 1719.
(340) Baker, P. K.; Campbell, M. J. M.; White, M. V. Inorg. Chim. Acta **1985**, *98*, L27.
(341) Campbell, M. J. M.; Morrison, E.; Rogers, V.; Baker, P. K. Transition-Met. Chem. **1986**, *11*, 381.
(342) Campbell, M. J. M.; Houghton, E.; Rogers, V.; Baker, P. K. Synth. React. Inorg. Met.-Org. Chem. **1986**, *16*, 1237.
(343) Miguel, D.; Pérez-Martínez, J. A.; Riera, V.; García-Granda, S. Organometallics **1994**, *13*, 1336.
(344) Curtis, M. D.; Fotinos, N. A. J. Organomet. Chem. **1984**, *272*, 43.
(345) Graham, A. J.; Fenn, R. H. J. Organomet. Chem. **1969**, *17*, 405.
(346) Graham, A. J.; Fenn, R. H. J. Organomet. Chem. **1970**, *25*, 173.
(347) Graham, A. J.; Akrigg, D.; Sheldrick, B. Cryst. Struct. Commun. **1976**, *5*, 891.
(348) Graham, A. J.; Akrigg, D.; Sheldrick, B. Cryst. Struct. Commun. **1977**, *6*, 253.
(349) Drew, M. G. B.; Brisdon, B. J.; Edwards, D. A.; Paddick, K. E. Inorg. Chim. Acta **1979**, *35*, L381.
(350) Drew, M. G. B.; Brisdon, B. J., Cartwright, M. Inorg. Chim. Acta **1979**, *36*, 127.
(351) Graham, A. J.; Akrigg, D.; Sheldrick, B. Acta Crystallogr., Sect. C **1983**, *39*, 192.
(352) Graham, A. J.; Akrigg, D.; Sheldrick, B. Acta Crystallogr., Sect. C **1985**, *41*, 995.
(353) Cotton, F. A.; Luck, R. L. Acta Crystallogr., Sect. C **1990**, *46*, 138.
(354) Jordanov, J.; Behm, H.; Beurskens, P. T. J. Cryst. Spectrosc. Res. **1991**, *21*, 657.
(355) Curtis, M. D.; Eisenstein, O. Organometallics **1984**, *3*, 887.
(356) Faller, J. W.; Haitko, D. A.; Adams, R. D.; Chodosh, D. F. J. Am. Chem. Soc. **1979**, *101*, 865.
(357) Brisdon, B. J.; Griffin, G. F.; Pierce, J.; Walton, R. A. J. Organomet. Chem. **1981**, *219*, 53.
(358) Brisdon, B. J.; Conner, K. A.; Walton, R. A. Organometallics **1983**, *2*, 1159.

(359) Brisdon, B. J.; Enger, S. K.; Weaver, M. J.; Walton, R. A. *Inorg. Chem.* **1987,** *26,* 3340.
(360) Brisdon, B. J.; Floyd, A. J. *J. Organomet. Chem.* **1985,** *288,* 305.
(361) Trost, B. M.; Lautens, M. *J. Am. Chem. Soc.* **1982,** *104,* 5543.
(362) Trost, B. M.; Lautens, M. *J. Am. Chem. Soc.* **1983,** *105,* 3343.
(363) Trost, B. M.; Hung, M.-H. *J. Am. Chem. Soc.* **1983,** *105,* 7757.
(364) Maitlis, P. M.; Games, (Mrs.) M. L. *Chem. Ind. (London)* **1963,** 1624.
(365) Maitlis, P. M.; Efraty, A. *J. Organomet. Chem.* **1965,** *4,* 172.
(366) Efraty, A. *Can. J. Chem.* **1969,** *47,* 4695.
(367) Mathew, M.; Palenik, G. J. *Can. J. Chem.* **1969,** *47,* 705.
(368) Davidson, J. L.; Sharp, D. W. A. *J. Chem. Soc., Dalton Trans.* **1975,** 2531.
(369) Davidson, J. L.; Green, M.; Stone, F. G. A.; Welch, A. J. *J. Chem. Soc., Dalton Trans.* **1976,** 738.
(370) Davidson, J. L.; Green, M.; Stone, F. G. A.; Welch, A. J. *J. Chem. Soc., Dalton Trans.* **1977,** 287.
(371) Davidson, J. L. *J. Chem. Soc., Chem. Commun.* **1980,** 113.
(372) Davidson, J. L. *J. Organomet. Chem.* **1991,** *419,* 137.
(373) Hughes, R. P.; Reisch, J. W.; Rheingold, A. L. *Organometallics* **1984,** *3,* 1761.
(374) Davidson, J. L.; Vasapollo, G. *J. Organomet. Chem.* **1983,** *241,* C24.
(375) Davidson, J. L.; Vasapollo, G. *J. Chem. Soc., Dalton Trans.* **1985,** 2231.
(376) Cotton, F. A.; Meadows, J. H. *Inorg. Chem.* **1984,** *23,* 4688.
(377) Carlton, L.; Davidson, J. L.; Vasapollo, G.; Douglas, G.; Muir, K. W. *J. Chem. Soc., Dalton Trans.* **1993,** 3341.
(378) Davidson, J. L. *Personal Communication.*
(379) Davidson, J. L.; Vasapollo, G.; Nobile, C. F.; Sacco, A. *J. Organomet. Chem.* **1989,** *371,* 297.
(380) King, R. B.; Efraty, A.; Douglas, W. M. *J. Organomet. Chem.* **1973,** *60,* 125.
(381) Giannotti, C.; Merle, G. *J. Organomet. Chem.* **1976,** *105,* 97.
(382) Laine, R. M.; Ford, P. C. *Inorg. Chem.* **1977,** *16,* 388.
(383) Scott, S. L.; Espenson, J. H.; Bakac, A. *Organometallics* **1993,** *12,* 1044.
(384) Hooker, R. H.; Mahmoud, K. A.; Rest, A. J. *J. Chem. Soc., Dalton Trans.* **1990,** 1231.
(385) Sterzo, C. L. *J. Organomet. Chem.* **1991,** *408,* 253.
(386) Song, L.-C.; Zhang, L.-Y.; Hu, Q.-M.; Huang, X.-Y. *Inorg. Chim. Acta* **1995,** *230,* 127.
(387) Curtis, M. D.; Klingler, R. J. *J. Organomet. Chem.* **1978,** *161,* 23.
(388) Saadeh, C.; Colbran, S. B.; Craig, D. C.; Rae, A. D. *Organometallics* **1993,** *12,* 133.
(389) King, R. B.; Bisnette, M. B. *Inorg. Chem.* **1965,** *4,* 475.
(390) Salzer, V. A.; Werner, H. *Z. Anorg. Allg. Chem.* **1975,** *418,* 88.
(391) Chaiwasie, S.; Fenn, R. H. *Acta Crystallogr., Sect. B* **1968,** *24,* 525.
(392) Mawby, A.; Pringle, G. E. *J. Inorg. Nucl. Chem.* **1972,** *34,* 525.
(393) Hart-Davis, A. J.; White, C.; Mawby, R. J. *Inorg. Chim. Acta* **1970,** *4,* 431.
(394) White, C.; Mawby, R. J. *J. Chem. Soc. (A)* **1971,** 940.
(395) King, R. B.; Gimeno, J. *Inorg. Chem.* **1978,** *17,* 2396.
(396) Blumer, D. J.; Barnett, K. W.; Brown, T. L. *J. Organomet. Chem.* **1979,** *173,* 71.
(397) Alway, D. G.; Barnett, K. W. *Inorg. Chem.* **1980,** *19,* 1533.
(398) Coville, N. J. *J. Organomet. Chem.* **1980,** *190,* C84.
(399) Fick, H.-G.; Beck, W. *J. Organomet. Chem.* **1983,** *252,* 83.
(400) Faller, J. W.; Anderson, A. S.; Jakubowski, A. *J. Organomet. Chem.* **1971,** *27,* C47.
(401) Alway, D. G.; Barnett, K. W. *J. Organomet. Chem.* **1975,** *99,* C52.
(402) Xia, W.; Hill, R. H. *Polyhedron* **1992,** *11,* 1319.
(403) Beach, D. L.; Barnett, K. W. *J. Organomet. Chem.* **1975,** *97,* C27.

(404) Todd, L. J.; Wilkinson, J. R.; Hickey, J. P.; Beach, D. L.; Barnett, K. W. *J. Organomet. Chem.* **1978,** *154,* 151.
(405) Lau, Y. Y.; Gipson, S. L. *Inorg. Chim. Acta* **1989,** *157,* 147.
(406) Lau, Y. Y.; Huckabee, W. W.; Gipson, S. L. *Inorg. Chim. Acta* **1990,** *172,* 41.
(407) Snow, M. R.; Stiddard, M. H. B. *Chem. Commun.* **1965,** 580.
(408) Stiddard, M. H. B.; Townsend, R. E. *J. Chem. Soc. (A)* **1969,** 2355.
(409) Snow, M. R.; Pauling, P.; Stiddard, M. H. B. *Aust. J. Chem.* **1969,** *22,* 709.
(410) Barbati, A.; Calderazzo, F.; Poli, R.; Zanazzi, P. F. *J. Chem. Soc., Dalton Trans.* **1986,** 2569.
(411) Calderazzo, F.; Poli, R.; Zanazzi, P. F. *Gazz. Chim. Ital.* **1988,** *118,* 583.
(412) Barbati, A.; Calderazzo, F.; Poli, R. *Gazz. Chim. Ital.* **1988,** *118,* 589.
(413) Calderazzo, F.; Poli, R.; Zanazzi, P. F. *Gazz. Chim. Ital.* **1988,** *118,* 595.
(414) Cotton, F. A.; Poli, R. *Inorg. Chem.* **1987,** *26,* 1514.
(415) Beddoes, R. L.; Davies, E. S.; Whiteley, M. W. *J. Chem. Soc., Dalton Trans.* **1995,** 3231.

ADVANCES IN ORGANOMETALLIC CHEMISTRY, VOL. 40

Substituent Effects as Probes of Structure and Bonding in Mononuclear Metallocenes

MELANIE L. HAYS and TIMOTHY P. HANUSA

Department of Chemistry
Vanderbilt University
Nashville, Tennessee

I

INTRODUCTION AND SCOPE OF REVIEW

The discovery of bis(cyclopentadienyl)iron (ferrocene) in 1951[1,2] and the subsequent development of its substitution reactions[3] provided the first evidence for the seemingly inexhaustible chemistry that exists between metals and cyclopentadienyl rings. During the past four decades, virtually every metal and metalloid in the periodic table has been used to form thermally stable compounds with the cyclopentadienyl (Cp) ligand, and nearly 80% of all organometallic compounds are now estimated to contain one or more Cp rings.[4] An enormous number of variations on the classic metallocene framework have been made by using Cp ligands with one or more organic, inorganic, or organometallic substituents.[5] In addition, linked Cp rings have been used to produce metallocenophanes and *ansa*-metallocenes,[6–9] and Cp relatives such as indene and fluorene have been incorporated into many sandwich compounds.[10–15]

A continuing motivation for the study of substituent effects in metallocenes has been to modify and consequently illuminate the interactions between metals and cyclopentadienyl rings. Two examples from the recent literature illustrate this point. Main-group metallocenes possess a surprisingly large range of structural types; "calcocene" $[CaCp_2]_n$ (**1**) (Fig. 1), for instance, forms a three-dimensional coordination polymer in the solid state, which is considered a consequence of highly ionic bonding, i.e., $\{[Ca]^{2+}[Cp]_2^-\}_n$.[16] In contrast, crystalline stannocene (**2**) contains monomeric $[SnCp_2]$ units; the canting of the rings in this presumably more covalent molecule has been ascribed to the operation of a stereochemically active pair of valence electrons.[17] The permethylated metallocenes $[M(C_5Me_5)_2]$ of both elements (**3**, **4**) are bent monomers, however, with virtually identical angles between the ring planes.[18,19] The structural similarities between these and related metallocene pairs have prompted a reevaluation of the bending forces in Group 2 and Group 14 metallocenes and have led to the proposal that ligand interactions dominate the geometries of many main-group metallocenes.[20]

A second example concerns the relative orientation of the Cp rings in transition-metal metallocenes, which have not always been easy to rationalize. Ferrocene itself is disordered in the solid state at room temperature,[21] but it has an eclipsed geometry in the gas phase,[22] as do rutheno-

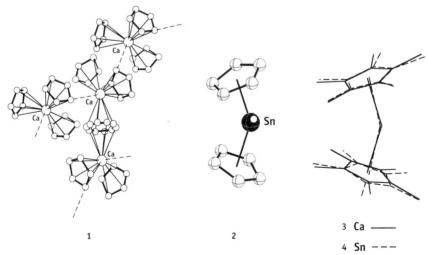

1   2

3 Ca ———
4 Sn – – –

FIG. 1.   Solid-state structures of polymeric $[CaCp_2]_n$ (**1**) and monomeric $[SnCp_2]$ (**2**). The decamethylmetallocenes of each metal (**3** and **4**) are virtually isostructural.

cene $[RuCp_2]^{23}$ and osmocene $[OsCp_2]^{24}$ in the solid state. Decamethylfer-
rocene $[Fe(C_5Me_5)_2]$ is perfectly staggered in the gas phase[25] and as a
solid.[26] Crystalline decachlororuthenocene $[Ru(C_5Cl_5)_2]$, in contrast, is
eclipsed.[27] Molecular mechanics calculations[28] indicate that an electronic
preference for eclipsed geometries exists in all these compounds, which
can be overcome when steric repulsions are high (as in $[Fe(C_5Me_5)_2]$), but
which reasserts itself when such interactions are reduced (as in
$[Ru(C_5Cl_5)_2]$).

These examples illustrate how substituent effects are closely coupled
to fundamental features of structure and bonding in metallocenes. This
review is focused on such relationships, especially with the structures
and physical properties of main-group metallocenes, and the energetics
and electronic properties of the transition metal compounds. For our
purposes, we define a *metallocene* as a mononuclear bis(cyclopentadienyl)
metal compound with independent rings (i.e., $[MCp'Cp'']$, where Cp',
$Cp'' = \pi\text{-}C_5R_nH_{5-n}$). Excluded by this definition are, for example, tris
(cyclopentadienyl)lanthanide "metallocenes" $[LnCp_3']$ and bridged metallocenes (metallocenophanes) or bimetallocenes (e.g., $[CpM(C_5H_4)-$
$(C_5H_4)M'Cp]$). Also excluded are various Group 4 "metallocenes" of the
form $[MCp_2'X_2]$ and their ions ($[MCp_2'X]^+$). Such organometallic complexes
have proven valuable in the stereospecific polymerization of olefins, and
the effects of substituents in controlling the properties of the resulting
polymers have received considerable attention.[29-32]

Because metallocenes themselves can serve as functional groups, many
compounds containing the MCp_2' unit that might be considered as substituted metallocenes are not discussed here. For example, the compound
$[Fe\{(C_5H_4)PPh_2\}_2]$ (**5**)[33] could be regarded as a phosphino-substituted ferrocene (**5a**) (Fig. 2) or as a ferrocenyl-substituted diphosphine (**5b**) (Fig. 2).

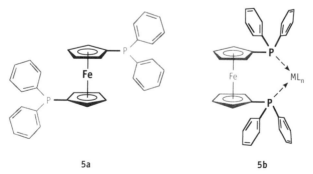

5a   5b

FIG. 2.   Depending on the part of the molecule of interest, $[Fe\{(C_5H_4)PPh_2\}_2]$ (**5**) could
be viewed as a substituted ferrocene (**5a**) or as a substituted phosphine (**5b**); it is used as a
chelating phosphine in synthesis.

In fact, the interest in this complex is as a chelating phosphine, and the metal–ring interaction is generally of little concern.[34] Similarly, metallocenyl fragments have been incorporated as electron donor units into nonlinear optical materials (e.g., [FeCp(C_5H_4)CH=CH(4-C_6H_4-1-NO_2)] (**6**),[35,36] but the metallocenyl units are valued for their contribution to molecular hyperpolarizabilities; **6** is not principally studied as an alkenyl substituted metallocene. In the same way, the redox properties of the ferrocenyl group in *cis*-[Pt(C_6H_4X)C≡C{FeCp(C_5H_4)}(dppe)] (**7**) are believed to be responsible for the facile conversion of **7** by oxidatively induced reductive elimination to the coupled compound (**8**). The novel chemistry occurs around the Pt(II) center, however, and not at the ferrocenyl moiety.[37] Compounds such as **5–8** are thus placed outside the scope of this survey.

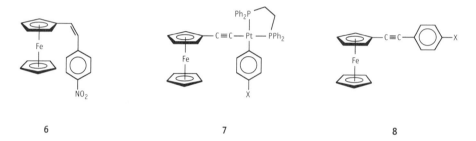

6   7   8

Even with a narrowed definition of substituted metallocenes, the extent of the literature is such that exhaustive coverage is not possible within the limits of this review. Recent reviews are available for some areas of substituted metallocene research; we have noted these and correspondingly omitted or restricted our coverage of such subjects here [e.g., solid-state NMR studies[38]; synthetic methods (see Section II)]. We have attempted to highlight major developments of the past 10 years; older material is cited when required to provide context.

II

SUBSTITUENT EFFECTS ON METALLOCENE SYNTHESIS AND STABILITY

The preparation of most monosubstituted and many polysubstituted metallocenes (especially ferrocene derivatives) is now routine, and extensive reviews of synthetic methods have been published.[4,5] In brief, one of four approaches is used to produce derivatized metallocenes: (1) a

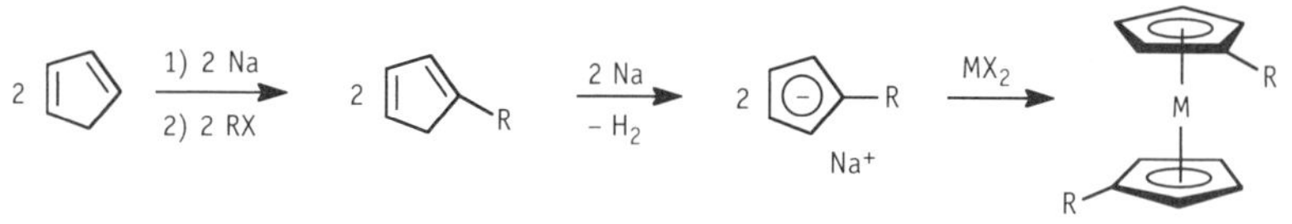

SCHEME 1

substituted cyclopentadiene is deprotonated and then allowed to react with a metal halide or amide (Scheme 1); (2) the rings of an existing metallocene are substituted *in situ* (e.g., Scheme 2[39]); (3) an organometallic precursor complex is altered so that the coordinated organic ligand becomes a cyclopentadienyl ligand (e.g., Scheme 3[40]); or (4) cyclopentadienyl ring metathesis is performed with an existing metallocene (e.g., [Eq. (1)]).[41]

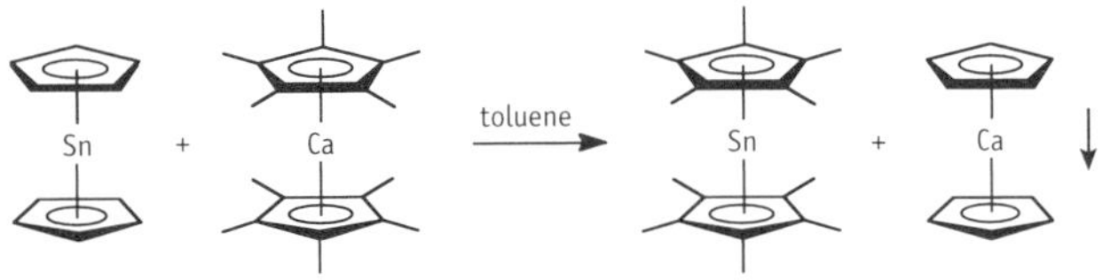

(1)

There are several types of highly substituted metallocenes, however, that are difficult enough to prepare that questions have been raised about the stability of their metal–ring interactions[42]; these are discussed here.

A. *Octaphenylmetallocenes of Main-Group Metals*

Although octaphenylmetallocenes $[M(C_5Ph_4H)_2]$ of several transition metals (M = V, Cr, Co, Ni,[43] Fe,[44] Ru[45]) have been known for some time, similar compounds have not been prepared for the lanthanide metals, and the main-group analogues display surprising instability. The octaphenyl-

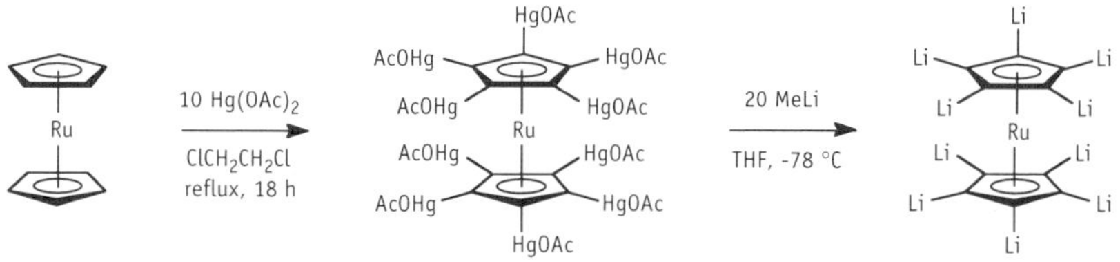

SCHEME 2

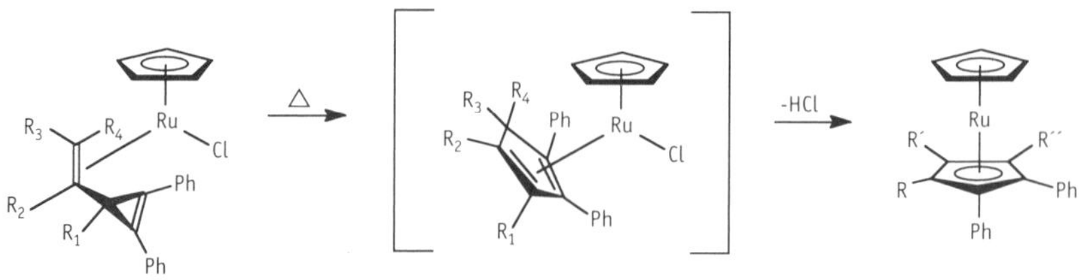

R	a	b	c	d	e
1	Ph	H	H	Ph	Ph
2	H	H	Me	H	H
3	H	H	H	Me	H
4	H	H	H	H	Me

SCHEME 3

metallocenes of Ge, Sn, and Pb have been made by the reaction of the lithium salt with the metal dichloride[46] or acetate (Pb only)[47] [Eq. (2)].

$$2 \, Li[C_5Ph_4H] \; + \; MCl_2 \; \longrightarrow \; [M(C_5Ph_4H)_2] \; + \; 2 \, LiCl \downarrow \qquad (2)$$

These compounds are less thermally and kinetically stable than either their unsubstituted or their decaphenylmetallocene analogues; they decompose near 200°C in the solid state (cf. 350°C for the decaphenylmetallocene derivatives $[M(C_5Ph_5)_2]$; M = Ge, Sn, Pb[47]) and in minutes in THF at room temperature.[46] A yellow barium complex has been prepared by the reaction of two equivalents of $C_5Ph_4H_2$ with the soluble barium amide $[Ba\{N(SiMe_3)_2\}_2(thf)_2]$ in THF or toluene [Eq. (3)].[41]

$$2 \, C_5Ph_4H_2 + [Ba\{N(SiMe_3)_2\}_2(thf)_2] \; \longrightarrow \; [Ba(C_5Ph_4H)_2(thf)] \; + \; 2 \, HN(SiMe_3) \qquad (3)$$

As with the Group 14 octaphenylmetallocenes, $[Ba(C_5Ph_4H)_2(thf)]$ exhibits a surprising lack of thermal stability. Attempted sublimation at 120°C and 10^{-6} torr for 3 hr or storage at room temperature for 12 days causes decomposition. The Lewis acidity of the metals, open metal coordination sites, and a lack of efficient electron donation from the $[C_5Ph_4H]^-$ ligand have been suggested as contributors to the instability of the main-group octaphenylmetallocenes.[4] Considering the high thermal and chemical stability of the related decaphenylmetallocenes, the open coordination sphere of the Group 14 octaphenyl analogues is probably the most critical of these. A thf ligand fills a coordination site in the barium complex, although it may be loosely bound.

B. *Polyhalogenated Metallocenes*

The effect of multiply halogenated cyclopentadienyl rings on the electro-chemistry and ionization energies of their associated metallocenes has received increased attention in recent years (see Section V). Some haloge-nated (Cl, Br, I) ferrocenes and ruthenocenes have been prepared by repetitive metalation exchange/halogenation reactions (Scheme 4).[48] Other highly halogenated derivatives have been synthesized using permercura-tion followed by perhalogenation of the parent metallocene (Scheme 5),[49–51] and from the reaction of diazotetrachlorocyclopentadiene with $[RuCl_2(C_5Me_5)]$ [Eq. (4)].[52]

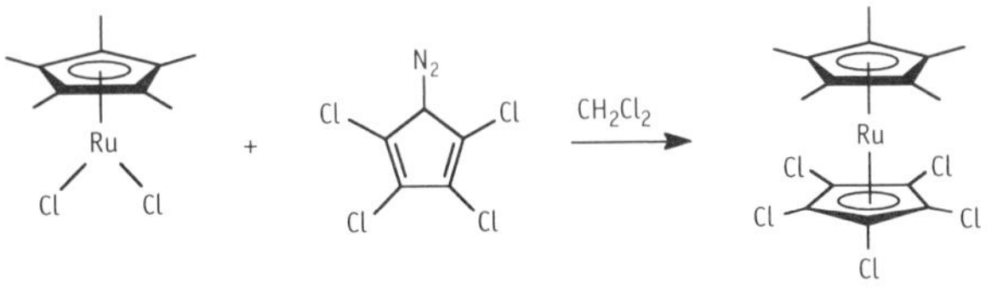

$$(4)$$

Unlike the compounds obtained with the heavier halogens, polyfluori-nated metallocene derivatives cannot be prepared from the permercurated metallocenes. Thermally stable pentafluorinated ruthenocenes have been made from the (oxocyclohexadienyl)(cyclopentadienyl)metal complexes by flash vacuum pyrolysis (Scheme 6).[53,54] Decafluorometallocenes, $[M(C_5F_5)_2]$, are still unknown; potential routes to these compounds starting from fluorinated cyclopentadiene (C_5F_5H)[55] or from the decachlorometal-locenes[48] have been unsuccessful. However, there seems to be no funda-mental steric or electronic barrier to their eventual preparation.

C. *Decaphenylferrocene*

A yellow solid described as decaphenylferrocene, $[Fe(C_5Ph_5)_2]$ (9) (Scheme 7), was originally reported as the product from the reaction of $[Fe(CO)_2Br(C_5Ph_5)]$ (10) with $[Li(C_5Ph_5)]$ in refluxing xylene.[56,57] Later

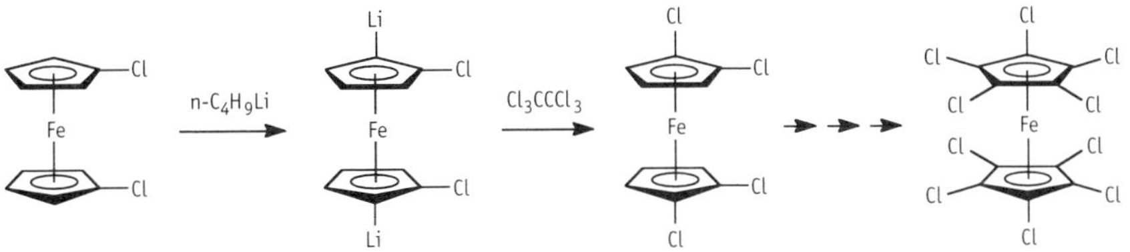

SCHEME 4

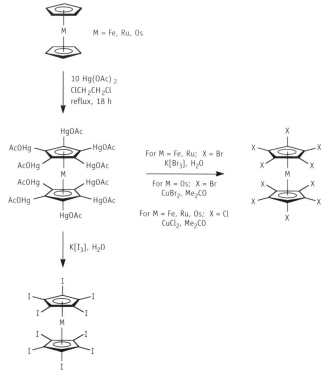

SCHEME 5

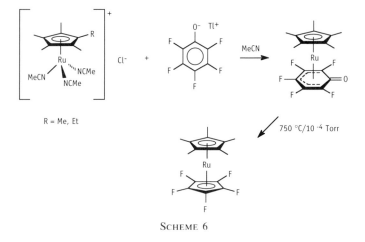

SCHEME 6

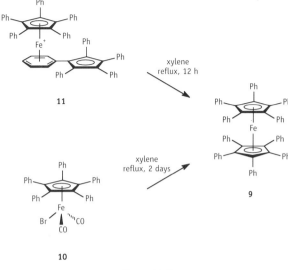

SCHEME 7

attempts to reproduce this synthesis were unsuccessful,[4] and an attempt to prepare **9** from the reaction of [Fe(CO)$_5$], Zn dust, and C$_5$Ph$_5$Br led instead to a deep-blue solid characterized by NMR, mass spectroscopy, and electrochemical analysis as the linkage isomer [(η^5-C$_5$Ph$_5$)$_2$Fe(η^6-Ph-C$_5$Ph$_4$)] (**11**) [Eq. (5)].[58] Although the question was raised whether a sym-

$$\text{(5)}$$

11

metrical sandwich structure for **9** (i.e., [Fe(η^5-C$_5$Ph$_5$)$_2$]) was too sterically encumbered to exist, two groups independently reported the synthesis of pink or maroon **9** by thermal treatment of either **10**[59] or **11**[60] (Scheme 7). A single-crystal X-ray diffraction study of **9** confirmed its sandwich structure.[59] Decaphenylferrocene was oxidized to the [Fe(η^5-C$_5$Ph$_5$)$_2$]$^+$ cation using either Br$_2$[60] or NOBF$_4$[59]; both salts were structurally characterized with X-ray diffraction.

D. Decaisopropylated Metallocenes

Although 1,2,3,4,5-pentaisopropylcyclopentadiene can be prepared by successive isopropylations starting from sodium cyclopentadienide,[61] the reaction of $Na[C_5(i\text{-}Pr)_5]$ (12) with $FeCl_2$ did not produce the desired decaisopropylferrocene, but instead the unusually stable pentaisopropylcyclopentadienyl radical (13) [Eq. (6)].[62] A direct synthesis of 13 from bromine

oxidation of 11 proceeds in 70% yield; the product has been characterized by ESR/ENDOR, NMR, mass spectroscopy, and X-ray diffraction.[63] In the crystal lattice, the rings stack with an interplanar spacing of 5.82 Å, a distance that may be sterically imposed and that has been suggested as the reason two such rings would be prohibited from coordinating to an iron center. Although the steric interactions between two $\cdot C_5(i\text{-}Pr)_5$ rings could be substantial, it should be noted that the closely related tetraisopropylcyclopentadiene also crystallizes with stacked rings (interplanar spacing = 5.11 Å)[64] and readily forms a ferrocene derivative.[65]

The formation of 13 may hinder some attempts to generate decaisopropylmetallocenes by solution methods, but the stable decaisopropylrhodocenium cation $[Rh(C_5(i\text{-}Pr)_5)_2]^+[PF_6]^-$ (14) was isolated in 55% yield as the product of a novel methylation reaction starting from $[Rh(C_5Me_5)_2]^+[PF_6]^-$ [Eq. (7)].[66] A similar reaction was used to produce the chiral mixed ring complexes $[Co\{C_5(i\text{-}Pr)_5\}Cp]^+[PF_6]^-$ and $[Co\{C_5(i\text{-}CHEt_2)_5\}Cp]^+[PF_6]^-$.[67]

14

III

PHYSICAL AND CHEMICAL PROPERTIES AFFECTED BY LIGAND SUBSTITUTION

A. *Melting Points*

Melting points of organometallic solids vary with the strength of intermolecular interactions, which in turn are affected by ligand bulk and lattice packing. If strictly isostructural metallocenes are being compared, their melting points do not vary systematically with the metal. The classic transition metal metallocenes $[MCp_2]$ (M = V—Ni), for example, have melting points within the range 167–176°C (Table I). Monosubstitution significantly lowers this value, as the melting points for the 1,1′-dimethyl-metallocenes are all below 65°C (Table I). The reduced melting points for these compounds indicate the extent to which asymmetry in the Cp′ ring can interfere with lattice packing. Similar comparisons are not possible with the main-group metallocenes, as the parent compounds are not all isostructural; some are polymeric (e.g., $[CaCp_2]_n$, $[PbCp_2]_n$), whereas others are monomeric ($[GeCp_2]$, $[SnCp_2]$) (Table I).

The melting points of main-group metallocenes with bulky Cp rings are more easily compared, as they are generally monomeric. Such highly substituted metallocenes have elevated melting points, a consequence of increased molecular weight and often more symmetrical structures, which facilitate crystal packing. A set of unusual exceptions to this generalization has been discovered with the main-group isopropyl-substituted metallocenes. The $[M(C_5(i\text{-}Pr)_4H_2]$ complexes are crystalline solids with melting points above 120°C, whereas the $[M(C_5(i\text{-}Pr)_3H_2)_2]$ compounds are oils or low-melting, easily supercooled solids (Table I). It is possible to trace the differences between the two types of metallocenes to the effect that the removal of an isopropyl group from the $[C_5(i\text{-}Pr)_4H]^-$ ring has on the remaining substituents. Molecular modeling calculations suggest that reduced steric congestion between the isopropyl groups in $[C_5(i\text{-}Pr)_3H_2]^-$ lowers the energy difference between various ring conformations (Fig. 3).[68] Such increased flexibility may translate into greater difficulty in achieving a unique packing arrangement, and ultimately into reduced lattice stabilization and melting temperature.

Owing to shorter M–C distances, increased intramolecular contact between the $[C_5(i\text{-}Pr)_3H_2]^-$ anions in the transition metal $[M(C_5(i\text{-}Pr)_3H_2)_2]$ complexes should decrease the orientational freedom of the isopropyl groups. The melting points of the complexes are in fact higher ($\geq 105°C$) (Table I).

TABLE I

MELTING POINTS OF METALLOCENES, DIMETHYLMETALLOCENES, AND ISOPROPYLATED METALLOCENES

Compound	Mol. wt. (g/mol)	Aggregation[a]	mp (°C)	Ref.
VCp_2	181	Monomeric	167	204,205
$CrCp_2$	182	Monomeric	173	206,207
$MnCp_2$	185	Monomeric (becomes polymeric below 160°C)	173	165,208
$FeCp_2$	186	Monomeric	173	1,21
$CoCp_2$	189	Monomeric	173	102,209
$NiCp_2$	189	Monomeric	173	209,210
$BeCp_2$	139	Monomeric	59–60	124,211
$MgCp_2$	154	Monomeric	176	112,212
$CaCp_2$	170	Polymeric	>265	16,213
$SrCp_2$	218	(Polymeric)	>360	213
$BaCp_2$	258	(Polymeric)	>420	213
$ZnCp_2$	196	Polymeric	150	131,214
$CdCp_2$	243	Unknown	250 (dec)	214
$GeCp_2$	203	Monomeric	78	215
$SnCp_2$	249	Monomeric	105	17,216
$PbCp_2$	337	Polymeric	140	216,217
$V(C_5MeH_4)_2$	209	(Monomeric)	26–28	218
$Cr(C_5MeH_4)_2$	210	(Monomeric)	41	219
$Mn(C_5MeH_4)_2$	213	(Monomeric)	61–63	220
$Fe(C_5MeH_4)_2$	214	Monomeric	38	101,221
$Co(C_5MeH_4)_2$	217	(Monomeric)	35–36	222
$Ni(C_5MeH_4)_2$	217	(Monomeric)	36–38	223
$Mg(C_5MeH_4)_2$	183	(Monomeric)	29–30	212
$Sn(C_5MeH_4)_2$	277	(Monomeric)	<25	216
$Pb(C_5MeH_4)_2$	365	(Monomeric)	(Low melting solid, readily supercools)	216
$Mn(C_5(i\text{-}Pr)_3H_2)_2$	438	Monomeric	116–118	179
$Fe(C_5(i\text{-}Pr)_3H_2)_2$	439	Monomeric	173	65
$Co(C_5(i\text{-}Pr)_3H_2)_2$	442	Monomeric	136–138	75
$Ni(C_5(i\text{-}Pr)_3H_2)_2$	441	Monomeric	107–110	68
$Mg(C_5(i\text{-}Pr)_3H_2)_2$	407	(Monomeric)	92–100	224
$Ca(C_5(i\text{-}Pr)_3H_2)_2$	423	Monomeric	40–45	224
$Sr(C_5(i\text{-}Pr)_3H_2)_2$	470	(Monomeric)	<30	225
$Ba(C_5(i\text{-}Pr)_3H_2)_2$	520	(Polymeric)	92–94	225
$Zn(C_5(i\text{-}Pr)_3H_2)_2$	448	(Monomeric)	ca. 30	78
$Sn(C_5(i\text{-}Pr)_3H_2)_2$	501	(Monomeric)	ca. −18	79
$Pb(C_5(i\text{-}Pr)_3H_2)_2$	590	Monomeric	34–35	75
$Mn(C_5(i\text{-}Pr)_4H)_2$	522	(Monomeric)	226–228	179
$Fe(C_5(i\text{-}Pr)_4H)_2$	523	(Monomeric)	250 (dec.)	65
$Co(C_5(i\text{-}Pr)_4H)_2$	526	(Monomeric)	148–150	75
$Sm(C_5(i\text{-}Pr)_4H)_2$	617	Monomeric	162–163	75
$Mg(C_5(i\text{-}Pr)_4H)_2$	491	(Monomeric)	229–231	75

TABLE I (*continued*)

Compound	Mol. wt. (g/mol)	Aggregation[a]	mp (°C)	Ref.
Ca(C$_5$(*i*-Pr)$_4$H)$_2$	507	Monomeric	196–200	226
Sr(C$_5$(*i*-Pr)$_4$H)$_2$	554	(Monomeric)	151–153	75
Ba(C$_5$(*i*-Pr)$_4$H)$_2$	604	Monomeric	149–150	226
Zn(C$_5$(*i*-Pr)$_4$H)$_2$	532	Monomeric	123–125	78
Sn(C$_5$(*i*-Pr)$_4$H)$_2$	586	Monomeric	162–163	79
Pb(C$_5$(*i*-Pr)$_4$H)$_2$	674	(Monomeric)	151–152	75

[a] Probable states of aggregation that have not been structurally authenticated are in parentheses.

B. *Volatility*

Like its melting point, the volatility of a compound depends on a mix of intermolecular forces, which are affected by molecular weight and geometry, and in solids, lattice structure.[69] Control over these forces is

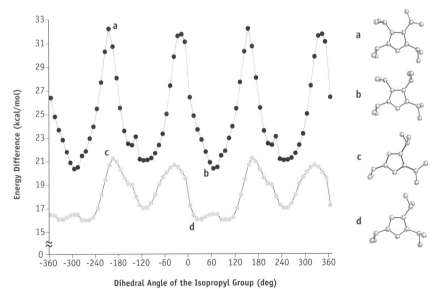

FIG. 3. Variations in energy caused by isopropyl group rotation in the [C$_5$(*i*-Pr)$_4$H]$^-$ ring (top trace) and the [(C$_5$(i-Pr)$_3$H$_2$)]$^-$ ring (bottom trace). On the right, from top to bottom, are the highest- and lowest-energy conformations of [C$_5$(*i*-Pr)$_4$H]$^-$, and the highest- and lowest-energy conformations of [(C$_5$(*i*-Pr)$_3$H$_2$)]$^-$. The greater energy difference for conformational changes in [(C$_5$(*i*-Pr)$_3$H$_2$)]$^-$ stems from the increased steric crowding between the isopropyl groups.

important in the use of these compounds as sources of the metals in chemical vapor deposition applications.[70,71] The changes that ligands can exert on the volatilities of organoalkaline-earth metallocenes by modifying their nuclearity have been reviewed,[72] and an example is presented in Fig. 4. The sublimation temperatures of strontocenes with increasing numbers and bulk of substituents drop by almost 300°C on progressing from $[SrCp_2]$ (probably polymeric) to $[Sr(C_5(t\text{-}Bu)H_4)_2]$ (partially associated[73]) to $[Sr(C_5Me_5)_2]$ (probably monomeric[74]), although their molecular weights increase by more than 60% during that interval. The drop in sublimation temperatures reflects the weakening of the intermolecular forces in the solids, which offsets the increase in molecular weight. Among the mononuclear complexes, however, molecular weight is a dominant controller of volatility, and sublimation temperatures rise on moving from $[Sr(C_5Me_5)_2]$ (358 g/mol) to $[Sr(C_5(i\text{-}Pr)_4H)_2]$ (554 g/mol).[75]

C. Oxygen Sensitivity

The chemical reactivity of metallocenes varies widely with the metal and the substituents on the rings; a general discussion of such properties is outside the scope of this review. The particular case of air (oxygen) sensitivity, however, can be analyzed with reference to the metal electron

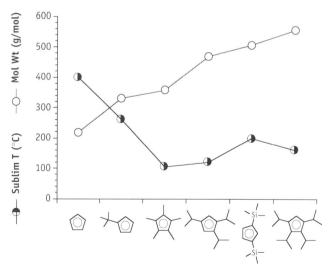

FIG. 4.   Sublimation temperatures of base-free strontocenes as a function of molecular weight and steric bulk of the cyclopentadienyl ligands.

configuration and to the presence of ring substituents. Except for the 18-electron metallocenes [MCp$_2'$] (M = Fe, Ru, Os) and [MCp$_2'$]$^+$ (M = Co, Rh, Ir), all transition metal and main-group metallocenes display some sensitivity to oxygen.[76] Such reactivity can be moderated, sometimes dramatically, by blocking physical access to the metal center with sterically bulky groups. In contrast to the high air-sensitivity of [CoCp$_2$], for example, octaphenylcobaltocene [Co(C$_5$Ph$_4$H)$_2$][43] and octaisopropylcobaltocene [Co(C$_5$(i-Pr)$_4$H)$_2$][75] are air-stable in the solid state. In the main-group series [M(C$_5$(i-Pr)$_4$H)$_2$] (M = Mg–Ba), air stability varies from several minutes (Mg) to seconds (Ba). In this case, the greater spacing between the rings in the barocene evidently increases the access to the metal center by oxygen. An interesting exception to this trend is provided by [Zn(C$_5$(i-Pr)$_4$H)$_2$], which despite having a metal center whose radius is similar to that of Mg^{2+} (0.58–0.60 Å[77]) decomposes almost instantaneously on exposure to air.[78] This reactivity reflects the exposure of the metal in [Zn(C$_5$(i-Pr)$_4$H)$_2$], which stems from its "slipped-sandwich" geometry (Fig. 5).

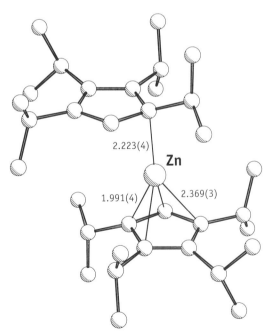

FIG. 5.   Structure of the "slipped-sandwich" metallocene, [Zn(C$_5$(i-Pr)$_4$H)$_2$][78]; selected distances are in angstroms.

There is also an electronic component to the oxygen stability of some main-group metallocenes. The relatively high air stability of the metallocenes $[Sn(C_5(i\text{-}Pr)_4H)_2]$ (weeks)[79] and $[Pb(C_5(i\text{-}Pr)_4H)_2]$ (days)[75] is partially a consequence of the small electronegativity differences between the Group 14 metals and carbon, which reduces the polarity of their bonds and consequently makes them less susceptible to attack by oxygen. A high degree of air stability in the solid state also has been reported for the decaphenyl- and decabenzylmetallocene derivatives of tin and lead,[47,80] although steric protection of the metal center is undoubtedly a major contributor to their stability.

IV

EFFECTS OF SUBSTITUENTS ON BONDING AND STRUCTURE

The extent to which the geometries of metallocenes are affected by the presence of substituents on the cyclopentadienyl rings varies widely with the metal type. With a few exceptions discussed later, transition-metal metallocenes with independent cyclopentadienyl rings (i.e., not *ansa*-metallocenes or metallocenophanes) have "linear" structures with parallel rings (Fig. 6a). In contrast, all structurally characterized divalent lanthanide metallocenes are "bent" (i.e., with canted rings) in the solid state and the gas phase (Fig. 6c). Main-group metallocenes display a wider range of metallocene geometries than do either the transition metal or lanthanide counterparts, and encompass linear, highly slipped (Fig. 6b), and bent forms (Figs. 6c and 6d). Since descriptions of structure are closely tied to questions of bonding, we discuss here both molecular orbital and molecular mechanical interpretations of metallocene structures.

A. *Transition Metal Metallocenes*

1. *Molecular Orbital Calculations and Spectra*

Transition metal metallocenes, especially ferrocene, have repeatedly been the subject of calculations. Until recently, these have had limited success in reproducing details of metallocene structure or in correlating energy levels with photoemission data.[81] Even with large basis sets, *ab initio* LCAO-SCF calculations of $[FeCp_2]$ and $[Fe(C_5Me_5)_2]$ at the Hartree–Fock level place the Fe–ring centroid distance at 1.89 Å, which is 15% too large.[82] Such errors appear to reflect fundamental limitations of the Hartree–Fock method with transition-metal compounds. Park and Almlöf

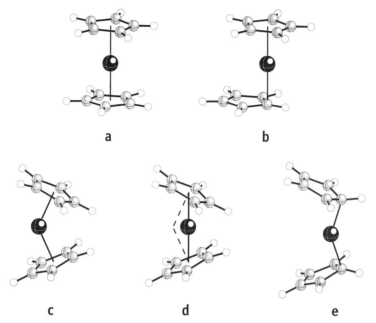

a   b

c   d   e

FIG. 6.   The range of structural variations documented in metallocenes.

demonstrated that a large-scale MCPF (modified couple pair functional) calculation in which all 66 valence electrons of ferrocene were correlated and ring–ring dispersion interactions were explicitly addressed produced an Fe–ring centroid distance of 1.65–1.67 Å,[83] in agreement with the experimental value (1.65 Å[21]). Calculations of comparable quality have not been performed for substituted ferrocenes or other metallocenes, and as the size of such calculations scale as N^4 (N = basis set size), they will become prohibitive for large molecules. A density functional approach with nonlocal corrections to correlation and exchange has provided accurate geometries of ferrocene (metal–ring distance = 1.648 Å[84]; calculations using this approach should be more computationally cost-effective than traditional post-Hartree–Fock treatments.

A widely used MO scheme for ferrocene and other first-row metallocenes was originally developed from a consideration of group theory and spectroscopic data (Fig. 7)[85]; it is qualitatively useful for the parent metallocenes, but substitution of the Cp rings will change the energy and mixing of various orbitals. For example, the symmetry of the Cp⁻ ring is lowered from D_{5h} to C_{2v} on monosubstitution, with concomitant splitting of the e_1 and e_2 orbitals. Such splitting is not always large enough to be observed

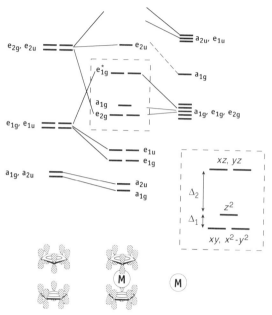

FIG. 7. A qualitative MO scheme for ferrocene and other first-row transition-metal metallocenes. The inset highlights the ordering of the d-orbitals in the frontier orbital region and provides the definition for Δ_1 and Δ_2 (see the text).

spectroscopically, however. An early attempt to rationalize trends in the absorption spectra of a series of monosubstituted ferrocenes [Fe(C$_5$XH$_4$)Cp] (X = CH$_3$, C$_2$H$_4$, Cl, F, NH$_2$) with CNDO and INDO-level calculations either predicted energy levels inconsistent with the experimental data or failed to identify a consistent pattern in the energies of metal–ring interactions across the series.[86] CNDO/2 calculations of various 1,1′-disubstituted ferrocenes [Fe(C$_5$XH$_4$)$_2$] (X = OCH$_3$, C$_2$H$_5$, CH$_3$, Cl, COOCH$_3$, COCH$_3$, CN) and the monosubstituted compound [Fe(C$_5$ClH$_4$)Cp] have been used to analyze trends in ionization energies derived from photoelectron spectra[87]; the absolute values of the calculated Cp′ π levels do not match those derived from experiment, although the general trends in the energies do.

Ligand field parameters for [Fe(C$_5$XH$_4$)$_2$] (X = H, CH$_3$, Br, Cl) have been derived from photoabsorption spectra.[88] The ligand field splittings of the Fe d orbitals follow those expected from the electron donating properties of the substituents; e.g., chlorine or bromine substitution increases the Δ_1 (e_{2g}–a_{1g}) splitting (by ~0.16 and ~0.28 eV, respectively) and decreases the Δ_2 (e_{1g}–a_{1g}) splitting (by ~0.22 and 0.32 eV) (see Fig.

7). The increase in Δ_1 means that the charge transfer from the ring to the halide is accompanied by greater mixing of the Fe $d_{x^2-y^2,xy}$ orbitals and $Cp(e_{2g})\pi$ orbitals. Although the spectrum of $[Fe(C_5MeH_4)_2]$ could not be unambiguously assigned, methyl substitution of the ring appeared to have the opposite effect.

2. Molecular Mechanics Calculations

In contrast to the difficulty in achieving reliable structural results with traditional molecular orbital calculations on metallocenes, molecular mechanics (MM) methods have been able to reproduce some features of their geometry with good accuracy. Bosnich and co-workers have developed a force field potential for transition metal metallocenes and used it to analyze structures and ring orientations.[28] Bond lengths and angles are duplicated to within a few percent of the experimental values, although the agreement largely reflects the experimental data used to derive the force constants for the calculations. The stability of the ring orientations in $[FeCp_2]$, $[RuCp_2]$, and $[OsCp_2]$ also was examined, and MM calculations successfully reproduce the 0.7–0.9 kcal/mol difference in energy between the eclipsed and staggered forms of ferrocene. A torsional (i.e., electronic) term is identified as the primary reason for the preference of the eclipsed form. Interestingly, the van der Waals interactions between the rings are essentially the same for both conformations and are *attractive* in nature, rather than repulsive, as sometimes assumed. For $[RuCp_2]$, a barrier to rotation of 6.8 kcal/mol was calculated, which compares favorably with the value of 8.1 kcal/mol estimated from thermal motion data of crystalline $[RuCp_2]$.[89]

$[Fe(C_5Me_5)_2]$ is perfectly staggered in the gas phase and the solid state (Fig. 8), although the barrier to rotation is placed at only 1.0 ± 0.3 kcal/mol.[25] MM calculations[28] find that the staggering is only partial (angle =

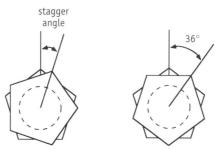

Fig. 8. Definition of the stagger angle in linear metallocenes. Perfect staggering occurs at an angle of 36°.

$18°$) for the isolated molecule. If [Fe(C$_5$Me$_5$)$_2$] is placed in a crystalline environment, however, perfect staggering is preferred, although there are some differences in the methyl group orientations from those found in the crystal. Calculations on [Fe(C$_5$Me$_4$H)$_2$] predict a partially staggered structure (angle = $18°$) in contrast to the perfectly staggered geometry found in the solid state.[90] The gearing of the methyl groups is also not exactly reproduced by the MM calculations. Calculations were also performed for [Ru(C$_5$Cl$_5$)(C$_5$Me$_5$)], [Ru(C$_5$Me$_5$)$_2$], and [Os(C$_5$Me$_5$)$_2$]. The perfectly eclipsed structure of [Ru(C$_5$Cl$_5$)(C$_5$Me$_5$)] is duplicated in the MM calculations. Stagger angles of $11.16°$ and $10°$ are calculated for [Ru(C$_5$Me$_5$)$_2$] and [Os(C$_5$Me$_5$)$_2$], respectively; because of disorder in the X-ray structures for both compounds,[91] however, accurate comparisons with the calculated geometries cannot be made.

3. Structural Distortions Induced by Bulky Substituents

The linearity of most transition metal metallocenes is a consequence of strong d_π–p_π metal ring bonding; although substituents on the metallocenes can change the relative orientations of the rings, the angle between them is more difficult to affect (Table II). Except in *ansa*-metallocenes or metallocenophanes, multiple ring substitution with bulky groups is required before bending is observed, and it may not occur even then. For example, the six bulky cyclohexyl rings in [Fe(C$_5$(C$_6$H$_{11}$)$_3$H$_2$)$_2$] do not distort the parallel sandwich geometry and leave the average Fe–C distance at 2.06(1) Å,[92] indistinguishable from the value in ferrocene itself.

a. *Ferrocene derivatives.*   Slight lengthening of the Fe–C distances in a neutral ferrocene derivative beyond the 2.05 Å value of ferrocene itself[21] has been observed in octaphenylferrocene [Fe(C$_5$Ph$_4$H)$_2$].[44] The Fe–C bond lengths for the unsubstituted carbons [2.054(3) Å] are shorter than those of the phenyl-substituted carbons [2.088(3) and 2.099(3) Å]. Greater lengthening is present in decaphenylferrocence [Fe(C$_5$Ph$_5$)$_2$]; although the structure determination is of limited precision, an average Fe–C distance of 2.11 Å was found.[59]

The 17-e^- ferrocenium salts appear to be slightly more amenable to distortion than their neutral counterparts. The Fe–C distances in unsubstituted ferrocenium compounds, [FeCp$_2$]X, range from 2.05 to 2.08 Å[93]; the average distance is ~2.09 Å for several decamethylferrocenium complexes, [Fe(C$_5$Me$_5$)$_2$]X.[94–98] In the structure of [Fe(C$_5$(*i*-Pr)$_4$H)$_2$]BF$_4$, the average Fe–C distance is 2.14(1) Å; the bond lengths to the unsubstituted carbons [2.10(1) Å] are shorter than those to the isopropyl-substituted carbons, which range from 2.13(1) to 2.18(1) Å.[99] The longest Fe–ring

TABLE II

TRANSITION METAL METALLOCENES WITH DISTORTED STRUCTURES

| | Distances (Å) | | Angles (deg) | | |
| | | | Ring centroid– M–ring centroid | Ring normal– M–ring normal | |
Metallocene	M–C(ring) (avg., range)	Displacement of *ipso*-carbons			Ref.
Fe(C$_5$(Me$_3$Si)$_3$H$_2$)$_2$	2.08(2), 2.050(1)–2.113(1)		Not given	173.9	100
Fe(C$_5$Ph$_4$H)$_2$	2.09(2), 2.054(3)–2.099(3)	0.123–0.248	180	180	44
Fe(C$_5$Ph$_5$)$_2$	2.11	Not given	180	180	59
[Fe(C$_5$(SiMe$_3$)H$_4$)$_2$][AlCl$_4$]	2.090(15), 2.067(6)–2.115(5)		Not given	175.1	101
[Fe(C$_5$(*i*-Pr)$_3$H$_2$)$_2$]BF$_4$ 2 independent molecules)	2.08(2), 2.07(1)–2.10(1) 2.10(2), 2.04(1)–2.13(1)	0.019–0.19 0.068–0.12	180 180	180 180	99
[Fe(C$_5$(*i*-Pr)$_4$H)$_2$]BF$_4$	2.14(1), 2.10(1)–2.18(1)	0.067–0.33	174.3	170.7	99
Co(C$_5$Ph$_4$H)$_2$	2.152(4), 2.091(4)–2.188(5)	0.110–0.219	180	180	43
[Co(C$_5$(*i*-Pr)$_3$H$_2$)$_2$]BPh$_4$	2.05(1), 2.02(1)–2.08(1)	0.17–0.32	177.9	174.7	75
[Co(C$_5$(*i*-Pr)$_4$H)$_2$]PF$_6$	2.09(1), 2.033(6)–2.138(6)	0.038–0.36	174.8	171.5	75

interaction in a ferrocenium complex observed to date is in the decaphenylferrocenium complex [Fe(C$_5$Ph$_5$)$_2$]BF$_4$; the steric repulsion of 10 phenyl substituents leads to an average Fe–C distance of 2.16(1) Å.[59] The *ipso* carbons of the phenyl substituents are displaced 0.23 Å from the ring plane.

Actual canting of the rings is rare in ferrocene and ferrocenium complexes; even [Fe(C$_5$Ph$_5$)$_2$][59] and [Fe(C$_5$Ph$_5$)$_2$]BF$_4$,[75] with Cp ligands bulky enough to lengthen the Fe–C bond distances, have linear geometries. Slight nonparallelism between the rings is occasionally observed in solid-state structures of neutral ferrocenes; this does not normally exceed an interplanar angle of 4°. Excluding *ansa*-bridged metallocenes and metallocenophanes, the only structurally authenticated neutral ferrocene complex with substantially bent rings is [Fe(C$_5$(SiMe$_3$)$_3$H$_2$)$_2$],[100] which has a ring normal–Fe–ring normal angle of 173.9°; the bent angle in this complex stems from the steric repulsion between two nearly eclipsed SiMe$_3$ groups on the rings. The C$_5$(*i*-Pr)$_4$H rings in [Fe(C$_5$(*i*-Pr)$_4$H)]BF$_4$ are distinctly nonparallel, with ring centroid–Fe–ring centroid and ring normal–Fe–ring normal angles of 174.3° and 170.7°, respectively. The $C_5(i\text{-Pr})_4H$ ligands are bent toward the two unsubstituted carbons of the ring.[99] The two rings in $[Fe(C_5(SiMe_3)H_4)_2][AlCl_4]$ adopt a partially staggered conformation (stagger angle = 17.7(2)°) presumably to lessen steric crowding.[101] The angle between the rings is 175.1°.

b. *Cobaltocene derivatives.* The average Co–C distance in neutral cobaltocene, $[CoCp_2]$, is 2.096(8) Å.[102] The only substituted cobaltocene for which a significant lengthening of the Co—C bonds has been reported is octaphenylcobaltocene, which has an average Co–C distance of 2.152(4) Å.[43] Probably because cobaltocenium compounds are 18-electron complexes, only a few have distorted bond lengths. The average Co–C value of 2.013(7) Å observed for the unsubstituted cobaltocenium cation in $[CoCp_2]^+[MoEt(NO)_2Cp]^{-\,103}$ lengthens to 2.05 Å in several $[Co(C_5Me_5)_2]X$ complexes,[97,104] and in $[Co(C_5(i\text{-Pr})_3H_2)_2]BPh_4$.[75]

The average Co–C distance in $[Co(C_5(i\text{-Pr})_4H)_2]PF_6$ is 2.09(1) Å, the longest such value observed for a cobaltocenium complex.[75] The $C_5(i\text{-Pr})_4H$ rings are nonparallel, with ring centroid–Co–ring centroid and ring normal–Co–ring normal angles of 174.8° and 171.5°, respectively.

c. *Other derivatives.* Neutral metallocenes in which electrons have been removed from bonding orbitals (e.g., $[VCp_2']$, $[CrCp_2']$, $[MnCp_2']$) or in which electrons have been added to antibonding orbitals (e.g., $[CoCp_2']$, $[NiCp_2']$) have longer M—C bonds than does ferrocene.[85] The steric effects of bulky ligands should consequently diminish, and in fact the difference in the average V–C distance between $[VCp_2]$ and $[V(C_5Ph_4H)_2]$, for example, is only 0.02 Å.[43] Even in 18-electron compounds with sufficiently long M—C bonds, distortions become insignificant; e.g., the average Ru–C distances in $[RuCp_2]$ and $[Ru(C_5Ph_4H)_2]$ are indistinguishable at 2.19 Å[23] and 2.20(1) Å,[45] respectively.

B. *Main-Group and Lanthanide Metallocenes*

Metallocenes of the main-group elements possess a much wider range of geometries than do those of the transition metals and include polymeric, linear, and bent geometries. In some cases substituents on the rings have substantial effects on the nuclearity of the compounds. Owing to the similarity in ionic radii [i.e., 1.00 Å (Ca^{2+}), 1.02 (Yb^{2+}); 1.18 Å (Sr^{2+} and Sm^{2+}) for CN = 6][77] and to a lack of involvement of the $4f$ electrons in bonding,[105] divalent lanthanide metallocenes are structurally related to the Group 2 analogues; consequently, they are included in this section.

1. *Oligomeric Metallocenes*

Three structurally characterized main-group metallocenes are oligomeric ([CaCp$_2$],[16] [Ba(C$_5$Me$_5$)$_2$],[19,106] and [PbCp$_2$][107]), and others, such as [SrCp$_2$] and [BaCp$_2$],[108] are presumed to be polymeric because of their low solubility and volatility. Increasing the steric bulk of the cyclopentadienyl rings will generate mononuclear complexes. For calcium and lead, the decamethylcyclopentadienyl derivatives, among others, are monomeric.[17,19] Because of its larger metal center, [Ba(C$_5$Me$_5$)$_2$] forms a one-dimensional coordination polymer; the use of the tetraisopropylcyclopentadienyl ligand will produce a monomeric barocene, however (Fig. 9).[109]

Among lanthanide metallocenes, both [Eu(C$_5$(SiMe$_3$)$_2$H$_3$)$_2$] and [Yb(C$_5$(SiMe$_3$)$_2$H$_3$)$_2$] are polymeric, despite the steric bulk of their cyclopentadienyl rings.[110] Monomeric metallocenes of each metal are known with pentamethylcyclopentadienyl ligands, however.[111]

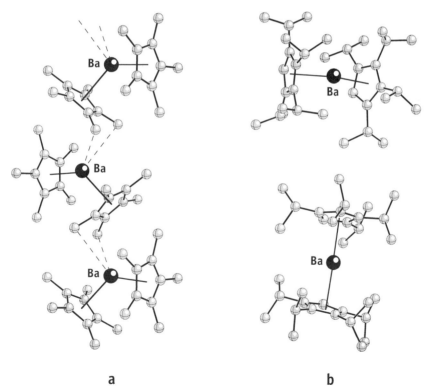

a   b

FIG. 9.   An oligomeric chain of [Ba(C$_5$Me$_5$)$_2$] in the solid state (a). On the right (b), the crystallographically independent molecules of [Ba(C$_5$(*i*-Pr)$_4$H)$_2$].[109]

2. Linear Metallocenes

The known linear, nonslipped main-group metallocenes are tabulated in Table III. Except for $[MgCp_2]^{112}$ and the lithocene anion $[LiCp_2]^{-}$ [113], all the compounds contain substituted cyclopentadienyl rings, which through steric interactions may contribute to their linear structures. The magnesocenes probably owe some of their linearity to covalency in the Mg—C bonds.[114] The importance of the valence electrons to the bending of Group 14 metallocenes is discussed later.

3. Bent Metallocenes

Bent metallocenes occur in a variety of forms. In "simple bent" compounds (Fig. 6c), the planes of the cyclopentadienyl rings are perpendicular to the metal–ring centroid vectors. Monomeric Group 2 metallocenes and several lanthanide complexes are typically found with this arrangement. In metallocenes with "slipped bent" geometries (Fig. 6d), the rings are tilted relative to the metal–ring centroid vectors. Group 14 metallocenes are the most common representatives of this structural type. A "slipped bent" structure can be regarded as an intermediate geometry on the way to a sigma-bound $[M(\eta^1\text{-Cp})_2]$ configuration (Fig. 6e) (e.g., found in $[HgCp_2]$),[115] which does not possess a sandwich structure.

One difficulty inherent in the analysis of bent metallocenes is the variety of structural parameters involved; these are diagrammed in Fig. 10. Note that the ring centroid–M–ring centroid angle (α) and ring normal–M–ring normal angle (β) can vary independently from each other. Failure to observe this distinction can produce misleading comparisons between

TABLE III

METAL–CARBON DISTANCES IN LINEAR MAIN-GROUP METALLOCENES

Complex	Average M–C distance (Å)	Method[a]	Ref.
$[PPh_4]^+[LiCp_2]^-$	2.318(4)	X	113
$[Li(12\text{-crown-}4)_2]^+$ $[Li(isodiCp)_2]^-$ (2 independent molecules)	2.319(16), 2.34(3)	X	227
$MgCp_2$	2.304(8)	X	112
$Mg(C_5Me_5)_2$	2.341(6)	E	135
$Si(C_5Me_5)_2{}^b$	2.42(1)	X	119
$Sn(C_5Ph_5)_2$	2.692(8)	X	228
$Pb(C_5(i\text{-Pr})_3H_2)_2$	2.748(8)	X	75

[a] E = gas electron diffraction; X = single crystal X-ray diffraction.
[b] A second conformer in the asymmetric unit has nonparallel rings.

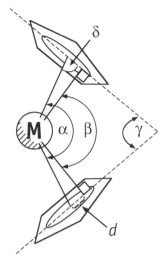

FIG. 10. Geometric parameters in a bent metallocene. The angles are as follows: α = ring centroid–M–ring centroid angle; β = angle between the normals to the ring planes; γ = angle between the ring planes; δ = angle between the ring plane and the metal–ring centroid vector. The distance d ("ring slippage") is the displacement of the ring centroid from the normal to the ring plane.

metallocenes (Table IV). For example, the stannocene $[Sn\{C_5[(i\text{-}Pr)_2N]_2 PH_4\}]^{116}$ and the barocene $[Ba(C_5Me_5)_2]^{117}$ have indistinguishable ring normal–M–ring normal (β) angles (133.4° and 133.3°, respectively). Their ring centroid–M–ring centroid angles (α) are substantially different, however, at 148.6° and 130.9°.

Metallocenes with large differences between α and β angles will have correspondingly large variations in M–C distances. In $[Ge(C_5Bz_5)_2]$ (α = 163.1°; β = 149°), for example, the M–C lengths range from 2.429 to 2.723 Å (Δ_{M-C} = 0.29 Å; 12% difference).[118] Such variability, however, does not correlate directly with the bulk of the substituents on the cyclopentadienyl rings (Fig. 11). There is as much variation in M–C distances in the decabenzylmetallocenes $[M(C_5Bz_5)_2]$ (M = Ge, Sn, Pb; Δ_{M-C} = 0.22 Å) (Table IV) as with the unsubstituted $[MCp_2]$ complexes (M = Ge, Sn; Δ_{M-C} = 0.21 Å). The amount of variation that can arise solely from crystal packing effects is illustrated by the two independent molecules of $[SnCp_2]$, which display differences in M–C bond lengths of 0.170 and 0.290 Å.[17] The differences for $[Si(C_5Me_5)_2]$ are even greater: the linear form has a small Δ_{M-C} of 0.033 Å, whereas the bent form (found in the same crystal), displays a Δ_{M-C} of 0.22 Å.[119]

TABLE IV

Solid-State Structural Data on Bent Monomeric Group 14, Group 2, and Lanthanide Metallocenes

	Angles (deg)		Distances (Å)			
	Ring centroid–M–ring centroid	Ring normal–M–ring normal	M–ring centroid	M–C(ring) (range)	δ	Ref.
Metallocene						
Group 14						
Si(C$_5$Me$_5$)$_2$[a]	167.4	154.7	2.120, 2.122	2.323–2.541	83.65	119
GeCp$_2$	Not given	129.6	Not given	2.347–2.730	n/a	215
Ge(C$_5$(SiMe$_3$)$_3$H$_2$)$_2$	169.48	159.23	2.261, 2.256	2.458–2.661	84.88	229
	171.77	165.10	2.252, 2.250	2.490–2.633	86.67	
Ge(C$_5$(PhCH$_2$)$_5$)$_2$	163.1	149	2.240, 2.288	2.429–2.723	82.95	118
Sn(C$_4$(t-Bu)$_2$H$_2$N)$_2$	142.5	114.4	2.428	2.485–2.943[a]	75.95	230
SnCp$_2$	143.7	134	2.38, 2.45	2.56–2.85	85.15	17
(2 independent molecules)	148	133	2.41, 2.41	2.58–2.75	82.50	
Sn{C$_5$[(i-Pr)$_2$N]$_2$PH$_4$}$_2$	148.6	133.4	2.38, 2.39	2.527–2.807	82.40	116
Sn(C$_5$Ph$_5$)Cp	151.1	136.1	2.391 (Cp)	2.541–2.792 (Cp)	82.50	120
			2.487 (Ph$_5$C$_5$)	2.63–2.92 (Ph$_5$C$_5$)	82.50	
Sn(C$_5$Me$_5$)$_2$	154.8	144.6	2.388, 2.399	2.585–2.77	84.90	18
(2 independent molecules)	154.9	143.6	2.400, 2.399	2.57–2.776	84.35	
Sn(C$_5$(PhCH$_2$)$_5$)$_2$	155.9	147.2	2.439, 2.415	2.615–2.818	85.65	80
Sn(C$_5$(SiMe$_3$)$_3$H$_2$)$_2$	162	164	Not given	Not given	n/a	231
Sn(C$_5$(i-Pr)$_4$H)$_2$	165.0	152.2	2.423, 2.424	2.575–2.821	83.60	79
Pb(C$_5$Me$_5$)$_2$	151	142.9	Not given	2.69–2.90	85.95	17
Pb(C$_5$(PhCH$_2$)$_5$)$_2$	153.4	146.7	2.507, 2.500	2.680–2.871	86.65	
				2.740–2.818	86.65	
Group 2						
Mg(C$_5$(SiMe$_3$)$_3$H$_2$)$_2$	Not given	172.2	2.02, 2.04	Not given	n/a	232
(2 independent molecules)		172.2	2.02, 2.03			
Ca(C$_5$Me$_5$)$_2$	147.7	146.5	2.358, 2.333	2.597–2.653	89.40	19
(2 independent molecules)	146.3	144.0	2.359, 2.354	2.612–2.681	88.85	
Ca(C$_5$(i-Pr)$_3$H$_2$)$_2$	169.7	170.2	2.33 (avg.)	2.587–2.643	90.25	224
Ca(C$_5$(i-Pr)$_4$H)$_2$	162.3	158.9	2.349, 2.352	2.594–2.682	88.30	109
Ba(C$_5$(i-Pr)$_4$H)$_2$	154.3	153.2	2.679, 2.676	2.921–2.960	89.45	109
(2 independent molecules)	154.1	153.7	2.681, 2.683		89.80	
Lanthanide						
Sm(C$_5$Me$_5$)$_2$	140.1	140.0	Not given	2.775–2.815	89.95	111
Eu(C$_5$Me$_5$)$_2$	140.3	139.8	Not given	2.765–2.822	89.75	111
Yb(C$_5$Me$_5$)$_2$	145.7	145.1	Not given	2.665(4) (avg.)	89.70	134
(2 independent molecules)	145.0	143.0			89.00	

[a] A second conformer in the asymmetric unit has parallel rings.

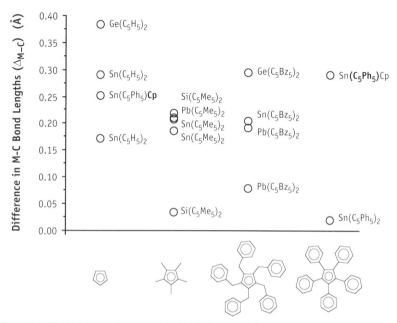

FIG. 11.   Variations in the spread of M–C distances (Δ_{M-C}) in main-group metallocenes with cyclopentadienyl rings of different steric bulk.

Another way the variation in bending angles can be evaluated is by a comparison of the tilt angle δ (Fig. 10) in various substituted metallocenes. This angle can be extracted from other angles by recognizing that the angles β and γ (the complement to β) form part of a quadrilateral, and thus

$$\beta + \gamma = 180° \quad \text{or} \quad \gamma = 180° - \beta.$$

Summing the angles around the quadrilateral involving the metal gives

$$\alpha + \gamma + 2\delta = 360°.$$

Substituting the first equation into the second yields

$$\alpha + (180° - \beta) + 2\delta = 360° \quad \text{or} \quad \delta = (180 + \beta - \alpha)/2.$$

Table IV contains calculated δ values for bent main-group metallocenes; there is not a strong correlation with the substituents on the ring. Thus, for the case of stannocenes, both the Cp and the Ph$_5$Cp ring in [Sn(C$_5$Ph$_5$)Cp][120] have $\delta = 82.5°$, and the highly substituted [Sn(C$_5$(i-Pr)$_4$H)$_2$][79] has $\delta = 83.6°$, whereas one of the two forms of [SnCp$_2$] has $\delta = 85.2°$.[17] Clearly, packing effects can influence the angle, as the other [SnCp$_2$] molecule has $\delta = 82.5°$.

4. Slipped Metallocenes

Besides the metallocenes with noncoplanar rings, there are several whose rings are "slipped"—i.e., they are coplanar or nearly so, but the ring centroids are displaced from the perpendicular projection of the metal atom on the least-square plane of the ring.[121] This structural type is represented by the highly asymmetric geometries (η^1; η^5) of beryllocene [BeCp$_2$] and the zincocenes [ZnCp$_2'$]; they have proven difficult to characterize unambiguously.

a. *Beryllocene.* A gas-phase electron diffraction study of [BeCp$_2$] identified an unsymmetrical sandwich geometry for the complex,[122,123] while a low-temperature single crystal X-ray investigation of [BeCp$_2$] revealed a slipped-sandwich structure in the solid state[124]; subsequent experiments confirmed that this is likely to be the ground state geometry for the complex in solution and in the gas phase.[125–127]

Although the preparation of substituted beryllocenes would be a natural target for exploring electronic and steric effects on its unusual structure, they have proven to be elusive. Steric overcrowding may interfere with the formation of decamethylberyllocene in solution.[72] The mixed ring species [Be(C$_5$Me$_5$)Cp] has been reported, but IR evidence suggests that the compound should be described as [Be(η^5-C$_5$Me$_5$)(η^1-Cp)].[128]

b. *Zincocenes.* Gas-phase electron diffraction studies of [Zn(C$_5$Me$_5$)$_2$][129,130] and [Zn(C$_5$(SiMe$_3$)H$_4$)$_2$][130] and X-ray studies of [Zn(C$_5$Me$_4$Ph)$_2$][121] and [Zn(C$_5$(i-Pr)$_4$H)$_2$][78] revealed that these monomeric metallocenes possess slipped-sandwich geometries similar to that of [BeCp$_2$]. The length of the η^1-Zn—C bond differs appreciably among the structures, ranging from 2.04(6) Å and 2.07(10) Å for the monohapto rings in the gas-phase structures of [Zn(C$_5$Me$_5$)$_2$] and [Zn(C$_5$(SiMe$_3$)H$_4$)$_2$], respectively, to 2.094(3) Å for the η^1-bonded cyclopentadienyl ligand in [Zn(C$_5$Me$_4$Ph)$_2$],[121] to 2.223(4) Å for the η^1-ring in [Zn(C$_5$(i-Pr)$_4$H)$_2$].[78] The latter complex also displays a wide range of Zn–C distances in the polyhapto ring, from 1.991(4) Å to 2.514(3) Å; the last distance is probably nonbonding.[131,132] Lengthening of the Zn—C bonds apparently occurs to relieve the steric pressure caused by the bulk of the [C$_5$(i-Pr)$_4$H]$^-$ ligand, as the corresponding range of Zn–C distances for the η^5-ring in [Zn(C$_5$Me$_4$Ph)$_2$] is much narrower (2.093(3) to 2.299(3) Å).

5. Structural Energetics in Organolanthanide and Group 2 Metallocenes

In the past 10 years, substituted main-group and lanthanide metal metallocenes have appeared whose structures are not easily rationalized when only differences in metal–ligand bonding are considered. For example, the crystal lattice of [Si(C_5Me_5)$_2$] contains both linear and bent forms of the molecule (ring normal–Si–ring normal angle = 154.7°)[119]; the energetic differences between the two conformations cannot be greater than that supplied by crystal packing forces. In another set of cases, the decamethyl-metallocenes of Sm,[133] Eu,[111] Yb,[134] Ca,[19] and Ba[106] are found with bent structures in the solid state[117]; electron diffraction studies reveal that the Yb, Ca, Sr, and Ba compounds are also bent in the gas phase.[74,135] As the *f*-electrons are well shielded in the lanthanide elements and are generally thought to play little role in the bonding of their compounds,[105,136] and the Group 2 compounds have no valence electrons, both electrostatic and steric considerations suggest that structures with parallel ring structures should be preferred.

Study of these "anomalous" structures has led to a reinvestigation of bonding energetics in main-group and lanthanide metallocenes, and to the recognition of underlying energetic similarities between the two classes of compounds.[20] The gas-phase electron diffraction (GED) data of the alkaline-earth metallocenes and Group 14 metallocenes have large-amplitude ring–metal–ring bending vibrations, which are believed to reflect shallow energy potentials for bending. The energy required to bend the Group 2 decamethylmetallocenes by 20° is estimated at only 2 to 3.5 kJ/mol,[74,137] and similar results are found for the Group 14 compounds [M(C_5MeH_4)$_2$] (M = Ge and Sn),[138] [PbCp$_2$],[139] and [Si(C_5Me_5)$_2$],[119] and the lanthanide complex [Yb(C_5Me_5)$_2$].[135] In all these cases, deciding whether the *equilibrium* structure is actually bent or not is difficult, as structures determined by GED can appear bent for statistical reasons.

Apparently owing to the small energy differences involved, molecular orbital calculations of unsubstituted lanthanide and Group 2 species have had difficulty confirming an energetic preference for bent structures. The compounds [SmCp$_2$] (extended Hückel level),[140] [(Eu, Yb)Cp$_2$] (SW-X$_\alpha$),[141] and [CaCp$_2$] (SCF Hartree–Fock level)[137] have been predicted to be linear or "quasilinear." Calculations taking into account electron correlation (MP2 level) on [(Ca, Sr, Ba)Cp$_2$] also predict a linear structure for [CaCp$_2$] and a quasilinear geometry for [SrCp$_2$]. [BaCp$_2$] and [SmCp$_2$] are predicted to be bent, with ring centroid–M–ring centroid angles of 147° and 170°, respectively; however, only 1.5 kJ/mol (0.2 kJ/mol for [SmCp$_2$]) separates the bent and linear forms.[142]

For the Group 14 metallocenes, a simplified rationalization for the bent geometries uses valence bond hybridization arguments.[143] The metal atom is considered approximately *sp*^{2} hybridized, with the lone pair of valence electrons occupying a stereochemically distinct region around the metal center (Fig. 12). Support for this scheme came from low-level MO calcula-

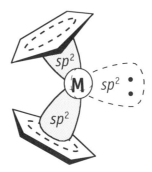

FIG. 12. Valence bond interpretation of the bending in Group 14 metallocenes. The bent structure stems from a stereochemically active lone pair of electrons, which is approximately sp^2 hybridized.

tions (e.g., extended Hückel, MNDO), which found large energy differences between the bent and linear forms. Such calculations are sensitive to the starting geometries used, however, and sandwich geometries are not always predicted to be the most stable. For example, an early MNDO calculation on $[SiCp_2]$ favored a linear sandwich over a bent one by 35 kJ/mol (Fig. 13),[144] but a later MNDO result starting with the geometry found experimentally for $[Si(C_5Me_5)_2]$ favored the bent form by 99 kJ/mol.[119] The same calculation found that an asymmetric $[Si(\eta^5\text{-}Cp)(\eta^1\text{-}Cp)]$ configuration was more stable than either sandwich geometry, however. With *ab initio* calculations, the energetic difference between the bent and linear forms is less; at the DZP–MP2 level, the difference decreases to 11.7 kJ/mol (Fig. 13).[145]

An extended Hückel calculation found the bent form of $[SnCp_2]$ to be 100 kJ/mol more stable than the linear,[18] but at the AM1 level the preference for the bent form decreases to only 4.0 kJ/mol.[146] Other calculations on $[GeCp_2]$,[138] $[PbCp_2]$,[147] and the $[TlCp_2]^-$ anion[148] (isoelectronic with the Group 14 metallocenes) favor small (<10 kJ/mol) energy differences between the bent and linear forms; an MNDO calculation for lead identified the $[Pb(\eta^1\text{-}Cp)_2]$ sandwich as having the lowest energy. A summary of the calculations reported for main-group and lanthanide metallocenes is given in Table V.

6. *Molecular Mechanics Calculations*

Molecular mechanics calculations have been performed for Group 2, Group 14, and lanthanide compounds with substituted rings; they indicate small energetic differences between the bent and linear forms. For the Group 2 and lanthanide decamethylmetallocenes, van der Waals attraction

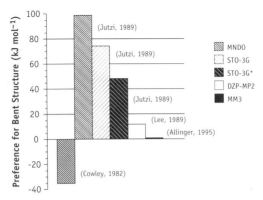

FIG. 13. Variation in estimated preference of bent over linear structures for [SiCp$_2$] as a function of the level of MO or MM calculation. Higher levels of MO theory favor smaller differences between the two geometries; MM finds the two structures to be essentially isoenergetic.[149]

between the methyl groups has been found to account for more than 90% of the stabilization of the bent forms; the energy involved is less than 10 kJ/mol.[114] A detailed study by Allinger using the MM3 force field found similar results, with no more than 2.0 kcal/mol (~9.6 kJ/mol) between the bent and linear forms of all [MCp$_2$] and [M(C$_5$Me$_5$)$_2$] complexes (M = Si, Ge, Sn, Pb, Ca, Ba, Sm, Eu, and Yb); the values for [SiCp$_2$] and [Si(C$_5$Me$_5$)$_2$] were only 0.2 and <0.1 kcal/mol, respectively. The staggered forms were found to have the lower energy in each case.

In these MM analyses, the valence electrons in the Group 14 metallocenes play no role in determining the bending angle. Their bent structures could arise from attractive van der Waals forces between the cyclopentadienyl rings similar to those proposed for the Group 2 metallocenes. That only small energy differences exist between the conformations of Group 2 and Group 14 metallocenes suggests that both are equally "floppy" systems. This conclusion is supported by the strong linear correlation that exists between the metal–ring normal distances and ring normal–metal–ring normal bending angles in the main-group and lanthanide decamethylmetallocenes (Fig. 14). The correlation reflects the fact that metallocenes with long ring-normal distances can bend proportionally farther back before encountering steric resistance between the methyl groups. Allinger has shown that, owing to steric interactions, metallocenes with bulkier rings (i.e., [M(C$_5$Bz$_5$)$_2$], M = Ge, Sn, Pb; [M(C$_5$(i-Pr)$_4$H)$_2$]$_2$, M = Ca, Ba) will have smaller bending angles than [MCp$_2$] or [M(C$_5$Me$_5$)$_2$] complexes. Note that this does not refute the point made earlier that the bulkiness of

TABLE V

MOLECULAR ORBITAL CALCULATIONS ON GROUP 2, GROUP 14,
AND LANTHANIDE METALLOCENES

Element	Method	Bent/linear preference (kJ/mol)	Global minimum	Comments	Ref.
Mg	SCF-HF		Linear sandwich		137
Ca	SCF-HF		Linear sandwich		137
	SCF-HF (MP2)		Linear sandwich		142
Sr	SCF-HF (MP2)		"Quasilinear" sandwich		142
Ba	SCF-HF (MP2)	Bent (1.5)	Bent sandwich		142
Si	MNDO	Linear (35)	Linear sandwich		144
	MNDO		$Si(\eta^1\text{-}Cp)_2$		233
	MNDO	Bent (99)	$Si(\eta^5\text{-}Cp)(\eta^1\text{-}Cp)$		119
	STO-3G	Bent (20.1)	Bent sandwich	Si–C distances in $Si(C_5Me_5)_2$ were used	119
	STO-3G	Bent (74.5)	Bent sandwich	Optimized geometry	119
	STO-3G*	Bent (48.5)	Bent sandwich	Optimized geometry	119
	DZP-MP2	Bent (11.7)	Bent sandwich	Optimized geometry	145
Ge	Extended Hückel	Bent (110)			138
	Ab initio	Linear (3)	Linear sandwich	Shallow potential well for ring orientations	138
Sn	Extended Hückel	Bent (100)			18
	SW-Xα			Used experimental geometry	144
	MNDO	Bent (6.3)	$Sn(\eta^1\text{-}Cp)_2$	Minimum is 154 kJ/mol below sandwiches	234
	AM1	Bent (4.0)	Bent sandwich		146
Pb	MNDO	Bent (2.6)	$Pb(\eta^1\text{-}Cp)_2$	Minimum is 140 kJ/mol below sandwiches	147
Tl	DZP-MP4	Bent (3.4)	Bent sandwich	$[TlCp_2]^-$ is isoelectronic with $SnCp_2$	148
Sm	Extended Hückel		Linear sandwich		140
	SCF-HF (MP2)	Bent (0.2)	Bent sandwich		142
Eu	SW-Xα		Linear sandwich		141
	SCF-HF (MP2)		"Quasilinear" sandwich		142
Yb	SW-Xα		Linear sandwich		141
	SCF-HF (MP2)		Linear sandwich		142

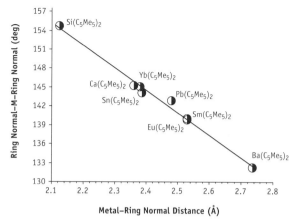

FIG. 14.   Bending angles in solid base-free main-group and lanthanide decamethylmetallo-cenes [M(C$_5$Me$_5$)$_2$] as a function of the metal to ring-normal distance. Both bent and linear conformers of [Si(C$_5$Me$_5$)$_2$] are found in the solid state[119]; only the value for the bent form is on the graph. For the eight compounds, $r^2 = 0.989$.

the ring does not directly correlate with the *spread* in M–C distances (Fig. 11).

The influence of the crystal field on the conformations of metallocenes has been explicitly studied with MM3 for [M(C$_5$Me$_5$)$_2$] (M = Sm, Yb, Ca, Ba).[149] The lattice environment is always found to influence the bending angle, leading to larger bending angles by an average of ca. 14°. Crystal packing forces are not, however, the primary source of the bending.

V

ELECTROCHEMICAL BEHAVIOR OF SUBSTITUTED METALLOCENES

Owing to its stability, solubility, and highly reproducible oxidation behavior, ferrocene has long been used as an electrochemical standard in nonaqueous solvents. Not surprisingly, the electron-donor or -acceptor properties of ring substituents in ferrocenes and other metallocenes have been repeatedly evaluated with electrochemical techniques. Measurements have been obtained using polarography,[150] cyclic voltammetry (CV),[151] chronopotentiometry,[152] photoelectron spectroscopy,[153] and Fourier transform ion cyclotron resonance mass spectrometry.[154] Extensive compilations of such data are available.[155,156] Historically, variations of oxidation potentials have been discussed almost solely in terms of the electronic properties of the substituents, but solvent and even steric effects also need to be considered when discussing electrochemical behavior.

The typical additive effect of alkyl substituents on half-wave potentials was illustrated in a study of the polarographic potentials of ferrocene derivatives.[157] For instance, three ethyl groups on ferrocene decrease its $E_{1/2}$ from 0.34 V to 0.19 V, a change of 0.05 V/(C_2H_5) group[157]; with eight ethyl groups, the potential is lowered to -0.05, a similar decrease of approximately -0.05 V/(C_2H_5) group. An additive effect in the opposite direction is observed when ferrocene is substituted with electron withdrawing phenyl groups (Fig. 15).[157]

Although additive electrochemical relationships hold in selected series of transition-metal metallocenes,[158,159] there are many exceptions to this behavior; substituents do not always shift potentials linearly, nor are the shifts necessarily in the expected direction. For example, the difference in potentials between 1,1'-dimethylferrocene (0.23 V) and 1,1'-diethylferrocene (0.37, 0.39 V)[150] indicate that the methyl groups produce a larger change than the ethyl groups, although the latter are normally considered to be better electron donors. As a second example, acetylferrocene has a half-wave potential of 0.62 V, which is appropriately higher than that of ferrocene itself (0.34 V)[150]; the potential increases even further (to 0.66 V) in 1-acetyl-1'-ethylferrocene, however, contrary to the expected effect of an added ethyl group.

CV data has been analyzed using the same additive concept. Solvent effects are often dismissed when analyzing CV data,[155] as they generally do not change the direction of potential shifts caused by the electron

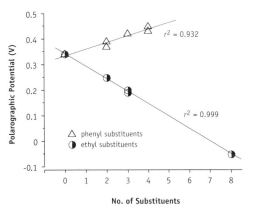

FIG. 15.   Typical additive effect of ethyl and phenyl substituents on the polarographic potential of ferrocene.[157]

donating/withdrawing properties of the ring substituents (Fig. 16). However, they still can be an important source of nonlinear electrochemical behavior. For example, successive substitution of benzyl groups in ferrocene produces a linear change in the half-wave potentials when measured in CH_3CN with CV.[158] When measured in CH_2Cl_2, however, the potentials of $[Fe(C_5Bz_nH_{5-n})_2]$ do not vary linearly,[158] and the shifts from solvent to solvent can be substantial ($E_{1/2}$ for $[Fe(C_5BzH_4)_2]$ in CH_3CN, CH_2Cl_2, and thf vs. SCE are 0.38, 0.45, and 0.54 V, respectively).[158]

The importance of solvent effects can be extracted from correlations of half-wave potentials determined by CV or polarographic methods with other measures of oxidizability, especially gas-phase free energies of ionization and ionization potentials. Within a series of *unsubstituted* metallocenes that differ only by the metal, the oxidation potentials of the compounds in solution and their gas-phase free energies of ionization vary directly (Fig. 17),[154,160] indicating that the effect of solvent is consistent from one metallocene to another.

In *substituted* metallocenes, however, oxidation potentials do not always correlate well with gas-phase ionization energies. For example, although there is a positive trend between the irreversible oxidation potentials of a series of substituted ruthenocenes and their gas-phase energies of ionization, the quantitative correlation between the two is poor (Fig. 18).[161] The weak correlation is believed to reflect changes in solvation that accompany the addition of substituents.

The potentials of substituted ferrocenes determined by chronopotentiometric techniques have been related to the Hammett para σ-con-

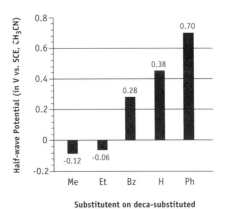

FIG. 16. $E_{1/2}$ values of decasubstituted ferrocenes[158]; the value for $[Fe(C_5Ph_5)_2]$ is taken from Ref. 59.

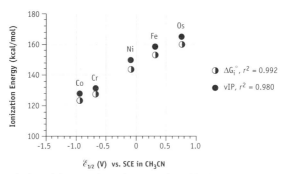

FIG. 17. Correlation of the gas-phase free energies of ionization and oxidation potentials for unsubstituted metallocenes. The vIP for all compounds are from Ref. 154; the sources of the $E_{1/2}$ and ΔG_i values are as follows: [CoCp$_2$] and [CrCp$_2$], Refs. 198, 199; [NiCp$_2$] and [FeCp$_2$], Refs. 198, 200; [OsCp$_2$], Refs. 201, 202.

stants.[152,162] The fact that good correlations have been found between these sets of data (e.g., Fig. 19) has been used to argue that inductive or inductive/resonance effects are responsible for the shifts in the metallocene potentials.[152,162] The $E_{1/4}$ potentials do not correlate with Hammett meta σ-constants, Taft's polar constants, or Brown's σ^+-constants,[152] although the gas-phase free energies of ionization of alkylferrocenes[160] and nickelocenes[163] have been found to compare with Taft σ_I parameters, which are a measure of polar effects exclusive of resonance. Nevertheless, the Taft σ_I parameters do not correlate with the gas-phase free energies of ionization of substituted ruthenocenes.[161] No single electronic parameter

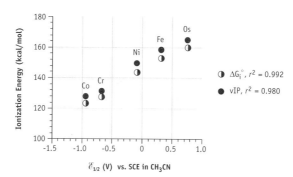

FIG. 18. Ionization energies and oxidation potentials for substituted ruthenocenes.[161] TTFMOSi = C$_5$(CF$_3$)$_4$OSiEt$_3$; TTFMH = C$_5$(CF$_3$)$_4$H.

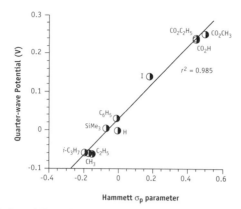

FIG. 19. Correlation of $E_{1/4}$ values of substituted ferrocenes with Hammett para σ-constants.[152,162]

scale has been identified that can be accurately used for predictions of substituent effects on the electrochemistry of all metallocenes.

Steric effects have recently been proposed as the source of some nonlinear electrochemical trends. The change in the oxidation potential of $[Fe(C_5(i\text{-}Pr)_3H_2)_2]$ (0.25 V) compared to ferrocene (0.45 V), for example, is equivalent to -0.033 V/i-Pr group,[92] but in $[Fe(C_5(i\text{-}Pr)_4H)_2]$ (0.06 V), the potential drops by 0.39 V, equivalent to -0.049 V/i-Pr group. The oxidation of $[Fe(C_5(i\text{-}Pr)_4H)_2]$ may be more favorable than expected because of the decreased steric crowding from the longer Fe—C bonds in the ferrocenium complex $[Fe(C_5(i\text{-}Pr)_4H)_2]^+$ relative to $[Fe(C_5(i\text{-}Pr)_3 H_2)_2]^+$.[99] The oxidation potentials of other ferrocenes may be similarly affected; e.g., the observed oxidation potential for $[Fe(C_5Ph_4H)_2]$ is only 0.03 V higher than that for ferrocene itself, in spite of the eight electron-withdrawing phenyl substituents on the rings.[44] Oxidation to the $[Fe(C_5Ph_4H)_2]^+$ cation may relieve steric congestion between the rings.

Steric effects may influence the potentials of some cobaltocenes as well. For $[Co(C_5(i\text{-}Pr)_3H_2)_2]BPh_4$, a reversible one-electron reduction occurs at -1.15 V (vs SCE) in CH_2Cl_2. The analogous reduction potential for $[Co(C_5(i\text{-}Pr)_4H)_2]BPh_4$ is virtually identical at -1.14 V, in spite of the two additional electron-donating isopropyl substituents. As with the $[Co(C_5(i\text{-}Pr)_4H)_2]^+$ cation, the mixed-ring cobaltocenium cation $[Co\{C_5(i\text{-}Pr)_5\}Cp]^+[PF_6]^-$ is not as reducing as expected given the presence of five isopropyl substituents; its redox potential in DMF is -0.96 V (vs SCE), only slightly more negative than that of $[CoCp_2]^+$ (-0.89 V) in the same solvent.[67] Reduction of the cobaltocenium complexes to 19-electron neutral metallocenes increases the Co—C bond distances; this process will be energetically favorable for $[Co(C_5(i\text{-}Pr)_4H)_2]^+$ and $[Co(i\text{-}Pr_5C_5)Cp]^+$ because it will relieve some steric strain present in the complexes. The relief of steric crowding may counteract the extra electron density provided by the additional isopropyl substituents, so that the reduction potentials for $[Co(C_5)_2(i\text{-}Pr)_4H)_2]^+$ and $[Co(i\text{-}Pr_5C_5)Cp]^+$ are not as negative as expected.

VI

MAGNETIC PROPERTIES OF SUBSTITUTED METALLOCENES

Neutral metallocenes of the transition and main-group metals normally exhibit an electronic ground state that cannot be easily perturbed. Some manganocenes are exceptions to this generalization in that their low and high spin states are nearly energetically degenerate and spin crossovers are often observed (Fig. 20). The position of equilibrium is dependent on the electron-donating ability and number of substituents on the ring, host matrix effects, and temperature.[164]

In the solid state at room temperature, manganocene itself forms a coordination polymer in which each manganese is coordinated to one Cp ring in an η^5 manner, and with η^3- and η^2-rings that bridge between manganese atoms (Fig. 21).[165] The average Mn—C bond length for the η^5-bonded ring is at 2.41 Å (the gas-phase value is 2.38 Å[166]). The spin state equilibrium for manganocene has been investigated by magnetic susceptibility,[167] ESR,[168,169] and He(I) and He(II) PES measurements.[170] Manganocene is essentially high-spin at room temperature and above (μ = 5.50 BM at 373 K),[171] crossing to low spin at reduced temperatures (μ = 1.99 BM at 193 K).[171] The long Mn—C bond lengths have been attributed to the high-spin state. Paramagnetic NMR spectroscopy has been used to determine the crossover enthalpy (21 ± 5 kJ/mol) and entropy (100 ± 20 J/mol · K).[171]

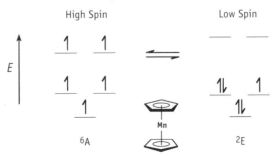

FIG. 20.   Changes in orbital occupancy accompanying spin-state crossovers in manganocene.

For gaseous 1,1'-dimethylmanganocene, He(I) PES data conclusively established the existence of a mixture of both low and high spin states in almost equal proportions.[172] ESR measurements on the compound are also consistent with the presence of a low-spin/high-spin equilibrium.[168,169] The spin crossover in d_8-toluene occurs at 303 K (Evans' method).[169] The enthalpy for the crossover from low to high spin was measured at 7.5 ± 0.4 kJ/mol, with an entropy of 24.3 ± 2.5 J/mol · K.[169] From gas-phase electron diffraction studies, the average Mn—C bond length for the low-spin state was found to be 2.14 Å.[173] The low-spin state was also found to have a large vibrational amplitude, possibly resulting from Jahn–Teller distortion involving ring tilting modes.[174]

Magnetic susceptibility measurements on solid $[Mn(C_5Me_5)_2]$ samples indicated Curie behavior from 4 to 116 K with an effective magnetic moment of 2.16 ± 0.1 BM, while measurements by Evans' method gave a magnetic moment of 1.97 ± 0.1 BM in d_8-toluene at 313 K.[175] The average Mn–C distance found by X-ray crystallography is 2.11 Å at ambient temperature, a length consistent with a low-spin manganocene complex.[176] In addition, the similarities between the solid-state structures of $[Mn(C_5Me_5)_2]$ and $[Fe(C_5Me_5)_2]$ discount dynamic Jahn–Teller distortion in the solid state for this manganocene, although a static distortion is evident in the C—C bond lengths of the ligand. ESR and variable temperature X-ray crystallography also indicate that the structure of $[Mn(C_5Me_5)_2]$ is independent of temperature, although variable-temperature X-ray crystallography detected a change in the crystallographic symmetry arising from greater ring slippage with increased temperature.[177] The inability of decamethylmanganocene to achieve a high-spin state even at elevated temperatures reflects the increase in stability of the low-spin state that accompanies deca-substitution of manganocene by electron-donating methyl groups. The temperature independence of the spin state of decamethylmangano-

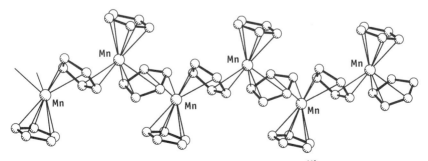

FIG. 21.   Structure of polymeric $[MnCp_2]$.[165]

cene has been used in studies of magnetic phase transitions.[178] The deca-methylmanganocenium cation has been employed as a donor component of a charge transfer salt with tetracyanoethylenide. This salt exhibits bulk ferromagnetic behavior below a Curie temperature of 6.2 K with a high coercive field of 3600 gauss.

Substituted manganocenes with sterically bulky groups have been used to investigate the conditions required for spin crossover. Hexaisopropyl-manganocene, $[Mn(C_5(i\text{-}Pr)_3H_2)_2]$, approaches spin crossover at 370 K.[179] It is evident from a series of alkylated manganocenes that with more electron donation from the ligand, the energy difference between the high- and low-spin states increases until the high-spin state is inaccessible, as with decamethylmanganocene (Fig. 22). This trend is further supported by the electronic state of 1,1'-bis(trimethylsilyl)manganocene, which is high-spin at all temperatures accessible by NMR.[180] Its high-spin state can be attributed to the electron-withdrawing behavior of the substituents that destabilize the low-spin state.

Steric effects may also influence magnetic properties of metallocenes. An example of this occurs with octaisopropylmanganocene, $[Mn(C_5(i\text{-}Pr)_4H)_2]$, which exhibits Curie behavior with a high-spin magnetic moment of 5.73 BM.[179] This anomalous behavior is believed to result from the large steric bulk of the eight isopropyl substituents, which sterically force the cyclopentadienyl ligands to remain at longer distances from the metal center, thereby generating a weaker ligand field.

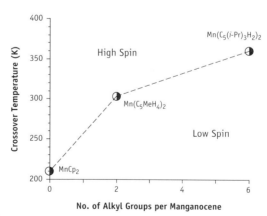

FIG. 22. Spin crossover temperatures for manganocenes as a function of alkyl groups. Data for $[MnCp_2]$, $[Mn(C_5MeH_4)_2]$, and $[Mn(C_5(i\text{-}Pr)_3H_2)_2]_2$ from Refs. 203, 169, and 179, respectively.

VII

SUBSTITUENT EFFECTS ON MÖSSBAUER SPECTRA

The two metal nuclei that can be studied by Mössbauer spectroscopy and are relevant to metallocenes are ^{57}Fe and ^{119}Sn. Spectra from both nuclei have been used to examine the electronic effects of substitution on metallocenes.

A. ^{57}Fe Mössbauer Spectra

^{57}Fe Mössbauer spectroscopy has been used to study ferrocene, ferrocene derivatives, and ferrocenium complexes; molecular orbital interpretations have been given for the trends in the spectra.[181–183] An increase in the s-electron density at the iron center leads to a negative change in the isomer shift. For ferrocene and its derivatives, changes in ring substitution do little to affect the s-electron density; thus, the I.S. usually falls within a range of 0.50–0.55 mm/s for ferrocenes referenced to iron metal (Table VI). Two recently reported exceptions to this generalization are the shifts for $[Fe(C_5Ph_4H)_2]$ and $[Fe(C_5Ph_5)_2]$.[60] In $[Fe(C_5Ph_4H)_2]$, the I.S. varies from 0.47 mm/s at 298 K to 0.60 mm/s at 15 K; it has been suggested that a solid-state thermal isomerization (toward a $[Fe(\eta^5\text{-}C_5Ph_4H)(C_5(\eta^6\text{-}$

TABLE VI

REPRESENTATIVE ^{57}Fe MÖSSBAUER DATA FOR FERROCENES AND FERROCENIUM SALTS

Complex	I.S. (mm/s) (T, K)	Q.S. (mm/s) (T, K)	Ref.
$[FeCp_2]$	0.52 (80)	2.37 (80)	183
$[Fe(C_5MeH_4)_2]$	0.53 (80)	2.39 (80)	183
$[Fe(C_5CH_4OMe)_2]$	0.52 (80)	2.15 (80)	183
$[Fe(C_5H_4CN)_2]$	0.55 (80)	2.29 (80)	183
$[Fe(C_5Me_5)_2]$	0.53 (80)	2.50 (80)	183
$[Fe(C_5Ph_4H)_2]$	0.47 (298)	2.37 (298)	60
	0.60 (15)	2.37 (15)	
$[Fe(C_5Ph_5)_2]$	0.60 (298)	2.55 (298)	60
	0.71 (13.5)	2.58 (13.5)	
$[Fe(\eta^5\text{-}C_5Ph_5)(\eta^6\text{-}(C_6H_5)C_5Ph_4)]$	0.51 (298)	1.78 (298)	60
	0.63 (15)	1.80 (15)	
$[FeCp_2][BF_4]$	0.84 (80)	Not observed	235
$[FeCp_2]Cl$	0.84 (80)	Not observed	235
$[Fe(C_5Me_5)_2][PF_6]$	0.75 (85)	Not observed	236

$C_6H_5)Ph_3H)$] form) may be causing this effect. The I.S. in decaphenylferrocene varies from 0.60 mm/s at 298 K to 0.71 mm/s at 15 K; both thermal isomerization and the increase in Fe–ring distance found in the room-temperature structure (see Section IV) may contribute to the high I.S. values.

Changes in the population of the d-orbitals can affect the electric field gradients around the metal, so that the quadrupolar splittings do change with substitution. Even here, however, the range of values is usually confined to 2.1–2.5 mm/s. Electron donating groups cause an increase in the Q.S.; for example, diethylferrocene displays a Q.S. that is +0.05 mm/s larger than in unsubstituted ferrocene,[184] and as the number of electron donating substituents is increased, the Q.S. rises proportionally (Fig. 23). It has been suggested that at certain levels of substitution, a saturation point is reached so that Q.S. no longer increases, possibly owing to a complete suppression of back-bonding.[185] Evidence for this effect is limited and based largely on a study of ethyl substituted ferrocenes for which no experimental errors in data are reported.[186] However, a saturation effect has been used to explain an anomalous Q.S. for the p-cyanophenyl ferrocenyl carbenium ion.[187]

Electron-withdrawing groups exert more complicated effects on the Q.S. than do electron donors. Those groups that withdraw electron density through conjugation, such as —CO_2Me, decrease Q.S. (e.g., ΔQ.S. = −0.07 mm/s for acetoxyferrocene); however, strongly electron-withdrawing groups such as —Br cause an *increase* in Q.S. (ΔQ.S. = +0.09 mm/s for bromoferrocene). Exceptions to these trends are found with protonated or bulky ligands that lead to conformations having poor orbital overlap with the Cp rings.[187]

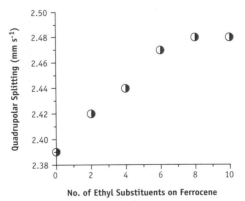

FIG. 23.   Effect of successive ethyl group substitution in ferrocene on the ^{57}Fe Mössbauer quadrupolar splitting values.[184]

The nonlinear effects of ring substituents on the Q.S. in ferrocenes have been rationalized by considering their influence on the orbitals of e_1 and e_2 symmetry, which are those most affected by changes in substitution.[183] The orbital with e_2 symmetry is mainly metal d-orbital in character and is responsible for metal-to-ligand back-bonding, whereas the orbital with e_1 symmetry is a combination of ligand p- ($\sim$54%) and metal d-orbitals ($\sim$40%) and is responsible for ring-to-metal donation.[188] The Q.S. is approximately twice as sensitive to changes in the electron density of the e_2 orbital than to similar changes in the e_1 orbital, although the latter is more extensively involved in the bonding of the complex.[183]

When ferrocene possesses electron-donating substituents, the increase in electron donation through e_1 lessens the need for electron donation from e_2. The decrease in back-bonding produces an increase in d-electron density with unequal electron populations in the d-orbitals, so that Q.S. increases.[182] If a group that withdraws electron density through conjugation is present on a ferrocene ring, the contribution from e_1 is lessened; in compensation, back donation from e_2 increases.[183,189] The resulting reduction in electron density donation to the metal tends to equalize the relative d-orbital populations, leading to a decrease in Q.S. When a stronger electron-withdrawing group is substituted on a ferrocene ring, the effect of reducing the contribution of e_1 is so great that the contribution from e_2 is maximized, and enough electron density is removed from the metal that Q.S. increases.[189]

Oxidation of ferrocene to the ferrocenium cation almost always results in the collapse of the Q.S.[182] This effect has been related to the redistribution of electron density in the d-orbitals upon oxidation, which leads to a more symmetrical electric field gradient. Consequently, observed Q.S.s

TABLE VII

REPRESENTATIVE ^{119}Sn MÖSSBAUER DATA FOR STANNOCENES
AND RELATED COMPOUNDS

Complex	I.S. (mm/s)	Q.S. (mm/s)	Ring normal–M– ring normal (°)	Ref.
SnCp$_2$	3.74 (6)	0.86 (6)	134	237
			133	
Sn(C$_5$MeH$_4$)$_2$	3.83	0.78	n/a	238
Sn(C$_5$Me$_5$)$_2$	3.53 (3)	0.99 (6)	144.6	191
			143.6	
Sn(C$_5$Ph$_5$)$_2$	3.74	0.58	180	192
[Sn(C$_5$Me$_5$)$_2$]$^+$[CF$_3$SO$_3$]$^-$	3.74	0.86	n/a	191
{[Sn(μ-η^5-C$_5$H$_5$)$_2$(η^5-C$_5$H$_5$)(thf)]$^+$ [BF$_4$]$^-$}$_n$	3.79 (6)	0.90 (6)	120.3	239

for ferrocenium cations at 78 K are only about 0.1–0.6 mm/s. Since the s-orbitals remain essentially unaffected by oxidation, the isomer shift usually decreases by only 0.1–0.2 mm/s.

The effect of substituents on the ^{57}Fe Mössbauer Q.S. of oxidized metal-locenes has been compared with their influence on quarter-wave poten-tials.[188] The correlation of the two parameters for 17 complexes produced an r^2 of 0.93, but the correlation did not apply to diketones or ferrocenyl carbenium ions. Because substituent effects are mainly transmitted through the more ligand-based e_1 orbital, and oxidation potentials are governed by what is thought to be the HOMO e_2 orbital, it is surprising that the correlation is as good as it is.[188] As substituent effects are ligand-based, correlations between Q.S. and Hammett-σ_p^+ parameters were good if the data for dichloroferrocene was excluded, indicating that resonance involves the iron center and the ring. The anomalous data for dichloroferro-cene could not be explained.[188] Q.S. values for multiply trimethylsilyl- and tert-butyl-substituted ferrocenes have been found to correlate moderately well with σ_R resonance parameters.[190]

Substituent effects of aryl ferrocenyl ketones and ferrocenyl carbenium ions have been studied.[187] Changes in the Q.S. of aryl ferrocenyl ketones do not correlate with the position of substituents on the phenyl group. Protonation of the ketones in acid normally decreases Q.S. by 0.03–0.24 mm/s; however, changes in substituents on the phenyl group have a more pronounced effect on the protonated species. Ferrocenyl carbenium ions formed by dissolving the corresponding alcohol in strong acid display increases in Q.S. of 0.16–0.29 mm/s. The increase in Q.S. for sterically bulky carbenium ions is not as large as expected, most likely owing to reduced orbital overlap from twisting of the carbenium ion sp^2 plane out of the plane of the cyclopentadienyl ring.

B. ^{119}Sn Mössbauer Spectra

Stannocenes are notable for their high values of ^{119}Sn isomer shift, well above the 2.1–2.4 mm/s range commonly taken as the dividing region between tetravalent and divalent tin species (all known stannocenes have I.S. > 3.2 mm/s) (Table VII).[191] The isomer shift reflects the s-electron density at the nucleus, $|\psi_s(0)|^2$, and is a function of the total of s-electron density for all the orbitals. Thus, it is fairly insensitive to the hybridization of individual orbitals and cannot be used in a quantitative way to estimate the amount of s-orbital character in a stannocene lone pair. It is noteworthy that the I.S. for the bent stannocenes [SnCp$_2$] and [Sn(C$_5$Me$_5$)$_2$] [3.74 and 3.83(3) mm/s, respectively] are as high as or higher than that for the linear complex $[Sn(C_5Ph_5)_2]$ (3.74 mm/s), although valence bond analysis (Section IV) gives the electrons in the bent and linear species significantly different degrees of hybridization.

Potentially more informative with regard to bonding is the quadrupole splitting, which depends on the partitioning of the s-electron density between various orbitals, with larger values of the Q.S. reflecting greater mixing of the s-electron density with p or d orbitals. In a valence bond approach to the bonding, changing the amount of p-character in the lone pair orbital should affect the ring–tin–ring angle and the Q.S. simultaneously. However, the average interplanar ring angles for $[SnCp_2]$ (133.5°), $[Sn(C_5Me_5)_2]$ (144.1°), and $[Sn(C_5Ph_5)_2]$ (180°) do not track with the Q.S. values [0.86(6), 0.99(6), and 0.58 mm/s, respectively]. Furthermore, the Mössbauer spectrum for a complex material such as $\{[Sn(\mu-\eta^5-C_5H_5)_2(\eta^5-C_5H_5)(thf)]^+[BF_4]^-\}_n$[193] (Fig. 24) consists of only one doublet with I.S. = 3.79(6) and Q.S. = 0.90(6) mm/s, values that are experimentally indistinguishable from those of $[SnCp_2]$. This is true even though the stannocene/stannocenium compound contains tin in two different coordination environments with approximately $[SnX(\mu-\eta^5-Cp)_3]$ coordination; the angle between the most closely bonded rings is 120.3°. It seems clear that the available ^{119}Sn Mössbauer data cannot address the question of the hybridization of the lone pair in stannocenes.

It is possible that the donor/acceptor properties of the substituents do affect the Q.S. in a consistent way, in that the Q.S. increases when electron-donating substituents are present {e.g., Q.S. = 0.86(6) for $[SnCp_2]$, and 0.99(6) for $[Sn(C_5Me_5)_2]$} and decreases with electron-withdrawing substituents {Q.S. = 0.58 mm/s for $[Sn(C_5Ph_5)_2]$, but the data are too few to permit dependable conclusions to be drawn.

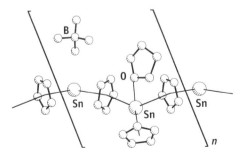

FIG. 24.   Portion of the polymeric species $\{[Sn(\mu-\eta^5-C_5H_5)_2(\eta^5-C_5H_5)(thf)]^+[BF_4]^-\}_n$.[193]

VIII

CONCLUSIONS AND FUTURE DIRECTIONS

The effects of ring substitution on the chemical, structural, and physical properties of compounds containing cyclopentadienyl rings can be profound. These effects have dramatically influenced the applications of such compounds in synthetic, catalytic, and materials chemistry. Despite the large quantity of literature covering what in some respects is a relatively mature field of investigation, there is still room for improvement in our understanding of the interactions of main-group and transition metals with cyclopentadienyl rings. The many empirical correlations that exist among structure, spectra, and electronic properties of metallocene chemistry could be better integrated with theoretical analysis.

Continuing advances in the theoretical treatment of metallocenes and substituted derivatives can be anticipated. Despite the historical difficulties with molecular orbital calculations on ferrocene, the problems are now recognized to stem from failure to account for electron correlation effects. New approaches to addressing this issue, coupled with inevitable increases in computing power, should make more metallocenes with substituted groups systems amenable to accurate calculation. More realistic predictions of donor ability, and thus better estimates of their effects in nonlinear optical (NLO) systems, will be possible.

It is becoming apparent that nonbonding interactions play a greater role than previously appreciated in determining the structures of both main-group and transition-metal metallocenes. Molecular mechanics are particularly efficient in describing such interactions and can be readily applied to much larger systems than traditional MO methods. The analysis of structural conformations of bis(cyclopentadienyl) compounds used in stereoselective synthesis[194] can be expected to benefit from wider applications of MM methods.

An area of experimental investigation that has not received much attention is the determination of basic thermodynamic data for substituted metallocenes (i.e., bond and vaporization enthalpies). Considering how important these values can be in the prediction and interpretation of reaction mechanisms in organometallic reactivity,[195] it is surprising that more data are not available. One would expect this to change in the future.

Metallocenyl fragments have been incorporated as electron donor units into various NLO materials. In most of the studies of these systems, the metal–ring interaction is not the focus of interest, and only a few have systematically examined the effects that ring substituents have on the second-order nonlinearities and related them to changes in metal–ring interactions.[36] For example, replacement of the Cp ring in $[FeCp(C_5H_4)\text{-}CH{=}CH(4\text{-}C_6H_4\text{-}1\text{-}NO_2)]$ (**6**) with C_5Me_5 produces an almost 30% increase in the molecular hyperpolarizability.[196] This represents an area of potentially large growth in applications.

Metallocenes have also played key roles in the development of "designer magnets," i.e., molecule-based systems that display enhanced electron-spin coupling and that could serve as a source of magnetic materials that can be fabricated at room temperature.[197] Compounds such as $[M(C_5Me_5)_2]^{\cdot+}[TCNE]^{\cdot-}$ (M = Cr, Mn, Fe) become ferromagnetic when cooled to sufficiently low temperatures, and changing the anion or metal can significantly change the magnetic properties (e.g., $[Fe(C_5Me_5)_2]^{\cdot+}$ $[TCNQ]^{\cdot-}$ and $[Ni(C_5Me_5)_2]^{\cdot+}[TCNE]^{\cdot-}$ display metamagnetic and antiferromagnetic properties, respectively). Changes in substituents on the metallocenes are obviously important ($[FeCp_2][TCNE]$ is a diamagnetic material), but more systematic investigations of substituent effects on magnetic properties are still to be done. A high degree of customizability in magnetic properties may result.

As members of the "first family" of organometallics, metallocenes and their substituted analogues will undoubtedly continue to influence all areas of inorganic and organometallic chemistry in the foreseeable future.

ACKNOWLEDGMENTS

I would like to thank the past and present members of my research group for their contributions to the work from my laboratory. Financial support has been received from the Army Research Office; the National Science Foundation; the donors of the Petroleum Research Fund, administered by the American Chemical Society; and the Natural Science Committee of Vanderbilt University.

REFERENCES

(1) Pauson, P. L.; Kealy, T. J. *Nature* **1951,** *168,* 1039.
(2) Miller, S. A.; Tebboth, J. A.; Tremaine, J. F. *J. Chem. Soc.* **1952,** 632.
(3) Woodward, R. B.; Rosenblum, M.; Whiting, M. C. *J. Am. Chem. Soc.* **1952,** *74,* 3458.
(4) Janiak, C.; Schumann, H. *Adv. Organomet. Chem.* **1991,** *33,* 291.
(5) Macomber, D. W.; Hart, W. P.; Rausch, M. D. *Adv. Organomet. Chem.* **1982,** *21,* 1.
(6) Rieckhoff, M.; Pieper, U.; Stalke, D.; Edelmann, F. T. *Angew. Chem.* **1993,** *105,* 1102.
(7) Erker, G.; Mollenkopf, C.; Grehl, M.; Froehlich, R.; Krueger, C.; Noe, R.; Riedel, M. *Organometallics* **1994,** *13,* 1950.
(8) Stehling, U.; Diebold, J.; Kirsten, R.; Roell, W.; Brintzinger, H. H.; Juengling, S.; Muelhaupt, R.; Langhauser, F. *Organometallics* **1994,** *13,* 964.
(9) Manners, I. *Adv. Organomet. Chem.* **1995,** *37,* 131.
(10) Cowley, A. H.; Mardones, M. A.; Avendano, S.; Roman, E.; Manriquez, J. M.; Carrano, C. *Polyhedron* **1993,** *12,* 125.
(11) O'Hare, D.; Murphy, V. J.; Kaltsoyannis, N. *J. Chem. Soc., Dalton Trans.* **1993,** 383.

(12) O'Hare, D.; Green, J. C.; Marder, T.; Collins, S.; Stringer, G.; Kakkar, A. K.; Kaltsoyannis, N.; Kuhn, A.; Lewis, R.; Mehnert, C.; Scott, S.; Kurmoo, M.; Pugh, S. *Organometallics* **1992**, *11*, 48.

(13) Evans, W. J.; Gummersheimer, T. S.; Boyle, T. J.; Ziller, J. W. *Organometallics* **1994**, *13*, 1281.

(14) Viebrock, H.; Abeln, D.; Weiss, E. Z. *Naturforsch., B: Chem. Sci.* **1994**, *49*, 89.

(15) Mösges, C.; Hampel, F.; Schleyer, P. v. R. *Organometallics* **1992**, *11*, 1769.

(16) Zerger, R.; Stucky, G. *J. Organomet. Chem.* **1974**, *80*, 7.

(17) Atwood, J. L.; Hunter, W. E.; Cowley, A. H.; Jones, R. A.; Stewart, C. A. *J. Chem. Soc., Chem. Commun.* **1981**, 925.

(18) Jutzi, P.; Kohl, F.; Hofmann, P.; Krueger, C.; Tsay, Y. H. *Chem. Ber.* **1980**, *113*, 757.

(19) Williams, R. A.; Hanusa, T. P.; Huffman, J. C. *Organometallics* **1990**, *9*, 1128.

(20) Burkey, D. J.; Hanusa, T. P. *Comments Inorg. Chem.* **1995**, *17*, 41.

(21) Seiler, P.; Dunitz, J. D. *Acta Crystallogr., Sect. B* **1979**, *35*, 1068.

(22) Haaland, A.; Nilsson, J. E. *Acta Chem. Scand.* **1968**, *22*, 2653.

(23) Seiler, P.; Dunitz, J. D. *Acta Crystallogr., Sect. B* **1980**, *36*, 2946.

(24) Boeyens, J. C. A.; Levendis, D. C.; Bruce, M., I.; Williams, M. L. *J. Crystallogr. Spectrosc. Res.* **1986**, *16*, 519.

(25) Almenningen, A.; Haaland, A.; Samdal, S.; Brunvoll, J.; Robbins, J. L.; Smart, J. C. *J. Organomet. Chem.* **1979**, *173*, 293.

(26) Freyberg, D. P.; Robbins, J. L.; Raymond, K. N.; Smart, J. C. *J. Am. Chem. Soc.* **1979**, *101*, 892.

(27) Brown, G. M. *J. Chem. Soc., Chem. Commun.* **1972**, 5.

(28) Doman, T. N.; Landis, C. R.; Bosnich, B. *J. Am. Chem. Soc.* **1992**, *114*, 7264.

(29) Jordan, R. F. *Adv. Organomet. Chem.* **1991**, *32*, 325.

(30) Kaminsky, W. *Angew. Makromol. Chem.* **1994**, *223*, 101.

(31) Kaminsky, W. *Catal. Today* **1994**, *20*, 257.

(32) Horton, A. D. *Trends Polym. Sci.* **1994**, *2*, 158.

(33) de Lang, R. J.; van Soolingen, J.; Verkruijsse, H. D.; Brandsma, L. *Synth. Commun.* **1995**, *25*, 2989.

(34) Blake, A. J.; Harrison, A.; Johnson, B. F. G.; McInnes, E. J. L.; Parsons, S.; Shephard, D. S.; Yellowlees, L. J. *Organometallics* **1995**, *14*, 3160.

(35) Green, M. L. H.; Marder, S. R.; Thompson, M. E.; Bandy, J. A.; Bloor, D.; Kolinsky, P. V.; Jones, R. J. *Nature* **1987**, *330*, 360.

(36) Kanis, D. R.; Ratner, M. A.; Marks, T. J. *Chem. Rev.* **1994**, *94*, 195.

(37) Sato, M.; Mogi, E.; Kumakura, So. *Organometallics* **1995**, *14*, 3157.

(38) Braga, D. *Chem. Rev.* **1992**, *92*, 633.

(39) Bretschneider-Hurley, A.; Winter, C. H. *J. Am. Chem. Soc.* **1994**, *116*, 6468.

(40) Hughes, R. P.; Robinson, D. J. *Organometallics* **1989**, *8*, 1015.

(41) Tanner, P. S.; Burkey, D. J.; Hanusa, T. P. *Polyhedron* **1995**, *14*, 331.

(42) It should be emphasized that the failure to isolate a compound by a particular synthetic procedure is not a legitimate criterion for the molecule's stability. This error in reasoning has a long history in chemistry (cf. the "instability" of the noble gas halides or the perbromate ion) and also occurs in the metallocene literature.

(43) Castellani, M. P.; Geib, S. J.; Rheingold, A. L.; Trogler, W. C. *Organometallics* **1987**, *6*, 1703.

(44) Castellani, M. P.; Wright, J. M.; Geib, S. J.; Rheingold, A. L.; Trogler, W. C. *Organometallics* **1986**, *5*, 1116.

(45) Hoobler, R. J.; Adams, J. V.; Hutton, M. A.; Francisco, T. W.; Haggerty, B. S.; Rheingold, A. L.; Castellani, M. P. *J. Organomet. Chem.* **1991**, *412*, 157.

(46) Schumann, H.; Janiak, C.; Zuckerman, J. J. *Chem. Ber.* **1988,** *121,* 207.
(47) Janiak, C.; Schumann, H.; Stader, C.; Wrackmeyer, B.; Zuckerman, J. J. *Chem. Ber.* **1988,** *121,* 1745.
(48) Hedberg, F. L.; Rosenberg, H. *J. Am. Chem. Soc.* **1973,** *95,* 870.
(49) Han, Y.-H.; Heeg, M. J.; Winter, C. H. *Organometallics* **1994,** *13,* 3009.
(50) Winter, C. H.; Han, Y.-H.; Heeg, M. J. *Organometallics* **1992,** *11,* 3169.
(51) Boev, V. I.; Dombrovskii, A. V. *Zh. Obshch. Khim.* **1977,** *47,* 727.
(52) Gassman, P. G.; Winter, C. H. *J. Am. Chem. Soc.* **1988,** *110,* 6130.
(53) Curnow, O. J.; Hughes, R. P. *J. Am. Chem. Soc.* **1992,** *114,* 5895.
(54) Hughes, R. P.; Zheng, X.; Ostrander, R. L.; Rheingold, A. L. *Organometallics* **1994,** *13,* 1567.
(55) Paprott, G.; Lehmann, S.; Seppelt, K. *Chem. Ber.* **1988,** *121,* 727.
(56) Slocum, D. W.; Johnson, S.; Matusz, M.; Duraj, S.; Cmarik, J. L.; Simpson, K. M.; Owen, D. A. *Polym. Mater. Sci. Eng.* **1983,** *49,* 353.
(57) Slocum, D. W.; Duraj, S.; Matusz, M.; Cmarik, J. L.; Simpson, K. M.; Owen, D. A. In *Metal Containing Polymeric Systems;* J. E. C. Sheats, Charles E., Jr.; Pittman, Charles U., Jr., Eds.; Plenum: New York, 1985, p. 59.
(58) Brown, K. N.; Field, L. D.; Lay, P. A.; Lindall, C. M.; Masters, A. F. *J. Chem. Soc., Chem. Commun.* **1990,** 408.
(59) Schumann, H.; Lentz, A.; Weimann, R.; Pickardt, J. *Angew. Chem. Int. Ed. Engl.* **1994,** *33,* 1731.
(60) Field, L. D.; Hambley, T. W.; Humphrey, P. A.; Lindall, C. M.; Gainsford, G. J.; Masters, A. F.; St. Pierre, T. G.; Webb, J. *Aust. J. Chem.* **1995,** *48,* 851.
(61) Sitzmann, H. *Z. Naturforsch. [B]* **1989,** *44,* 1293.
(62) Sitzmann, H.; Boese, R. *Angew. Chem. Int. Ed. Engl.* **1991,** *30,* 971.
(63) Sitzmann, H.; Bock, H.; Boese, R.; Dezember, T.; Havlas, Z.; Kaim, W.; Moscherosch, M.; Zanathy, L. *J. Am. Chem. Soc.* **1993,** *115,* 12003.
(64) Burkey, D. J.; Alexander, E. K.; Hanusa, T. P. *Organometallics* **1994,** *13,* 2773.
(65) Sitzmann, H. J. *J. Organomet. Chem.* **1988,** *354,* 203.
(66) Buchholz, D.; Gloaguen, B.; Fillaut, J.-L.; Cotrait, M.; Astruc, D. *Chem. Eur. J.* **1995,** *1,* 374.
(67) Gloaguan, B.; Astruc, D. *J. Am. Chem. Soc.* **1990,** *112,* 4607.
(68) Overby, J. S.; Schoell, N. S.; Hanusa, T. P. Unpublished results. **1995.**
(69) Wulfsberg, G. *Principles of Descriptive Inorganic Chemistry;* Brooks/Cole: Monterey, CA, 1987, pp. 95–100.
(70) Benac, M. J.; Cowley, A. H.; Jones, R. A.; Tasch, A. F., Jr. *Chem. Mater.* **1989,** *1,* 289.
(71) Kirlin, P. S.; Brown, D. W.; Gardiner, R. A. U.S. Patent No. 5,225,561, 1993.
(72) Hanusa, T. P. *Chem. Rev.* **1993,** *93,* 1023.
(73) Gardiner, M. G.; Raston, C. L.; Kennard, C. H. L. *Organometallics* **1991,** *10,* 3680.
(74) Andersen, R. A.; Blom, R.; Burns, C. J.; Volden, H. V. *J. Chem. Soc., Chem. Commun.* **1987,** 786.
(75) Burkey, D. J. Ph.D. thesis, Vanderbilt University, August 1995.
(76) The 18-electron anion [MnCp$_2'$]$^-$ is air-sensitive.
(77) Shannon, R. D. *Acta Crystallogr., Sect. A* **1976,** *32,* 751.
(78) Burkey, D. J.; Hanusa, T. P. *J. Organomet. Chem.* **1996,** *512,* 165.
(79) Burkey, D. J.; Hanusa, T. P. *Organometallics* **1995,** *14,* 11.
(80) Schumann, H.; Janiak, C.; Hahn, E.; Kolax, C.; Loebel, J.; Rausch, M. D.; Zuckerman, J. J.; Heeg, M. J. *Chem. Ber.* **1986,** *119,* 2956.
(81) Driscoll, D. C.; Dowben, P. A.; Boag, N. M.; Grade, M.; Barfuss, S. *J. Chem. Phys.* **1986,** *85,* 4802.

(82) Lüthi, H. P.; Ammeter, J. H.; Almlöf, J.; Faegri, K., Jr. *J. Chem. Phys.* **1982,** *77,* 2002.
(83) Park, C.; Almlöf, J. *J. Chem. Phys.* **1991,** *95,* 1829.
(84) Fan, L.; Ziegler, T. *J. Chem. Phys.* **1991,** *95,* 7401.
(85) Haaland, A. *Acc. Chem. Res.* **1979,** *12,* 415.
(86) Armstrong, D. R.; Fortune, R.; Perkins, P. G. *Rev. Roum. Chim.* **1978,** *23,* 1179.
(87) Vondrák, T. *J. Organomet. Chem.* **1986,** *306,* 89.
(88) Dowben, P. A.; Driscoll, D. C.; Tate, R. S.; Boag, N. M. *Organometallics* **1988,** *7,* 305.
(89) Bodenheimer, J. S.; Low, W. *Spectrochim. Acta* **1973,** *29A,* 1733.
(90) Struchkov, Y. T.; Andrianov, V. G.; Salnikova, T. N.; Lyatifov, I. R.; Materikova, R. B. *J. Organomet. Chem.* **1978,** *145,* 213.
(91) Albers, M. O.; Liles, D. C.; Robinson, D. J.; Shaver, A.; Singleton, E.; Wiege, M. B.; Boeyens, J. C. A.; Levendis, D. C. *Organometallics* **1986,** *5,* 2321.
(92) Burman, J. A.; Hays, M. L.; Burkey, D. J.; Tanner, P. S.; Hanusa, T. P. *J. Organomet. Chem.* **1994,** *479,* 135.
(93) Martinez, R.; Tiripicchio, A. *Acta Crystallogr., Sect. C* **1990,** *46,* 202.
(94) Gebert, E.; Reis, A. H., Jr.; Miller, J. S.; Rommelmann, H.; Epstein, A. J. *J. Am. Chem. Soc.* **1982,** *104,* 4403.
(95) Dixon, D. A.; Calabrese, J. C.; Miller, J. S. *J. Am. Chem. Soc.* **1986,** *108,* 2582.
(96) Miller, J. S.; Zhang, J. H.; Reiff, W. M.; Dixon, D. A.; Preston, L. D.; Reis, A. H., Jr.; Gebert, E.; Extine, M.; Troup, J.; Epstein, A. J.; Ward, M. D. *J. Phys. Chem.* **1987,** *91,* 4344.
(97) Miller, J. S.; Calabrese, J. C.; Harlow, R. L.; Dixon, D. A.; Zhang, J. H.; Reiff, W. M.; Chittipeddi, S.; Selover, M. A.; Epstein, A. J. *J. Am. Chem. Soc.* **1990,** *112,* 5496.
(98) Pickardt, J.; Schumann, H.; Mohtachemi, R. *Acta Crystallogr., Sect. C* **1990,** *46,* 39.
(99) Hays, M. L.; Burkey, D. J.; Duderstadt, R. E.; Hanusa, T. P. Submitted.
(100) Okuda, J.; Herdtweck, E. *Chem. Ber.* **1988,** *121,* 1899.
(101) Foucher, D. A.; Honeyman, C. H.; Lough, A. J.; Manners, I.; Nelson, J. M. *Acta Crystallogr., Sect. C* **1995,** *51,* 1795. Note that in a ferrocenium derivative (TCNQ salt) in which the cyclopentadienyl rings are fused to bicyclic norbornane-like fragments (Gallucci, J. C.; Opromolla, G.; Paquette, L. A.; Pardi, L.; Schirch, P. F. T.; Sivik, M. R.; Zanello, P. *Inorg. Chem.* **1993,** *32,* 2292) the ring-normal–Fe–ring normal angle is 172.9°.
(102) Bünder, W.; Weiss, E. *J. Organomet. Chem.* **1975,** *92,* 65.
(103) Legzdins, P.; Wassink, B.; Einstein, F. W. B.; Jones, R. H. *Organometallics* **1988,** *7,* 477.
(104) Dixon, D. A.; Miller, J. S. *J. Am. Chem. Soc.* **1987,** *109,* 3656.
(105) Evans, W. J. *Adv. Organomet. Chem.* **1985,** *24,* 131.
(106) Williams, R. A.; Hanusa, T. P.; Huffman, J. C. *J. Chem. Soc., Chem. Commun.* **1988,** 1045.
(107) Panattoni, C.; Bombieri, G.; Croatto, U. *Acta Crystallogr.* **1966,** *21,* 823.
(108) Fischer, E. O.; Stölzle, G. *Chem. Ber.* **1961,** *94,* 2187.
(109) Williams, R. A.; Tesh, K. F.; Hanusa, T. P. *J. Am. Chem. Soc.* **1991,** *113,* 4843.
(110) Hitchcock, P. B.; Howard, J. A. K.; Lappert, M. F.; Prashar, S. *J. Organomet. Chem.* **1992,** *437,* 177.
(111) Evans, W. J.; Hughes, L. A.; Hanusa, T. P.; Doedens, R. J. *Organometallics* **1986,** *5,* 1285.
(112) Bünder, W.; Weiss, E. *J. Organomet. Chem.* **1975,** *92,* 1.
(113) Harder, S.; Prosenc, M. H. *Angew. Chem., Int. Ed. Engl.* **1994,** *33,* 1744.

(114) Hollis, T. K.; Burdett, J. K.; Bosnich, B. *Organometallics* **1993**, *12*, 3385.
(115) Fischer, B.; van Mier, G. P. .; Boersma, J.; van Koten, G.; Smeets, W. J. J.; Spek, A. L. *Recl. Trav. Chim. Pays-Bas* **1988**, *107*, 259.
(116) Cowley, A. C.; Lasch, J. G.; Norman, N. C.; Stewart, C. A.; Wright, T. C. *Organometallics* **1983**, *2*, 1691.
(117) Williams, R. A.; Hanusa, T. P.; Huffman, J. C. *J. Am. Chem. Soc.* **1990**, *112*, 2454.
(118) Schumann, H.; Janiak, C.; Hahn, E.; Loebel, J.; Zuckerman, J. J. *Angew. Chem., Int. Ed. Engl.* **1985**, *24*, 773.
(119) Jutzi, P.; Holtmann, U.; Kanne, D.; Krueger, C.; Blom, R.; Gleiter, R.; Hyla, K. I. *Chem. Ber.* **1989**, *122*, 1629.
(120) Heeg, M. J.; Herber, R. H.; Janiak, C.; Zuckerman, J. J.; Schumann, H.; Manders, W. F. *J. Organomet. Chem.* **1988**, *346*, 321.
(121) Fischer, B.; Wijkens, P.; Boersma, J.; van Koten, G.; Smeets, W. J. J.; Spek, A. L.; Budzelaar, P. H. M. *J. Organomet. Chem.* **1989**, *376*, 223.
(122) Almenningen, A.; Bastiansen, O.; Haaland, A. *J. Chem. Phys.* **1964**, *40*, 3434.
(123) Haaland, A. *Acta Chem. Scand.* **1968**, *22*, 3030.
(124) Nugent, K. W.; Beattie, J. K.; Hambley, T. W.; Snow, M. R. *Aust. J. Chem.* **1984**, *37*, 1601.
(125) Lustyk, J.; Starowieyski, K. B. *J. Organomet. Chem.* **1979**, *170*, 293.
(126) Nugent, K. W.; Beattie, J. K. *Inorg. Chem.* **1988**, *27*, 4269.
(127) Beattie, J. K.; Nugent, K. W. *Inorg. Chim. Acta* **1992**, *198–200*, 309.
(128) Pratten, S. J.; Cooper, M. K.; Aroney, M. J., *J. Organomet. Chem.* **1990**, *381*, 147.
(129) Blom, R.; Haaland, A.; Weidlein, J. *J. Chem. Soc., Chem. Commun.* **1985**, 266.
(130) Blom, R.; Boersma, J.; Budzelaar, P. H. M.; Fischer, B.; Haaland, A.; Volden, H. V.; Weidlein, J. *Acta Chem. Scand.* **1986**, *A40*, 113.
(131) Budzelaar, P. H. M.; Boersma, J.; van der Kerk, G. J. M.; Spek, A. L.; Duisenberg, A. J. M. *J. Organomet. Chem.* **1985**, *281*, 123.
(132) Budzelaar, P. H. M.; Boersma, J.; van der Kerk, G. J. M.; Spek, A. L. *Organometallics* **1984**, *3*, 1187.
(133) Evans, W. J.; Hughes, L. A.; Hanusa, T. P. *J. Am. Chem. Soc.* **1984**, *106*, 4270.
(134) Burns, C. J. Ph.D. thesis, University of California, Berkeley, 1987.
(135) Andersen, R. A.; Blom, R.; Boncella, J. M.; Burns, C. J.; Volden, H. V. *Acta Chem. Scand.* **1987**, *A41*, 24.
(136) Freeman, A. J.; Watson, R. E. *Phys. Rev.* **1962**, *127*, 2058.
(137) Blom, R.; Faegri, K., Jr.; Volden, H. V. *Organometallics* **1990**, *9*, 373.
(138) Almlöf, J.; Fernholt, L.; Faegri, K. J.; Haaland, A.; Schilling, B. E. R.; Seip, R.; Taugboel, K. *Acta Chem. Scand., Ser. A* **1983**, *37*, 131.
(139) Almenningen, A.; Haaland, A.; Motzfeldt, T. *J. Organomet. Chem.* **1967**, *7*, 97.
(140) Ortiz, J. V.; Hoffman, R. *Inorg. Chem.* **1985**, *24*, 2095.
(141) Green, J. C.; Hohl, D.; Rösch, N. *Organometallics* **1987**, *6*, 712.
(142) Kaupp, M.; Schleyer, P. v. R.; Dolg, M.; Stoll, H. *J. Am. Chem. Soc.* **1992**, *114*, 8202.
(143) Dave, L. D.; Evans, D. F.; Wilkinson, G. *J. Chem. Soc.* **1959**, 3684.
(144) Baxter, S. G.; Cowley, A. H.; Lasch, J. G.; Lattman, M.; Sharum, W. P.; Stewart, C. A. *J. Am. Chem. Soc.* **1982**, *104*, 4064.
(145) Lee, T. J.; Rice, J. E. *J. Am. Chem. Soc.* **1989**, *111*, 2011.
(146) Dewar, M. J. S.; Healy, E. F.; Kuhn, D. R.; Holder, A. J. *Organometallics* **1991**, *10*, 431.
(147) Dewar, M. J. S.; Holloway, M. K.; Grady, G. L.; Stewart, J. J. P. *Organometallics* **1985**, *4*, 1973.

(148) Armstrong, D. R.; Herbst-Irmer, R.; Kuhn, A.; Moncrieff, D.; Paver, M. A.; Russell, C. A.; Stalke, D.; Steiner, A.; Wright, D. S. *Angew. Chem., Int. Ed. Engl.* **1993**, *32*, 1774.
(149) Timofeeva, T. V.; Lii, J.-H.; Allinger, N. L. *J. Am. Chem. Soc.* **1995**, *117*, 7452.
(150) Gorton, J. E.; Lentzner, H. L.; Watts, W. E. *Tetrahedron* **1971**, *27*, 4353.
(151) Nagy, A. G.; Toma, S. *J. Organomet. Chem.* **1985**, *282*, 267.
(152) Little, W. F.; Reilley, C. N.; Johnson, J. D.; Sanders, A. P. *J. Am. Chem. Soc.* **1964**, *86*, 1382.
(153) Gassman, P. G.; Deck, P. A.; Winter, C. H.; Dobbs, D. A.; Cao, D. H. *Organometallics* **1992**, *11*, 959; Gassman, P. G.; Mickelson, J. W.; Sowa, J. R., Jr. *J. Am. Chem. Soc.* **1992**, *114*, 6942.
(154) Ryan, M. F.; Richardson, D. E.; Lichtenberger, D. L.; Gruhn, N. E. *Organometallics* **1994**, *13*, 1190. Electrospray mass spectrometry has also been used to study the electrochemistry of metallocenes. See: Xu, X.; Nolan, S. P.; Cole, R. B. *Anal. Chem.* **1994**, *66*, 119.
(155) Stretlets, V. V. *Coord. Chem. Rev.* **1992**, *114*, 1.
(156) Zanello, P. In *Ferrocenes, Homogeneous Catalysis, Organic Synthesis, Materials Science;* Togni, A., and Hayashi, T., Eds.; VCH: Weinheim, Germany, 1995, Chapter 7.
(157) Sabbatini, M. M.; Cesarotti, E. *Inorg. Chim. Acta* **1977**, *24*, L9.
(158) Zanello, P.; Cinquantini, A.; Mangani, S.; Opomolla, G.; Pardi, L.; Janiak, C.; Rausch, M. D. *J. Organomet. Chem.* **1994**, *471*, 171.
(159) Castellani, M. P.; Wright, J. M.; Geib, S. J.; Rheingold, A. L.; Trogler, W. C. *Organometallics* **1986**, *5*, 1116.
(160) Ryan, M. F.; Eyler, J. R.; Richardson, D. E. *J. Am. Chem. Soc.* **1992**, *114*, 8611.
(161) Ryan, M. F.; Siedle, A. R.; Burk, M. J.; Richardson, D. E. *Organometallics* **1992**, *11*, 4231.
(162) Slocum, D. W.; Ernst, C. R. *Adv. Organomet. Chem.* **1972**, *10*, 79.
(163) Richardson, D. E.; Ryan, M. F.; Dhan, M. N. I.; Maxwell, K. A. *J. Am. Chem. Soc.* **1992**, *114*, 10482.
(164) Ammeter, J. H.; Zoller, L.; Bachmann, J.; Baltzer, P.; Gamp, E.; Bucher, R.; Deiss, E. *Helv. Chim. Acta* **1981**, *64*, 1063.
(165) Bünder, W.; Weiss, E. *Z. Naturforsch. [B]* **1978**, *33*, 1235.
(166) Almenningen, A.; Haaland, A.; Motzfeld, T. In *Selected Topics in Structure Chemistry;* Universititsforlaget: Oslo, Norway, 1975, p. 107.
(167) Reynolds, L. T.; Wilkinson, G. *J. Inorg. Nucl. Chem.* **1959**, *2*, 95.
(168) Ammeter, J. H.; Bucher, R.; Oswald, N. *J. Am. Chem. Soc.* **1974**, *96*, 7833.
(169) Switzer, M. E.; Wang, R.; Rettig, M. F.; Maki, A. H. *J. Am. Chem. Soc.* **1974**, *96*, 7669.
(170) Cauletti, C.; Green, J. C.; Kelly, M. R.; Powell, P.; Van Tilborg, J.; Robbins, J.; Smart, J. *J. Elect. Spec. Relat. Phenom.* **1980**, *19*, 327.
(171) Cozak, D.; Gauvin, F. *Organometallics* **1987**, *6*, 1912.
(172) Rabalais, J. W., Werme, L. O.; Bergmark, T.; Karlsson, L.; Hussain, M.; Siegbahn, K. *J. Chem. Phys.* **1972**, *57*, 1185.
(173) Almenningen, A.; Samdal, S.; Haaland, A. *J. Chem. Soc., Chem. Commun.* **1977**, 14.
(174) Almenningen, A.; Haaland, A.; Samdal, S. *J. Organomet. Chem.* **1978**, *149*, 219.
(175) Smart, J. C.; Robbins, J. L. *J. Am. Chem. Soc.* **1978**, *100*, 3936.
(176) Freyberg, D. P.; Robbins, J. L.; Raymond, K. N.; Smart, J. C. *J. Am. Chem. Soc.* **1979**, *101*, 892.
(177) Augart, N.; Boese, R.; Schmid, G. *Z. Anorg. Allg. Chem.* **1991**, *595*, 27.
(178) Broderick, W. E.; Thompson, J. A.; Day, E. P.; Hoffman, B. M. *Science* **1990**, *249*, 401.
(179) Hays, M. L.; Burkey, D. J.; Hanusa, T. P. Submitted.

(180) Hebendanz, N.; Köhler, F. H.; Müller, G.; Riede, J. *J. Am. Chem. Soc.* **1986,** *108,* 3281.

(181) Bagus, P. S.; Walgren, U. I.; Almlof, J. *J. Chem. Phys.* **1976,** *64,* 2324.

(182) Ernst, R. D.; Wilson, D. R.; Herber, R. H. *J. Am. Chem. Soc.* **1984,** *106,* 1646.

(183) Houlton, A.; Miller, J. R.; Roberts, R. M. G.; Silver, J. *J. Chem. Soc., Dalton Trans.* **1990,** 2181.

(184) Houlton, A.; Miller, J. R.; Robert, R. M. G.; Silver, J. *J. Chem. Soc., Dalton Trans.* **1991,** 467.

(185) Brown, R. A.; Houlton, A.; Howe, S. D.; Roberts, R. M. G.; Silver, J. *J. Chem. Soc., Dalton Trans.* **1993,** 3329.

(186) Iijima, S. I.; Motoyama, I.; Sano, H. *Bull. Chem. Soc. Jpn.* **1980,** *53,* 3180.

(187) Neshvad, G.; Roberts, R. M. G.; Silver, J. *J. Organomet. Chem.* **1984,** *260,* 319.

(188) Roberts, R. M. G.; Silver, J. *J. Organomet. Chem.* **1984,** *263,* 235.

(189) Neshvad, G.; Roberts, R. M. G.; Silver, J. *J. Organomet. Chem.* **1982,** *236,* 237.

(190) Okuda, J.; Albach, R. W.; Herdtweck, E.; Wagner, F. E. *Polyhedron* **1991,** *10,* 1741.

(191) Dory, T. S.; Zuckerman, J. J. *J. Organomet. Chem.* **1984,** *264,* 295.

(192) Williamson, R. L.; Hall, M. B. *Organometallics* **1986,** *5,* 2142.

(193) Dory, T. S.; Zuckerman, J. J.; Barnes, C. L. *J. Organomet. Chem.* **1985,** *281,* C1.

(194) Halterman, R. L. *Chem. Rev.* **1992,** *92,* 965.

(195) *Bonding Energetics in Organometallic Compounds;* Marks, T. J., Ed.; American Chemical Society: Washington, D.C., 1990, Vol. 428.

(196) Cheng, L.-T.; Tam, W.; Meredith, G. R.; Marder, S. R. *Mol. Cryst. Liq. Cryst.* **1990,** *189,* 137.

(197) Miller, J. S.; Epstein, A. J. *Angew. Chem., Int. Ed. Engl.* **1994,** *33,* 385.

(198) Holloway, J. D. L.; Geiger, W. E., Jr. *J. Am. Chem. Soc.* **1979,** *101,* 2038.

(199) Cauletti, C.; Green, J. C.; Kelly, M. R.; Powell, P.; Van Tilborg, J.; Robbins, J.; Smart, J. *J. Electron Spectrosc. Relat. Phenom.* **1980,** *19,* 327.

(200) Green, J. C. *Struct. Bonding (Berlin)* **1986,** *43,* 37.

(201) Gubin, S. P.; Smirnova, S. A.; Denisovich, L. I.; Lubovich, A. A. *J. Organomet. Chem.* **1971,** *30,* 243.

(202) Lichtenberger, D. L.; Copenhaver, A. S. *J. Chem. Phys.* **1989,** *91,* 663.

(203) Köhler, F.; Schlesinger, B. *Inorg. Chem.* **1992,** *31,* 2853.

(204) Weiss, E.; Fischer, E. O. *Z. Anorg. Allg. Chem.* **1955,** *278,* 219.

(205) Fischer, E. O.; Vigoureux, S. *Chem. Ber.* **1958,** *91,* 2205.

(206) Cotton, F. A.; Wilkinson, G. *Z. Naturforsch. [B]* **1954,** *9,* 417.

(207) Weiss, E.; Fischer, E. O. *Z. Anorg. Allg. Chem.* **1958,** *284,* 69.

(208) Wilkinson, G.; Cotton, F. A.; Birmingham, J. M. *J. Inorg. Nucl. Chem.* **1956,** *2,* 95.

(209) Wilkinson, G.; Pauson, P. L.; Cotton, F. A. *J. Am. Chem. Soc.* **1954,** *76,* 1970.

(210) Seiler, P.; Dunitz, J. D. *Acta Cryst., Sect. B* **1980,** *36,* 2255.

(211) Fischer, E. O.; Hofmann, H. P. *Chem. Ber.* **1959,** *92,* 482.

(212) Duff, A. W.; Hitchcock, P. B.; Lappert, M. F.; Taylor, R. G. *J. Organomet. Chem.* **1985,** *293,* 271.

(213) Fischer, E. O.; Stölzle, G. *Chem. Ber.* **1961,** *94,* 2187.

(214) Lorberth, J. *J. Organomet. Chem.* **1969,** *19,* 189.

(215) Grenz, M.; Hahn, E.; Du, M. W. W.; Pickardt, J. *Angew. Chem., Int. Ed. Engl.* **1984,** *96,* 61.

(216) Fischer, E. O.; Grubert, H. *Z. Anorg. Allg. Chem.* **1956,** *286,* 237.

(217) Panattoni, C.; Bombieri, G.; Croatto, U. *Acta Crystallogr.* **1966,** *21,* 823.

(218) Eberl, K.; Koehler, F. H.; Mayring, L. *Angew. Chem.* **1976,** *88,* 575.

(219) Rettig, M. F.; Drago, R. S. *J. Am. Chem. Soc.* **1969,** *91,* 1361.

(220) Reynolds, L. T.; Wilkinson, G. *J. Inorg. Nucl. Chem.* **1959,** *9,* 86.

(221) Pauson, P. L.; Sandhu, M. A.; Watts, W. E. *J. Chem. Soc., Org.* **1966,** 251.
(222) Riemschneider, R.; Helm, D. *Z. Naturforsch.* [B] **1959,** *14,* 811.
(223) Koehler, F. H. *J. Organomet. Chem.* **1976,** *110,* 235.
(224) Burkey, D. J.; Hanusa, T. P.; Huffman, J. C. *Adv. Mater. Opt. Electron.* **1994,** *4,* 1.
(225) Burkey, D. J.; Williams, R. A.; Hanusa, T. P. *Organometallics* **1993,** *12,* 1331.
(226) Williams, R. A.; Tesh, K. F.; Hanusa, T. P. *J. Am. Chem. Soc.* **1991,** *113,* 4843.
(227) Zaegel, F.; Gallucci, J. C.; Meunier, P.; Gautheron, B.; Sivik, M. R.; Paquette, L. A. *J. Am. Chem. Soc.* **1994,** *116,* 6466.
(228) Heeg, M. J.; Janiak, C.; Zuckerman, J. J. *J. Am. Chem. Soc.* **1984,** *106,* 4259.
(229) Jutzi, P.; Schlueter, E.; Hursthouse, M. B.; Arif, A. M.; Short, R. L. *J. Organomet. Chem.* **1986,** *299,* 285.
(230) Kuhn, N.; Henkel, G.; Stubenrauch, S. *J. Chem. Soc., Chem. Commun.* **1992,** 760.
(231) Cowley, A. H.; Jutzi, P.; Kohl, F. X.; Lasch, J. G.; Norman, N. C.; Schlüter, E. *Angew Chem., Int. Ed. Engl.* **1984,** *23,* 616.
(232) Morley, C. P.; Jutzi, P.; Krüger, C.; Wallis, J. M. *Organometallics* **1987,** *6,* 1084.
(233) Glidewell, C. *J. Organomet. Chem.* **1985,** *286,* 289.
(234) Dewar, M. J. S.; Grady, G. L.; Kuhn, D. R.; Merz, K. M., Jr. *J. Am. Chem. Soc.* **1984,** *106,* 6773.
(235) Birchall, T.; Drummond, I. *Inorg. Chem.* **1971,** *10,* 399.
(236) Herber, R. H.; Hanusa, T. P. Submitted.
(237) Harrison, P. G.; Zuckerman, J. J. *J. Am. Chem. Soc.* **1970,** *92,* 2577.
(238) Harrison, P. G.; Healy, M. A. *J. Organomet. Chem.* **1973,** *51,* 153.
(239) Amini, M. M.; Ng, S. W.; Fidelis, K. A.; Heeg, M. J.; Muchmore, C. R.; Van der Helm, D.; Zuckerman, J. J. *J. Organomet. Chem.* **1989,** *365,* 103.

ADVANCES IN ORGANOMETALLIC CHEMISTRY, VOL. 40

Reactions of 17- and 19-Electron Organometallic Complexes

SHOUHENG SUN and DWIGHT A. SWEIGART

Department of Chemistry
Brown University
Providence, Rhode Island

I

INTRODUCTION

Organometallic complexes with a 17- or 19-electron count about the metal frequently show greatly enhanced reactivity in comparison to their 18-electron analogues. This fact has been amply demonstrated in ligand substitution, migratory insertion, reductive elimination, and atom abstraction reactions. One may take advantage of this reactivity by the intentional generation of radical species to initiate stoichiometric and catalytic transformations in 18-electron complexes. For example, CO insertion into the Fe—Me bond in $CpFe(CO)(PPh_3)Me$ does not occur to a detectable extent after 5 days, yet is over within 2 minutes when a catalytic amount of oxidant is added.[1] Similarly, the normally slow CO substitution by a phosphine (L) in $[CpFe(CO)_3]^+$ to give $[CpFe(CO)_2(L)]^+$ becomes essentially instantaneous and quantitative when a trace of reducing agent is added (*vide infra*). There are today a variety of synthetic procedures that involve the initial oxidation or reduction of an 18-electron precursor; often this methodology provides a route to otherwise inaccessible molecules.

A number of good reviews deal with various aspects of organometallic radicals.[2–8] Herein, reactions of organometallic radicals that are of particular interest to the authors are discussed. No attempt has been made to be comprehensive. To avoid excessive overlap with existing reviews, much of the focus is on recently reported chemistry.

171

II

SYNTHESIS AND CHARACTERIZATION

A. Seventeen-Electron Radicals

A fairly large number of 17-electron organometallic complexes have been synthesized and characterized. A representative sample is given in Table I.[9-35] The stabilities of these radicals run the gamut from extremely reactive ($[Re(CO)_5]$, $[CpFe(CO)_2]$, etc.) to sufficiently unreactive to allow routine storage, e.g., trans-$[Cr(CO)_4(PPh_3)_2]^+$, $[CpCo(PPh_3)_2]^+$, and $[Fe(CO)_3(PPh_3)_2]^+$. The stability of 17-electron complexes is greatly influenced by changes in electronic or steric factors. Replacing CO ligands by more electron-releasing tertiary phosphines generally enhances stability, sometimes enough to permit isolation of the radical species, as exemplified by $[Fe(CO)_3(PPh_3)_2]^+$ and trans-$[Cr(CO)_4(PPh_3)_2]^+$. The dienyl manganese series, **1–6**, nicely illustrates the effect of increasing electron density at

$$CpMn(CO)_3^+ \quad CpMn(CO)_2PPh_3^+ \quad (MeCp)Mn(CO)(dppe)^+ \quad Cp^*Mn(CO)_2SePh$$

1   **2**   **3**   **4**

5   **6**

the metal center; all but complex **1** can be isolated. Complex **5** is especially noteworthy because it is a stable optically active 17-electron radical.[36] An X-ray study of **6** in conjunction with its neutral precursor, $[(C_6H_6Ph)Mn(CO)dppe]$, indicates that an 0.12 Å increase in the Mn—P bond lengths in **6** is the only substantial structural difference between the 18- and 17-electron redox pair.[26] A comparison of the X-ray structure of the 18-electron $[Cp^*Mn(CO)_2PMe_3]$ with that of the 17-electron $[Cp^*Cr(CO)_2PMe_3]$ shows that the OC—Mn—CO bond angle is a "normal" 93°, while the OC—Cr—CO angle is only 80°, a value consistent with theoretical calculations.[18]

Steric congestion can stabilize 17-electron complexes by inhibiting bimolecular reaction pathways that lead to nonradical products. For example, the radicals $[Mn(CO)_5]$, $[CpCr(CO)_3]$, and $[CpFe(CO)_2]$ rapidly dimerize by forming a M—M bond (vide infra), but dimerization does not occur with $[Mn(CO)_3(PBu_3)_2]^{[37]}$ and $[CpCr(CO)_2PPh_3]^{[15,16]}$ and is retarded with

TABLE I

SOME REPRESENTATIVE 17-ELECTRON ORGANOMETALLIC COMPLEXES

Compound	Ref.	Compound	Ref.
[V(CO)$_6$]	9	[Re(CO)$_5$]	23
[Cp$_2$V(CO)]	10	[CpMn(CO)$_3$PPh$_3$]$^+$	24
trans-[Cr(CO)$_4$(PPh$_3$)$_2$]$^+$	11	[Cp*Mn(CO)$_2$SePh]	25
[W(CO)$_3$(P^{i}Pr$_3$)$_2$SR]	12	[(C$_6$H$_6$Ph)Mn(dppe)CO]$^+$	26
[(C$_6$Me$_6$)Cr(CO)$_2$PPh$_3$]$^+$	13	[CpRe(PPh$_3$)$_2$(H)$_2$]$^+$	27
[(C$_6$Et$_6$)M(CO)$_3$]$^+$ (M = Cr, Mo, W)	14	[Fe(CO)$_3$(PPh$_3$)$_2$]$^+$	28
[CpM(CO)$_3$] (M = Cr, Mo, W)	15–17	[(C$_4$Ph$_4$)Fe(CO)$_2$PPh$_3$]$^+$	29
[CpCr(CO)$_2$PPh$_3$]	15, 16	[Cp$_2$Fe]$^+$	30
[Cp*Cr(CO)$_2$PMe$_3$]	18	[CpFe(CO)$_2$]	31
[CpCr(NO)(NH$_3$)$_2$]$^+$	19	[Cp*Fe(dppe)(CCPh)]$^+$	32
[CpCr(NO)(PPh$_3$)I]	20	[CpCo(PPh$_3$)$_2$]$^+$	33
[CpMo(PMe$_3$)$_2$I$_2$]	21	[Cp$_2$Co]$^{2+}$	34
[Mn(CO)$_5$]	22	[CpCo(1,5-COD)]$^+$	35

[(C$_5$Ph$_5$)Fe(CO)$_2$].[38] Similarly, the radical cations [(benzene)M(CO)$_3$]$^+$ (M = Cr, Mo, W) are unobservable under most experimental conditions because they suffer very rapid associative attack by even weak nucleophiles. However, placing bulky groups on the arene ring greatly retards this process, and the 17-electron cations can be observed, as has been demonstrated[14] for [(C$_6$Et$_6$)M(CO)$_3$]$^+$ (**7$^+$**) and [(1,3,5-C$_6$H$_3^t$Bu$_3$)M(CO)$_3$]$^+$ (**8$^+$**).

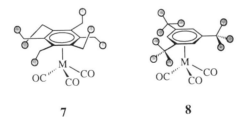

7   8

Although it is certainly true that in general 17-electron organometallic complexes are more reactive than nonradical analogues, there are a few exceptions to this pattern. For example, the 17-electron [CpMo(PMe$_3$)$_2$I$_2$] is more kinetically stable to iodide substitution than is the 16-electron cation (*vide infra*).[39] Similarly, the 17-electron [CpCr(NO)(PPh$_3$)I] is more inert to iodide substitution than is the 18-electron anionic complex.[20] A study of complexes of the form [CpCr(NO)L$_2$]$^{0,+}$ led to the conclusion that the 17-electron cation is preferred when ligand L is a σ-donor such as NH_3, and that the 18-electron neutral complex is more stable when L is a π-acceptor such as CO. With $P(OMe)_3$, a ligand that is both a σ-donor and a π-acceptor, both 17- and 18-electron species are stable.[19]

The generation of 17-electron complexes can be accomplished by a variety of methods. A classic procedure is the photolytic cleavage of metal–metal bonded dimers.[3,8] Some of the most thoroughly studied examples of this are given in Eqs. (1)–(3) for $[CpM(CO)_3]$,[17,40]

$$[CpM(CO)_3]_2 \xrightarrow[\text{(M = Cr, Mo, W)}]{hv} 2\,[CpM(CO)_3] \tag{1}$$

$$[CpFe(CO)_2]_2 \xrightarrow{hv} 2\,[CpFe(CO)_2] \tag{2}$$

$$[M(CO)_5]_2 \xrightarrow[\text{(M = Mn, Re)}]{hv} 2\,[M(CO)_5] \tag{3}$$

$[CpFe(CO)_2]$,[31,41,42] and $[M(CO)_5]$.[43,44] All of these 17-electron radicals can react rapidly in a number of ways, the most obvious being recombination back to the 18-electron dimers. The recombination rate constants for these radicals are extremely large and near the diffusion-controlled limit, except for $[CpCr(CO)_3]$, which is about a factor of 10 slower to dimerize. The radical $[Cp^*Cr(CO)_3]$ recombines about 40 times more slowly than $[CpCr(CO)_3]$. Replacement of CO with a P-donor ligand also reduces the recombination rate, e.g., in the series $[CpFe(CO)L]$ the relative rate as a function of L is CO (100), $P(OMe)_3$ (4.4), $P(OEt)_3$ (3.1), and $P(O^iPr)_3$ (0.2).[31] A similar trend holds for a series of $[Mn(CO)_4L]$ complexes.[44] *Thermal* dissociation of $[CpCr(CO)_3]_2$ and $[Cp^*Cr(CO)_3]_2$ into monomers occurs to a sufficient extent that photolytic activation is not required in order to study the chemistry of the 17-electron radicals.[15,16,45,46] The dimer dissociation constants have been determined to be 4×10^{-4} M and 8×10^{-2} M, respectively.[40] By comparison, the dissociation constants for $[CpMo(CO)_3]_2$ and $[CpW(CO)_3]_2$ are estimated[40] to be 7×10^{-18} M and 1×10^{-56} M (!), reflecting the stronger M—M bonds in these species.

The abstraction of hydrogen atoms by trityl radical from 18-electron metal hydrides has been used to generate neutral 17-electron complexes, as illustrated by Eqs. (4) and (5).[38,47]

$$[CpCr(CO)_3H] + Ph_3C \longrightarrow [CpCr(CO)_3] + Ph_3CH \tag{4}$$

$$[CpW(CO)_3H] + Ph_3C \longrightarrow [CpW(CO)_3] + Ph_3CH \tag{5}$$

Perhaps the most obvious way to generate 17-electron complexes is to start with an 18-electron precursor and perform an oxidation, either chemically or electrochemically. Indeed, chemical oxidation with reagents such as $Ag[BF_4]$, $NO[PF_6]$, and $[PhN_2][PF_6]$ has been used to prepare many 17-electron complexes, most of which are cationic. One of the best known is ferrocenium ion, which is easily synthesized[30] and finds use as a clean and mild oxidizing agent and as an electrochemical standard (the Fc^+/Fc couple). Similarly, by going to very positive potentials and with the use of ultramicroelectrode technology, cobaltocenium ion ($[Cp_2Co]^+$) can be oxidized in liquid $SO_2/[Bu_4N][AsF_6]$ to the 17-electron $[Cp_2Co]^{2+}$, which is stable on the cyclic voltammetry time scale at 25°C.[34] The heterogeneous charge transfer rate constant for the $[Cp_2Co]^{2+}/[Cp_2Co]^+$ couple suggests that the simple sandwich structure of $[Cp_2Co]^+$ is maintained upon oxidation.

The electrochemical oxidation of almost any 18-electron organometallic complex is readily accomplished. More often than not, however, the oxidation is chemically irreversible because the initial product of electron transfer reacts further, a process usually referred to as "decomposition." In order to explore the chemistry of 17-electron radicals produced by electrochemical oxidation, it is generally necessary that they be "observed," meaning that the electrochemical couple has some reversibility in the absence of intentionally added reactants or scavengers. To this end, the use of a non-nucleophilic solvent (e.g., CH_2Cl_2) and electrolyte ($[Bu_4N][PF_6]$, not $[Bu_4N][ClO_4]$) maximizes the chance of producing 17-electron radicals of sufficient lifetime that their chemistry can be explored. In addition, it is often possible to induce chemical reversibility by simply lowering the temperature.[48] Studies in CH_2Cl_2 can be conveniently carried out down to −90°C. For example, the one-electron oxidation of $[(C_6H_6)Cr(CO)_3]$ at room temperature is normally followed by rapid decomposition leading to Cr(II) or Cr(III) and free benzene. At low temperatures in CH_2Cl_2, the oxidation becomes a completely reversible one-electron process.

The generation of 17-electron radicals by electron transfer from or to an 18-electron precursor is summarized in more detail in Scheme 1. Oxida-

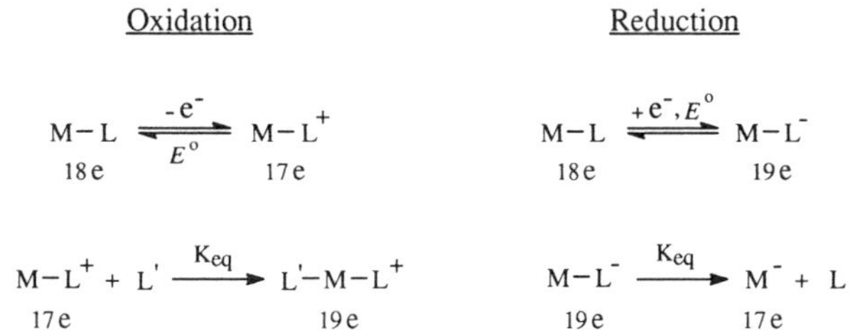

$$\text{Oxidation} \qquad\qquad \text{Reduction}$$

$$\underset{18e}{M-L} \underset{E^\circ}{\overset{-e^-}{\rightleftharpoons}} \underset{17e}{M-L^+} \qquad\qquad \underset{18e}{M-L} \overset{+e^-,E^\circ}{\rightleftharpoons} \underset{19e}{M-L^-}$$

$$\underset{17e}{M-L^+} + L' \overset{K_{eq}}{\longrightarrow} \underset{19e}{L'-M-L^+} \qquad\qquad \underset{19e}{M-L^-} \overset{K_{eq}}{\longrightarrow} \underset{17e}{M^-} + L$$

SCHEME 1.

tion of M—L gives M—L$^+$, which, because of its electron-deficient nature, may coordinate to another ligand L′ that happens to be available (possibly solvent). Obviously, in order for the predominant product to be M—L$^+$, K_{eq} for the 17e $\leftrightarrow$ 19e interconversion must be very small, and this is most likely to occur if M—L is fairly easy to oxidize (less positive $E°$). Even if K_{eq} is not especially large, there is an additional consideration that, as discussed in Section II.B, has led to serious misinterpretations of experimental data. Thus, the 19-electron adduct L′—M—L$^+$ is often very easily oxidized to the 18-electron L′—M—L^{2+}. This means that in an electrochemical experiment, L′—M—L$^+$ will be *spontaneously* oxidized at potentials positive enough to oxidize M—L, with the result that the only observed species will be L′—M—L^{2+} as in Eq. (6). In other

$$M-L \quad \xrightarrow[L']{-2\,e^-} \quad L'-M-L^{2+} \tag{6}$$

words, the disproportionation of L′—M—L$^+$ to L′—M—L^{2+} and M—L (plus L′) is exoergic and, therefore, radical species are not observed. Examples of this behavior are given in Eqs. (7)–(10).[14,49–53]

$$[(arene)M(CO)_3] \quad \xrightarrow[L]{-2\,e^-} \quad [(arene)M(CO)_3L]^{2+} \tag{7}$$

$$(M = Mo, W;\ L = MeCN, PR_3)$$

$$[Co(CO)_3PPh_3]^- \quad \xrightarrow[PPh_3]{-2\,e^-} \quad [Co(CO)_3(PPh_3)_2]^+ \tag{8}$$

$$[CpW(CO)_3Me] \quad \xrightarrow[MeCN]{-2\,e^-} \quad [CpW(CO)_3(MeCN)Me]^{2+} \tag{9}$$

$$[CpM(CO)_3]^- \quad \xrightarrow[MeCN]{-2\,e^-} \quad [CpM(CO)_3(MeCN)]^+ \tag{10}$$

$$(M = Mo, W)$$

As shown in Scheme 1, 17-electron radicals can in principle be generated by *reduction* of 18-electron precursors. For this to be successful, the 19e $\leftrightarrow$ 17e interconversion must have a large K_{eq}, which is most likely to occur if M—L has a very negative reduction potential ($E°$). However, a large K_{eq} does not guarantee a 17-electron radical because M$^-$ may dimerize or undergo a second spontaneous reduction to M^{2-}. Equation (11)

$$[Mn(CO)_6]^+ \quad \xrightarrow{+2\,e^-} \quad [Mn(CO)_5]^- + CO \tag{11}$$

provides an example of such behavior.[54] In practice, reduction of 18-electron complexes is not a generally useful route to 17-electron radicals.

An interesting electrochemical investigation[55] of a series of transition-metal hydrides (M—H) permitted an estimate of the acidities of the corresponding 17-electron cation radicals (M—H$^+$). A thermochemical cycle based on the pK_a of M—H, along with the potential of oxidation of M—H and its conjugate base M$^-$, produced pK_as for M—H$^+$ in MeCN. As examples, the pK_as of [CpW(CO)$_3$H]$^+$ and [CpW(CO)$_2$(PMe$_3$)H]$^+$ are -3 and $+5$, respectively. The neutral 18-electron [CpW(CO)$_3$H] has a pK_a of $+16$, showing that oxidation causes a very large increase in acidity, as would be expected. Prior to this study, the most acidic hydride reported was HCo(CO)$_4$ (pK_a = $+8.3$ in MeCN).

B. Nineteen-Electron Radicals

While a rather large number of 17-electron organometallic complexes have been characterized, there are relatively few examples of 19-electron complexes that have been identified as such, although they are believed to occur commonly as reaction intermediates (*vide infra*). The best-known 19-electron organometallic complex is the commercially available cobaltocene, which has the odd or 19th electron approximately equally distributed between the cobalt and the cyclopentadienyl ligands.[56] The rhodium analogue, Cp$_2$Rh, is known to dimerize very rapidly and must be generated and studied at low temperatures or on a short time scale.[57,58] Cobaltocene, in addition to having a rich chemistry of its own, is a convenient, generally useful moderate to mild reducing agent that often cleanly transfers an electron while being oxidized to the stable cobaltocenium ion. A more strongly reducing yet thermally stable 19-electron complex is [CpFe(arene)], obtained by treatment of the cation with Na/Hg. The "electron reservoir" radicals [CpFe(arene)], especially with C$_6$Me$_6$ as the arene, are fascinating compounds that initiate or promote a range of chemical reactions.[4,59] They are, for example, capable of reducing CO$_2$ to carbonate and CO.[60] The pK_a for proton loss from a methyl in [CpFe(C$_6$Me$_6$)]$^+$ is 29.2 in DMSO. With the use of a thermochemical cycle akin to that used for 17-electron metal hydrides,[55] the pK_a for [CpFe(C$_6$Me$_6$)] has been calculated to be 43.5.[61] This is the same as that for free toluene and provides a measure of the large decrease in acidity of ligands when the metal center is changed from 18 electrons to 19 electrons. Dimeric versions of [CpFe(arene)] such as **9** are known, which for z = $+1$ is a delocalized

$$(C_6Me_6)Fe - \boxed{} - \boxed{} - Fe(C_6Me_6) \rceil^{z = +1, 0}$$

9

metal-centered radical with each metal possessing about 40% of the odd electron.[62] The neutral complex **9** ($z = 0$) is a biradical with two 19-electron centers.

Although the unpaired spin density is predominantly localized on the metal in [CpFe(arene)], as well as in [Fe(C$_6$Me$_6$)$_2$]$^+$, there are many ostensibly 19-electron complexes in which the odd electron is mostly localized on the ligands. These are best described as 18-electron complexes with a radical ligand and are sometimes labeled as "18 + δ" compounds, in which δ indicates the (small) extent to which the odd electron is localized on the metal.[63] Necessarily, 18 + δ complexes possess a ligand of the "electron sink" variety as in [Co(CO)$_2$(NO)$_2$] and [M(CO)$_4$bipy]$^-$ (M = Cr, Mo, W).[64,65] The chelating phosphine 2,3-bis(diphenylphosphino)maleic anhydride is particularly good at accepting electron density and forming 18 + δ complexes. Examples are given in structures **10–12**.[66–69] Of course,

10   **11**   **12**

there are many times when the appropriate experimental data with which to determine the odd electron distribution is unavailable. Sometimes it is assumed that the desired information can be obtained by application of the Vlcek equation, which relates the amount of ligand (or metal) contribution in the HOMO to reduction potentials of appropriately related complexes and their ligands. As a note of caution, however, a recent analysis indicated that the Vlcek equation is likely to yield qualitatively erroneous results in many and perhaps most cases.[70]

Electrochemical and NMR studies of transition-metal sandwich compounds have provided information concerning interesting 19-electron complexes. The complexes [Cp*M(C$_6$Me$_6$)]$^{2+}$ (M = Rh, Ir) undergo a hapicity change upon two-electron reduction, as shown in Scheme 2. The monoreduced complexes retain the planar η^6-C$_6$Me$_6$ bonding and are best formulated as 19-electron. The second reduction is accompanied by η^6 to η^4 slippage of the arene, and the kinetic barrier to this structural change is manifested in a slow rate of heterogeneous charge transfer in the second reduction.[71–74] The [Cp*Co(C$_6$Me$_6$)]^z system differs from the others in having a planar η^6-arene in the neutral state (20-electron).[75] Chemistry similar to that described above obtains with [Ru(C$_6$Me$_6$)$_2$]$^{z+}$, which exists as a planar 19-electron complex when $z = +1$, but rearranges to η^6,η^4-bonding in the neutral state.[74,76]

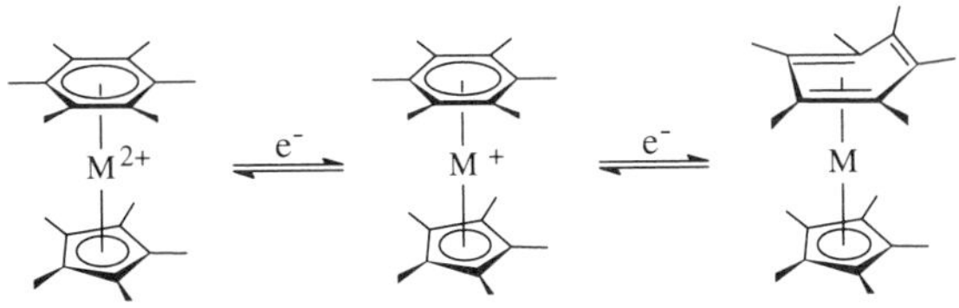

(M= Rh, Ir)

SCHEME 2.

A general characteristic of 19-electron complexes is that they are connected to 17-electron radicals by simple ligand dissociation. This 17e $\leftrightarrow$ 19e interconversion has been documented, with varying degrees of precision, and plays a dominant role in the chemistry of 19-electron systems, as illustrated throughout this review. Considering just one example at this time, [CpW(CO)$_3$] is reported to bind ligands (L) with the K_{eq} increasing in the order L = PPh$_3$ > py $\gg$ MeCN.[53,77]

Elegant studies by Tyler and co-workers have shown that 19-electron radicals play a prominent role in the disproportionation reactions of organometallic dimers such as [Cp$_2$Fe$_2$(CO)$_4$], [Mn$_2$(CO)$_{10}$], [Cp$_2$Mo$_2$(CO)$_6$], [Cp$_2$W$_2$(CO)$_6$], and [W$_2$(CO)$_{10}$]$^{2-}$. For example, photolysis of [Cp$_2$Mo$_2$(CO)$_6$] in the presence of certain phosphines and amines leads to ionic products, as illustrated in Scheme 3.[78,79] An analogous disproportionation of [Cp$_2$Mo$_2$(CO)$_6$] with solvent MeCN serving as the ligand was found to be catalyzed by the triphenylpyrylium ion/triphenylpyranyl radical redox pair.[80] The most prominent feature of the proposed mechanism of the

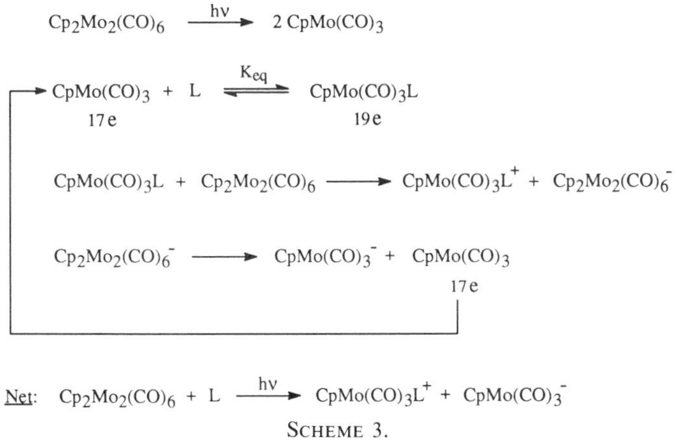

SCHEME 3.

reaction with P- and N-donors (Scheme 3) is the formation of a 19-electron adduct by ligand addition to the initially formed 17-electron radical $[CpMo(CO)_3]$. The 19-electron intermediate is extremely electron-rich and reduces the reactant dimer to yield product $[CpMo(CO)_3L]^+$ and dimer anion; the latter dissociates to produce $[CpMo(CO)_3]$, thus regenerating the original 17-electron radical. Termination of this chain process can occur by any of a variety of reactions, such as direct electron transfer from $[CpMo(CO)_3L]$ to $[CpMo(CO)_3]$, which yields products but consumes the radicals required to propagate the chain. Photolysis of $[Cp_2W_2(CO)_6]$ also leads to the 19-electron species $[CpW(CO)_3L]$, which transfers an electron to any reducible substrate or induces the disproportionation of the reactant dimer.[53,77,81-83]

Irradiation of $[W_2(CO)_{10}]^{2-}$ in solutions containing a N- or P-donor ligand (L) produces a transient 19-electron complex, $[W(CO)_5L]^-$, which is a powerful reducing agent and has been shown to reduce CO_2, $[Mn_2(CO)_{10}]$, $[Cp_2Co]^+$, benzophenone, and methylviologen.[84] Similarly, the photochemical disproportionation of $[Mn_2(CO)_{10}]$ in the presence of N-donors was shown to obey a mechanism analogous to that in Scheme 3, with the 19-electron intermediate formulated as $[Mn(CO)_3L_3]$.[85] The photochemically initiated thermal disproportionation of $[Cp_2Fe_2(CO)_4]$ with the bidentate ligand dppe in CH_2Cl_2 solvent has been studied in detail.[86,87] The reaction, Eq. (12), is postulated to follow a radical chain mechanism with the essential features similar to those in Scheme 3. The mechanism is thought to involve the 17e $\leftrightarrow$ 19e interconversion given in Eq. (13), with $k_1 \approx 3 \times 10^7 \ M^{-1}s^{-1}$ and $k_{-1} \approx 3 \times 10^7 \ s^{-1}$, so that K_{eq} is about unity.

$$[Cp_2Fe_2(CO)_4] + dppe + CH_2Cl_2 \longrightarrow [CpFe(CO)dppe]^+$$

$$+ \ [CpFe(CO)_2(CH_2Cl)] + Cl^- + CO \qquad (12)$$

$$[CpFe(CO)_2] + dppe \ \underset{k_{-1}}{\overset{k_1}{\rightleftharpoons}} \ [CpFe(CO)_2(\eta^1\text{-}dppe)] \qquad (13)$$

A very interesting recent study describes the intramolecular disproportionation of homodinuclear and heterodinuclear fulvalene complexes in the presence of PMe_3.[88] Equation (14) shows one of the six reactions reported. In this case, initiation of the radical chain process was accomplished with a catalytic amount of the 19-electron "reservoir" complex $[CpFe(C_6Me_6)]$, which reduces 13 to break the Ru—W bond and generate a 17-electron radical center (presumably at Ru). Addition of PMe_3 to the Ru is followed by electron transfer to the reactant (13) to afford the zwitterionic product 14 and regenerate the radical intermediate.

(14)

Chemical or electrochemical reduction of 18-electron precursors is, in principle, the most straightforward route to 19-electron radicals. The electrochemical reduction of stable organometallics has been studied for many years.[89] By reference to Scheme 1, one can see that ligand dissociation may accompany reduction of M—L to M—L$^-$, and to generate the 19-electron complex as the predominant species it is necessary that K_{eq} for the 19e $\leftrightarrow$ 17e interconversion be small. This is more likely to occur as $E°$ becomes less negative, i.e., an easily reduced (electron-poor) M—L is less prone to dissociate ligand L after reduction. Even if these conditions are met, the formation of *18-electron* M^{2-} may still be thermodynamically favored, and radical species may not be seen. Equations (11) and (15) are

$$[Co(CO)_3(PBu_3)_2]^+ \xrightarrow{+ 2 e^-} [Co(CO)_3PBu_3]^- + PBu_3 \qquad (15)$$

two examples of reductions leading to 18-electron products.[50,54] Scheme 1 also shows that oxidation of M—L can give 19-electron complexes by addition of ligand L' to the initially formed 17-electron M—L$^+$. For this to occur, K_{eq} for the 17e $\leftrightarrow$ 19e addition must be large, a condition favored when the precursor M—L has a very positive $E°$. As explained earlier, this is very often followed by a second spontaneous oxidation to the 18-electron L'—M—L^{2+}; examples of this behavior are illustrated by Eqs. (8)–(10). For this reason, it is normally preferable to utilize reduction chemistry to generate and study 19-electron complexes.

The electrochemistry of [(arene)W(CO)$_3$] nicely illustrates some of the foregoing principles. In a 1984 study, the voltammetric oxidation of [(C$_6$Me$_6$)W(CO)$_3$] (15) in MeCN was reported[90] to be a chemically irreversible one-electron process. It was concluded that the initially formed monocation rapidly adds solvent to yield the 19-electron species [(η^6-C$_6$Me$_6$)W(CO)$_3$(MeCN)]$^+$. A subsequent investigation of [(mesitylene)W(CO)$_3$] (16) gave essentially identical results except for a small and inconsequential methyl group substituent effect.[49,51] However, the interpretation presented was quite different. Figure 1 shows a cyclic voltammogram (CV) for the tungsten mesitylene complex. It can be seen that 16 is oxidized at wave

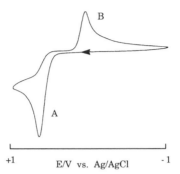

Fig. 1.   Cyclic voltammogram of 1.0 mM [(mesitylene)W(CO)$_3$] (**16**) in MeCN at 20°C. The scan rate was 0.20 V/s, and the working electrode was a 1.6-mm-diameter platinum disk.

A in a chemically irreversible manner to generate a product that is reduced at wave B (also chemically irreversible). Experiments with ultramicroelectrodes showed that waves A and B in the CV of **16** remain chemically irreversible even at a scan rate of 10^4 V/s. With complex **15**, wave B was assigned[90] to the reduction of the 19-electron [(C$_6$Me$_6$)W(CO)$_3$(MeCN)]$^+$. However, careful voltammetric measurements with **16** showed[49,51] that the oxidation at wave A involves two (not one) electrons. Furthermore, spectroelectrochemical and NMR experiments were consistent only with an electrogenerated W(II) 18-electron species. The obvious conclusion is that the species responsible for wave B (and hence "observed") is not a radical but merely the *18-electron* [(mesitylene)W(CO)$_3$(MeCN)]$^{2+}$, which is formed by the following sequence: oxidation of **16** by one electron to give **16**$^+$, coordination of MeCN, and spontaneous oxidation of the 19-electron adduct to [(mesitylene)W(CO)$_3$(MeCN)]$^{2+}$. Although the K_{eq} for the 17e $\leftrightarrow$ 19e interconversion is rather small (near unity), the spontaneous nature of the second oxidation means that the reactant **16** is entirely converted to the 18-electron dication. When P(OBu)$_3$ is present in a MeCN solution of **16**, oxidation affords the analogous dication [(mesitylene)W (CO)$_3$P(OBu)$_3$]$^{2+}$.

   The chromium analogue of **16**, [(mesitylene)Cr(CO)$_3$] (**17**) also undergoes a net two-electron oxidation in MeCN, but the chemistry involved is entirely different for the two complexes. With **17**$^+$, attack by solvent leads to rapid loss of arene and CO ligands, accompanied by further oxidation. (At fast scan rates, the **17**$^+$/**17** couple becomes a reversible one-electron process.) A comparison of the relative reactivity of the 17-electron complexes **16**$^+$ and **17**$^+$ toward associative attack by MeCN led to the rate order W $\gg$ Cr (ratio of ca. 10^4 : 1). It was suggested that this reactivity order reflects less steric congestion for nucleophilic attack at the third-row metal center.[49,51] The oxidation chemistry of $[(arene)M(CO)_3]$ in MeCN has been extended to $[CpM(CO)_3]^-$ (M = Cr, W) with very analogous results.[53] Thus, the tungsten complex is oxidized to $[CpW(CO)_3$ $(MeCN)]^+$ while the chromium complex gives rise to a reversible voltammetric wave.

The electrochemical reduction of $[Mn(CO)_5L]^+$ and related complexes leads to a variety of products, all of which are believed to originate from initially formed 19-electron radicals $[Mn(CO)_5L]$ (L = CO, PR_3, RNC, MeCN).[54,91,92] Cyclic voltammetry up to a half million volts per second produces only chemically irreversible waves, from which it may be concluded that the 19-electron intermediates have $t_{1/2} < 100$ ns with respect to ligand dissociation. The reduction of cis-$[Mn(CO)_2(dppe)_2]^+$ was interpreted in terms of the facile equilibrium shown in Eq. (16).[91,92] Note that,

$$(dppe)(OC)_2Mn\begin{matrix}Ph_2\\P\\\\P\\Ph_2\end{matrix}\underset{19e}{}\quad\underset{3.5\,s^{-1}}{\overset{10^6\,s^{-1}}{\rightleftharpoons}}\quad (dppe)(OC)_2Mn—P\underset{Ph_2\ Ph_2}{\quad P}\underset{17e}{}\qquad(16)$$

in spite of a "chelate effect" operating in the 19-electron complex, the 19e ↔ 17e equilibrium lies strongly on the side of the 17-electron radical ($K_{eq} \approx 3 \times 10^5$), a factor that can be attributed to the high electron density imparted by the dppe ligands.

III

LIGAND SUBSTITUTION AND ATOM ABSTRACTION REACTIONS

A. *Seventeen-Electron Radicals*

Although early work on the photogenerated $[Mn(CO)_5]$ radical suggested a dissociative mechanism for CO substitution,[93] there were hints in the literature concerning thermal reactions of $[Mn_2(CO)_{10}]$ that pointed to an associative pathway.[94,95] A few years later Pöe and co-workers studied the reaction of photogenerated $[Re(CO)_5]$ with CCl_4 in the presence of ligand L (PPh_3, PBu_3).[96,97] The products consisted of $[Re(CO)_5Cl]$ and $[Re(CO)_4(L)Cl]$, formed according to Eqs. (17)–(19). An analysis of the

$$[Re(CO)_5] + CCl_4 \longrightarrow [Re(CO)_5Cl] \qquad (17)$$

$$[Re(CO)_5] + L \longrightarrow [Re(CO)_4L] + CO \qquad (18)$$

$$[Re(CO)_4L] + CCl_4 \longrightarrow [Re(CO)_4(L)Cl] \qquad (19)$$

product ratio as a function of the concentration of CCl_4 and L showed that the CO substitution in Eq. (18) is strictly associative. Subsequent analogous experiments with $[Mn(CO)_5]$ verified that this 17-electron radical also followed an associative pathway.[98] This is true as well of the more stable radicals $[Mn(CO)_3L_2]$ (L = P^nBu_3, P^iBu_3), even with CO as the nucleophile.[99] In a study akin to the work with $[Re(CO)_5]$, the highly reactive radical $[CpW(CO)_3]$ was generated from $[CpW(CO)_3H]$ by hydrogen atom abstraction with trityl radical, Eq. (5), and then allowed to react competitively with Ph_3CCl and PR_3.[47] This afforded a product mixture consisting of $[CpW(CO)_3Cl]$ and $[CpW(CO)_2(PR_3)Cl]$, corresponding to chlorine atom abstraction before and after CO substitution, respectively. An analysis of the product ratios proved that the substitution reaction was associative.

Substitution of CO in the 17-electron $[V(CO)_6]$ with phosphine and phosphite nucleophiles follows a strictly second-order rate law.[100] Activation parameters and the rate dependence on the nature and basicity of the nucleophile established an associative mechanism. Substitution of a second CO to give $[V(CO)_4L_2]$ is also associative and at least 10^3 times slower than the first CO replacement. Similarly, N- and O-donors lead to associative CO substitution in $[V(CO)_6]$, although the products cannot be characterized before they disproportionate to $[VL_6][V(CO)_6]_2$. That all of these reactions are associative is not surprising since formation of a 19-electron intermediate (or transition state) results in a net gain of $\frac{1}{2}$ in bond order.[96] In conjunction with this is an enormous increase in lability in comparison to 18-electron systems. For example, the rate of CO replacement by $P(^nBu)_3$ in $[V(CO)_6]$ is $\approx 10^{10}$ times faster than the second-order (interchange) pathway in $Cr(CO)_6$. Likewise, $[V(CO)_6]^-$ is essentially inert to thermal ligand substitution. The work on 17-electron vanadium complexes was extended to $[Cp_2V(CO)]$ (**18**) and $[(pentadienyl)_2V(CO)]$ (**19**), with the discovery that CO exchange in **18** occurs fairly rapidly by an associative pathway. In contrast, steric factors inhibit attack on **19** and CO exchange occurs very slowly via a *dissociative* pathway.[101] A dissociative mechanism also appears to be important in halide substitution in the 17-electron $[CpMoI_2(PMe_3)_2]$.[21]

The dimer $[CpCr(CO)_3]_2$ has a weak metal–metal bond and dissociates in solution at room temperature to a measurable extent to generate the 17-electron $[CpCr(CO)_3]$. This permitted a detailed kinetic study of CO substitution in the radical, which established a rapid associative mechanism.[15,16] A comparison of these results with reactivity data[17,47] for $[CpW(CO)_3]$ suggests that CO substitution by PPh_3 in the 17-electron radicals is faster for tungsten than chromium by a factor of ca. 10^6. Photolysis of $[CpFe(CO)_2]_2$ in conjunction with time-resolved infrared spectroscopy was used to identify the monomer [CpFe(CO)$_2$] and measure its rate of reaction with P(OMe)$_3$ to give [CpFe(CO)P(OMe)$_3$]. The rate constant for this associative process is 9×10^8 M^{-1}s^{-1} at room temperature in heptane, fast enough to compete with the nearly diffusion-controlled dimerization back to [CpFe(CO)$_2$]$_2$.[31,42,102] In [(C$_5$Ph$_5$)Fe(CO)$_2$]$_2$, the phenyl groups are sufficiently bulky that partial dissociation into monomer occurs thermally, thus providing a mechanism for the observed rapid CO exchange in both monomer and dimer.[38]

N-donor induced disproportionation of [Fe(CO)$_3$(PR$_3$)$_2$]$^+$ (R = Me, Bu, Cy, Ph) as well as halide induced disproportionation of [M(CO)$_3$(PCy$_3$)$_2$]$^+$ (M = Fe, Ru, Os) has been interpreted in terms of nucleophilic attack being rate determining.[103,104] The rate data led to the conclusion that the reactivity of these 17-electron complexes is only weakly dependent on the metal, and the suggestion was made that periodic trends in 17-electron systems are generally attenuated in comparison to those for 18-electron analogues. However, it was noted previously that W $\gg$ Cr by ca. $10^6 : 1$ for substitution in [CpM(CO)$_3$]. A direct comparison of the rate of associative ligand substitution at a 17-electron center as a function of the metal for a complete triad (Cr, Mo, W) was reported for the reaction in Eq. (20).[14]

$$[(C_6Et_6)M(CO)_3]^+ \ + \ P(OBu)_3 \longrightarrow [(C_6Et_6)M(CO)_2P(OBu)_3]^+ \ + \ CO \qquad (20)$$

The starting complexes, [(C$_6$Et$_6$)M(CO)$_3$] (7), were oxidized to the cations in a chemically reversible one-electron process at a platinum electrode in CH$_2$Cl$_2$. A bulky arene was required in order to inhibit attack by solvent or electrolyte and provide sufficient stability that the radical cations could be observed on the cyclic voltammetry (CV) time scale in the absence of P(OBu)$_3$. Figure 2 shows CVs for the oxidation of 7 in the absence (A) and presence (B) of P(OBu)$_3$. In (B) the reversibility of the original oxidation wave is wholly (M = Mo, W) or partially (Cr) quenched because the radical 7$^+$ reacts rapidly according to Eq. (20). This gives rise to a new reversible couple, negative of the original oxidation, due to [(C$_6$Et$_6$)M(CO)$_2$P(OBu)$_3$]$^{+/0}$. With the chromium system the lack of complete quenching of the reduction wave for 7$^+$ allowed the rate of Eq. (20) to be determined by digital simulation of the CVs at various scan rates, nucleophile concentrations, and temperatures. The mechanism was established as associative with $\Delta H^{\ddagger} = 22 \pm 2$ kJ and $\Delta S^{\ddagger} = -130 \pm 20$ J K^{-1}.

In contrast to the behavior shown by 7$^+$(Cr), the lack of any wave due to 7$^+$(Mo, W) with excess P(OBu)$_3$ present, even up to 100 V/s, indicated that the rate of Eq. (20) is much larger for the heavier metals. Nevertheless, it was possible to determine these rates by voltammetry with a deficiency of P(OBu)$_3$ present. This procedure, developed by Parker,[105,106] leads to

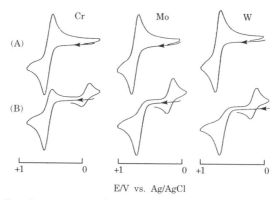

FIG. 2.   Cyclic voltammograms of 1 mM [(C$_6$Et$_6$)M(CO)$_3$] (**20**) in CH$_2$Cl$_2$ at 20°C with (A) no added P(OBu)$_3$ and (B) P(OBu)$_3$ present at 10 mM (Cr and W) and 2 mM (Mo). The scan rate was 0.5 V/s, and the working electrode was a 1.6-mm-diameter platinum disk.

a splitting of the oxidation wave for **7** into two waves, provided the second-order reaction of **7**$^+$ with nucleophile is very rapid. Figure 3 illustrates the theoretical splitting at a [nucleophile]/[complex] ([L]/[M]) ratio of 0.50. As the oxidation starts, a kinetic shift of the wave occurs, but the supply of nucleophile is quickly exhausted because it is in limited supply,

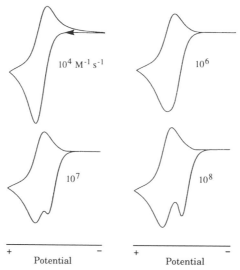

FIG. 3.   Theoretical cyclic voltammograms for the oxidation of a 1 mM solution of complex M in the presence of 0.5 mM nucleophile L at a scan rate of 0.5 V/s. The second-order reaction of M$^+$ and L occurs with the indicated rate constant.

and the remainder of the oxidation wave then appears at the normal unshifted potential. The result is a splitting into a kinetic prewave and the normal Nernstian wave. (If the nucleophile is present in excess, only a single kinetically shifted wave occurs.) The separation of the two waves is a sensitive function of the [L]/[M] ratio, the scan rate, and the rate of the homogeneous second-order reaction. The prewave method is applicable only to very rapid reactions; under ordinary conditions, the second-order homogeneous rate constant must be at least $10^6 \, M^{-1}s^{-1}$, or a prewave will not be seen.

The preceding analysis gives the following rate constants ($M^{-1}s^{-1}$) for Eq. (20) at 20°C: Cr (1×10^2), Mo (5×10^7), W (1×10^7). This translates into the relative reactivity order Mo > W ≫ Cr ($5 \times 10^5 : 10^5 : 1$). Because of the ethyl groups in 7, this series has a "steric" component, but it is likely[14] that the same qualitative order will always obtain because studies with [(mesitylene)M(CO)$_3$]$^+$ indicate[51] that W ≫ Cr by ca. $10^4 : 1$. In dissociative reactions of 18-electron chromium triad complexes, the "normal" reactivity order is Mo > Cr > W. Steric effects only serve to increase the rate for Cr compared to W.[107] The much greater rate of W compared to Cr for associative reactions at 17-electron centers is likely due to the greater size of W (and Mo). In any event, it is apparent that the reactivity of 17-electron complexes can be *very* dependent on the metal within a triad.

There appears to be a fundamental difference in the way P-donors react with the chromium triad complexes [(arene)M(CO)$_3$] and [(arene)M(CO)$_3$]$^+$: the former undergo arene displacement,[108,109] whereas the latter prefer CO substitution, Eqs. (21) and (22). Equation (21) is known

$$[(\text{arene})M(CO)_3] \; + \; 3\,PR_3 \; \longrightarrow \; [(PR_3)_3M(CO)_3] \; + \; \text{arene} \qquad (21)$$

$$[(\text{arene})M(CO)_3]^+ \; + \; 3\,PR_3 \; \longrightarrow \; [(\text{arene})M(CO)_2PR_3]^+ \; + \; CO \qquad (22)$$

to be associative and is thought to proceed via $\eta^6 \rightarrow \eta^4$ ring slippage when the first nucleophile binds to the metal, thus avoiding the necessity of forming a 20-electron intermediate. We suggest that [(arene)M(CO)$_3$] reacts with P-donors to lose arene while [(arene)M(CO)$_3$]$^+$ reacts to lose CO because ring slippage need not occur with substitution reactions of 17-electron complexes. Because there is no ring slippage, the metal–arene bond is not weakened when the nucleophile attacks the radical, so CO, rather than arene, is lost. Stated another way, 19-electron complexes are viable intermediates, but 20-electron ones are not. As shown earlier, there are a number of characterized 19-electron complexes containing a planar nonslipped arene ligand (e.g., [CpFe(arene)], [Cp*Rh(C$_6$Me$_6$)]$^+$, [Ru(C$_6$Me$_6$)$_2$]$^+$). This does not imply, of course, that slippage can never occur when a two-electron ligand is added to a 17-electron complex. For example, the reaction of 17-electron $[(\eta^5\text{-indenyl})_2V(CO)]$ with CO yields the adduct $[(\eta^5\text{-indenyl})(\eta^3\text{-indenyl})V(CO)_2]$.[110]

Elegant studies by Kochi and co-workers[111,112] established that rapid ligand substitution according to Eq. (23) occurs when a slight oxidizing

$$[(\text{MeCp})\text{Mn}(\text{CO})_2(\text{MeCN})] \ + \ PR_3 \ \longrightarrow \ [(\text{MeCp})\text{Mn}(\text{CO})_2 PR_3] \ + \ \text{MeCN} \qquad (23)$$

current is applied (or a small amount of a chemical oxidizing agent is added) to a MeCN solution of $[(\text{MeCp})\text{Mn}(\text{CO})_2(\text{MeCN})]$. Current efficiencies, defined as moles product per mole electrons passed, are as large as 1000. Detailed electrochemical investigation showed that the reaction occurs by a catalytic electron-transfer chain (ETC) process defined in Scheme 4. The catalytic cycle is initiated by oxidation of a small amount of reactant [Mn—NCMe]. The chain length depends on the relative reactivities of the cation radicals with respect to ligand substitution versus decomposition. Oxidation of [Mn—NCMe] yields the 17-electron $[\text{Mn—NCMe}]^+$, which rapidly substitutes PR_3 (L) for MeCN to give $[\text{Mn—L}]^+$. It happens that the reactant [Mn—NCMe] is more easily oxidized than is the product [Mn—L] $(E_1^0 < E_2^0)$, which ensures that $[\text{Mn—NCMe}]^+$ is continuously regenerated by electron transfer from [Mn—NCMe] to $[\text{Mn—L}]^+$. It is also possible for $[\text{Mn—L}]^+$ to be reduced directly at the electrode. Thus, the overall substitution in Eq. (23) is catalytic in electrons whether the electron transfer is homogeneous or heterogeneous. Based on activation parameters and on the rate dependence on the nature of the nucleophile, it was clearly established that substitution in $[\text{Mn—NCMe}]^+$ is associative.

Electrocatalytic ETC substitution initiated by oxidation has been reported with chromium triad carbonyl derivatives.[113] Several of the reactions studied, and ones that give little or no product when uncatalyzed, are shown in Eqs. (24)–(26). The mechanism described earlier for the reactions of $[(\text{MeCp})\text{Mn}(\text{CO})_2(\text{MeCN})]$ applies to these reactions. The

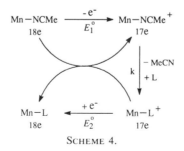

SCHEME 4.

$$fac\text{-}[(MeCN)_3W(CO)_3] \xrightarrow{\text{'BuNC}} fac\text{-}[(\text{'BuNC})_3W(CO)_3] \qquad (24)$$

$$cis\text{-}[(py)_2Mo(CO)_4] \xrightarrow{\text{'BuNC}} cis\text{-}[(BuNC)_2Mo(CO)_4] \qquad (25)$$

$$cis\text{-}[(MeCN)_2W(CO)_4] \xrightarrow{PPh_3} cis\text{-}[(MeCN)(PPh_3)W(CO)_4] \qquad (26)$$

success of electrocatalysis of ligand substitution initiated by oxidation depends on the stabilities of the cationic radical intermediates and usually requires that E^0 of the product be more positive than E^0 of the reactant. In other words, catalysis requires that the product be *harder* to oxidize than the reactant, which is rarely the case because departing ligands are usually replaced by ones that increase the electron density at the metal. Consequently, catalytic oxidative activation of organometallics to ligand substitution is quite rare. The same reasoning, however, implies that catalytic *reductive* activation should be common, and indeed it is (*vide infra*).

The complex $[(MeCp)Mn(CO)_3]$ (**20**) is extraordinarily inert to CO substitution.[114] For example, it does not react thermally with PPh_3 over 3 days at 140°C. However, **20** and analogues variously functionalized on

the cyclopentadienyl ring undergo substitution of one or two CO ligands within milliseconds at room temperature upon oxidation in the presence of $P(OEt)_3$.[115] The chemistry involved is illustrated in Scheme 5, and typical cyclic voltammograms are given in Fig. 4. The reversibility of the $20^+/20$ couple in Fig. 4A vanishes when $P(OEt)_3$ is in solution, and two new couples appear at more negative potentials, which are due to the products of single and double CO substitution. The extent of single versus double substitution is markedly influenced by steric congestion in the vicinity of the metal due to substituents on the cyclopentadienyl ring. Sufficiently bulky groups hinder access of the nucleophile to the metal center. A good example of this is complex **21**, the X-ray structure of which is shown in Fig. 5. It was found that CO replacement in 21^+ is extremely rapid ($k = 4 \times 10^7$ $M^{-1}s^{-1}$) and only slightly slower than that observed with 20^+. However, the second CO in 21^+ is replaced ca. 100 times more slowly than the second in 20^+. This difference is reflected in Fig. 4, which shows that double CO substitution in 21^+ is not nearly as prominent as

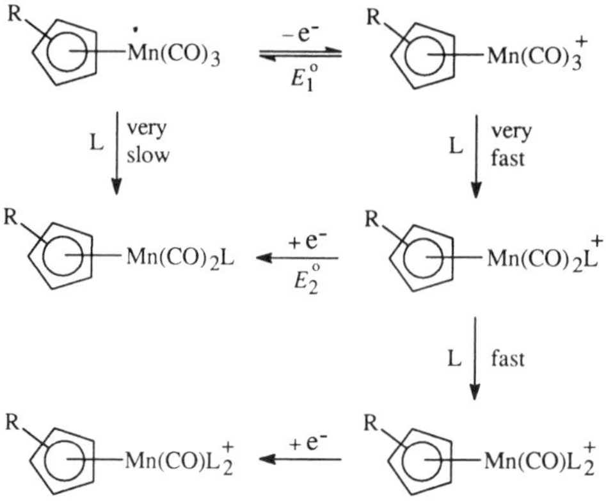

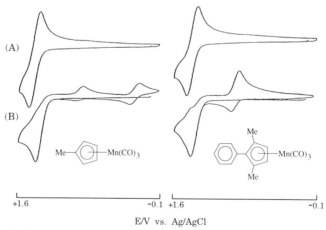

(L = P(OEt)$_3$)

Scheme 5.

it is in **20$^+$**. The reactivities of **20$^+$** and **21$^+$** show that they conform to the generalization that 17-electron organometallic complexes are vastly more reactive than their 18-electron analogues. In the case of **20$^+$**, it is estimated that the rate enhancement compared to **20** is 10^9 or more. It may be noted that the oxidatively promoted CO substitution in [(RCp)Mn(CO)$_3$] is *not* catalytic; the transformation [(RCp)Mn(CO)$_3$] $\rightarrow$ [(RCp)Mn(CO)$_2$L] requires an initial stoichiometric oxidation, followed by CO substitution in the 17-electron radical and then stoichiometric reduction.

(A)

(B)

| +1.6 | −0.1 | +1.6 | −0.1 |

E/V vs. Ag/AgCl

FIG. 4.   Cyclic voltammograms of 1.0 mM [(RCp)Mn(CO)$_3$] complexes **20** and **21** in CH$_2$Cl$_2$ at 25°C with (A) no added P(OEt)$_3$ and (B) P(OEt)$_3$ present at 1.2 mM. The scan rate was 0.5 V/s, and the working electrode was a 1-mm-diameter glassy carbon disk.

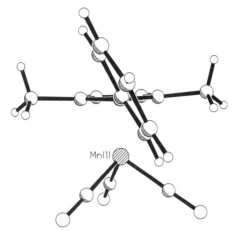

FIG. 5. Structure of complex **21**, showing the rotation of the phenyl ring from the cyclopentadienyl plane.

As indicated earlier, ligand substitution in 18-electron complexes can be "turned on" by oxidation. This can be a stoichiometric process, or it may be catalytic, depending on the relevant redox potentials and other factors. If the process is catalytic, a chain process occurs in which electron transfer is the major propagation step (e.g., Scheme 4). It is also possible for the overall substitution to occur via a radical chain process that involves not electron transfer, but atom transfer. Good organometallic examples of this are provided by CO substitution in [HRe(CO)$_5$] and [CpW(CO)$_3$H].[116-118] Considering the tungsten complex, it was found that the room temperature thermal reaction with PBu$_3$ proceeds very slowly, and at a rate difficult to reproduce, to afford a cis,trans mixture of [CpW(CO)$_2$(PBu$_3$)H]. Irradiation of [CpW(CO)$_3$H] and PBu$_3$ at 500 nm for an hour did not induce substitution, but irradiation in the presence of a small amount of photoinitiator [Cp$_2$W$_2$(CO)$_6$] led to rapid formation of [CpW(CO)$_2$(PBu$_3$)H]. The radical chain mechanism shown in Scheme 6 was proposed to account for the observations. The key features are (1) rapid CO substitution in the 17-electron [CpW(CO)$_3$] and (2) hydrogen atom abstraction from [CpW(CO)$_3$H] to propagate the chain. Based on quantum yield measurements, the chain length was estimated as ca. 2000. Termination likely occurs via radical dimerization.

An atom transfer chain mechanism very similar to that in Scheme 6 is thought to occur in some photoinitiated CO substitution reactions of organometallic halide complexes. For example, [Cp$_2$Mo$_2$(CO)$_6$] promotes the photosubstitution of CO by PR$_3$ in [CpMo(CO)$_3$X] (X = Br, I),[119] and

$$Cp_2W_2(CO)_6 \xrightarrow{h\nu} 2\ CpW(CO)_3$$

$$CpW(CO)_3 + L \longrightarrow CpW(CO)_2L + CO$$

$$CpW(CO)_2L + CpW(CO)_3H \longrightarrow CpW(CO)_2(L)H + CpW(CO)_3$$

Net: $CpW(CO)_3H + L \longrightarrow CpW(CO)_2(L)H + CO$

SCHEME 6.

$[Cp_2Fe_2(CO)_4]$ promotes photosubstitution of CO by RNC in $[CpFe(CO)_2I]$.[120] The propagation step in these reactions is halogen atom abstraction from reactant by the 17-electron radicals $[CpMo(CO)_2PR_3]$ and $[CpFe(CO)_2(CNR)]$, respectively. A radical chain pathway has also been proposed for the $[Cp_2Fe_2(CO)_4]$-promoted photosubstitution in Eq. (27).[121] In

$$[CpFe(CO)_2(\eta^1\text{-}Cp)] + P(OPh)_3 \xrightarrow[{[Cp_2Fe_2(CO)_4]}]{h\nu}$$

$$[CpFe(CO)(P(OPh)_3)(\eta^1\text{-}Cp)] + CO \qquad (27)$$

this case it is thought that the η^1-Cp group transfers to $[CpFe(CO)P(OPh)_3]$ by an addition–elimination process involving a bimetallic intermediate such as **22**.

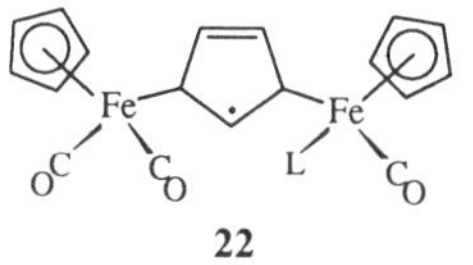

22

The ability of 17-electron radicals to abstract halogen atoms from organic and organometallic substrates has been investigated in some detail. A rough guide to relative reactivities is available from known rates of chlorine atom abstraction from CCl_4: $Re(CO)_5 > Mn(CO)_5 > CpMo(CO)_3 \approx CpW(CO)_3$.[8] A thorough kinetic study of the reaction in Eq. (28) has been reported for a series of non-β-hydrogen-containing organic halides, RX.[45,46] The $[CpCr(CO)_3]$ radical was formed by thermal homolysis of $[Cp_2Cr_2(CO)_6]$, and the rate law was found to be consistent with rate-determining halogen atom transfer from carbon to chromium according to the mechanism outlined by Eqs. (29) and (30). As would be expected, the reaction rates correlate inversely with the relative carbon–

$$2\,[CpCr(CO)_3] + RX \longrightarrow [CpCr(CO)_3X] + [CpCr(CO)_3R] \quad (28)$$

$$[CpCr(CO)_3] + RX \xrightarrow{\text{slow}} [CpCr(CO)_3X] + R \quad (29)$$

$$[CpCr(CO)_3] + R \xrightarrow{\text{fast}} [CpCr(CO)_3R] \quad (30)$$

halogen bond strengths. With organic halides containing one or more β-hydrogen atoms, the product distribution was found to be more complicated. For example, reaction with PhCHMeBr produces $[CpCr(CO)_3Br]$, $[CpCr(CO)_3H]$, styrene, and ethylbenzene. It is proposed that these products form because $[CpCr(CO)_3]$ can abstract a β-hydrogen from the radical PhCHMe to yield styrene plus $[CpCr(CO)_3H]$. The latter is capable of transferring hydrogen to PhCHMe to give ethylbenzene. The final product distribution reflects the relevant rate constants and time-dependent concentrations.

Laser flash photolysis of $[Cp_2M_2(CO)_6]$ (M = Mo, W) was used to generate the radicals $[CpM(CO)_3]$ and study their subsequent atom-transfer reactions with organic and inorganic halides.[17] The bimolecular rate constants for $[CpW(CO)_3]$ cover seven orders of magnitude, signaling a highly selective atom abstraction process. For example, k is 3.9×10^2 $M^{-1}s^{-1}$ with CH_2Br_2 and $1.3 \times 10^9\ M^{-1}s^{-1}$ with CBr_4 at 23°C. The reactivity of organic halides follows the trends RI > RBr > RCl and 3° > 2° > 1° > CH_3, consistent with rate-determining atom abstraction. The inorganic halides and pseudohalides $[XCo(NH_3)_5]^{2+}$ (X = Cl, Br, CN, SCN, N_3) also transfer atom X very rapidly to $[CpW(CO)_3]$, presumably by the same inner-sphere mechanism. A comparison of literature results for the reaction of $[CpM(CO)_3]$ with $BrCH_2CN$ indicates[17,46] that the relative reactivity with respect to bromine atom abstraction is in the order M = W $(10^6) \approx$ Mo $(10^6) \gg$ Cr (1). It may be noted that this order is approximately the same as that found[14,16,17] for associative ligand substitution in $[CpM(CO)_3]$ and $[(C_6Et_6)M(CO)_3]^+$.

B. Nineteen-Electron Radicals

In 1979, Butts and Schriver showed that the reaction of $[Fe(CO)_5]$ and PR_3 to give $[Fe(CO)_4PR_3]$ occurs cleanly, fairly rapidly, and in good yield when sodium benzophenone ketyl (BPK) is added to the solution *in a catalytic amount*.[122] This discovery was remarkable because $[Fe(CO)_5]$ normally requires rigorous thermal or photochemical conditions to afford even poor yields of the CO substitution product, $[Fe(CO)_4L]$, and is reported[123] to have a half-life of years for CO exchange at room temperature. BPK is known to reduce $[Fe(CO)_5]$ to a variety of polynuclear iron carbonyl anions, and these species, present in catalytic amounts, were shown to be responsible for the increased rate of ligand substitution in [Fe(CO)$_5$]. Although the mechanism of this *reductive activation* process was not understood at the time, the authors realized that the procedure had significant synthetic potential.

It was pointed out earlier that ligand substitution initiated by reduction is much more likely to be catalytic than is oxidatively induced substitution. The first such ETC-catalyzed process that was identified as such was reported by Rieger and co-workers in 1981.[124,125] In this case, an electrochemical study of the reduction of [R$_2$C$_2$Co$_2$(CO)$_6$] complexes in the presence of P-donor ligands showed that the [R$_2$C$_2$Co$_2$(CO)$_5$PR$_3$] products were formed by an ETC process. In the proposed mechanism, shown in Scheme 7, reduction leads to homolytic cleavage of the Co—Co bond, resulting in an intermediate bimetallic complex containing 18- and 17-electron centers. Rapid CO substitution by PR$_3$ at the 17-electron metal is followed by spontaneous electron transfer to the electrode or to [R$_2$C$_2$Co$_2$(CO)$_6$] reactant, thereby forming neutral product and propagating the reaction. The use of a small amount of BPK to initiate ETC catalysis of ligand substitution in [(CF$_3$)$_2$C$_2$Co$_2$(CO)$_6$] leads in most cases to virtually quantitative yields of [(CF$_3$)$_2$C$_2$Co$_2$(CO)$_5$L] (L = PR$_3$, P(OR)$_3$, MeCN) within a few minutes at room temperature.[126] The utility of this procedure in synthesis is nicely illustrated by Eq. (31), which proceeds in only 10%

$$[(CF_3)_2C_2Co_2(CO)_6] + MeCN \longrightarrow [(CF_3)_2C_2Co_2(CO)_5(MeCN)] + CO \qquad (31)$$

yield after 3 hr in boiling MeCN, but gives 100% product in less than 1 min upon the addition of a trace of BPK. In a similar fashion, BPK has been used to induce substitution of CO by PPh$_3$ and P(OEt)$_3$ in [(μ_3-RC)Co$_3$(CO)$_9$] (R = H, Ph).[127]

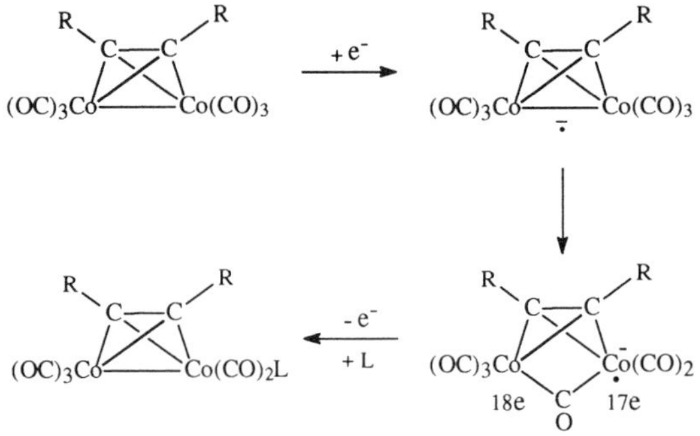

SCHEME 7.

In addition to the polynuclear cobalt complexes just described, ETC-catalyzed CO substitution initiated electrochemically or by use of BPK has been reported for the following polynuclear complexes: $[Fe_3(CO)_{12}]$, $[Ru_3(CO)_{12}]$, $[Os_3(CO)_{12}]$, $[Rh_6(CO)_{16}]$, $[H_4Ru_4(CO)_{12}]$, and $[Fe_3S_2(CO)_8 (C_3H_2S_2)]$.[128–131] The entering ligands were mostly of the tertiary phosphine type, although in some cases isonitriles, arsines, and $SbPh_3$ were used successfully. It was possible to control the number of COs replaced by variation in the number of equivalents of entering ligand. Mononuclear complexes have also been subjected to ETC substitution conditions. For example, reductive activation of $[CpFe(arene)]^+$ in the presence of $P(OMe)_3$ induces rapid catalytic formation of $[CpFe(P(OMe)_3)_3]^+$.[132] Similar activation of $[(OC)_3Mn(NCMe)_3]^+$ is reported to afford $[(OC)_3Mn (PR_3)_2(NCMe)]^+$.[133]

The chemical or electrochemical reduction of $[(arene)Mn(CO)_3]^+$ (**23**$^+$) can lead to bimolecular coupling through the arene rings, CO loss followed by dimerization, or arene ring slippage.[134] Which mode of decomposition is followed depends on many factors, especially the nature of the arene. In the presence of $P(OBu)_3$, these decomposition pathways are short circuited in favor of ETC substitution to give $[(arene)Mn(CO)_2P(OBu)_3]^+$. For example, Fig. 6 shows that application of a reducing current for a

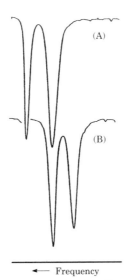

(A)

(B)

←— Frequency

FIG. 6.   IR spectrum of 1 mM $[(C_6Me_6)Mn(CO)_3]^+$ and 1.5 mM $P(OBu)_3$ in CH_2Cl_2. Spectrum (A) is before and (B) is after a reducing current was applied for a few seconds, corresponding to 0.03 equivalents of electrons. The spectrum in (b) is that of $[(C_6Me_6)Mn (CO)_2P(OBu)_3]^+$.

few seconds is sufficient to completely convert 23^+ (arene = C_6Me_6) to the substitution product. Voltammetric data showed that the reactivity of 23 with $P(OBu)_3$ depends on the arene in the order benzofuran > 1,3,5-$C_6H_3Me_3$ > 1,3,5-$C_6H_3Et_3$. There was no reaction with $P(OBu)_3$ when the arene was C_6Et_6, presumably because of the steric congestion provided by the alternating up and down ethyl groups (e.g., see structure 7). The behavior of $[(1,3,5-C_6H_3Me_3)Re(CO)_3]^+$ (24^+) was found to differ fundamentally from that of the manganese analogue. Thus, 24^+ is reduced in a chemically reversible two-electron step, which is not affected by the presence of CO or P-donor ligands. Very large peak-to-peak separations in the CVs indicate that 24^+ is reduced to the 19-electron 24, which then undergoes a slow and spontaneous second reduction as the arene slips to η^4 bonding. From this information, it is reasonable to infer that the rate of ligand substitution in the 19-electron complexes is in the order Mn $\gg$ Re. By way of contrast, it seems that the reactivity of the 17-electron complexes $[(Cp)M(CO)_3]^+$ is in the order Re $\gg$ Mn.[115]

 The question of mechanism of ligand substitution at 19-electron centers can be addressed at a number of levels. With the knowledge that 17-electron complexes follow an associative mechanism and that 19-electron ones are electron-rich, it is obvious that one would anticipate dissociative pathway(s) for reactions of the latter. With 19-electron polynuclear systems this can be accomplished by cleavage of an M—M bond to generate a 17-electron center at which the nucleophile can attack, leading ultimately to substitution (as in Scheme 7). Alternatively, the rate-determining step may be simple ligand dissociation. For example, ETC-catalyzed CO substitution in 25 is retarded by a CO atmosphere and the rate seems[135] to depend only slightly on the nature of the nucleophile. This suggests rate-determining CO dissociation from the 19-electron radical intermediate 25^-. Similarly, there is electrochemical evidence that ETC ligand substitution is dissociative in $[(\mu_3-MeC)Co_3(CO)_6(PR_2Ph)_3]^{136}$ and 26.[137]

     25   26

Quantitative studies of ligand substitution at 19-electron centers are hard to find. A study of CO substitution in the 19-electron complex 10 established[138] a dissociative mechanism and the reactivity order 10 $\gg$ 10^+.

However, ESR studies showed that **10** is predominantly an 18-electron complex with a radical ligand ("18 + δ" complex); only a few percent of the unpaired electron is localized on the $Co(CO)_3$ moiety.[67,68] Nevertheless, the slight delocalization of the extra electron onto the metal is believed to involve orbitals with Co—CO antibonding character. This weakens the Co—CO bond and accounts for the increase in lability of **10** compared to **10⁺**. Substitution of CO by PPh_3 in $[M(CO)_4bipy]$ (M = Cr, Mo, W) is ETC catalyzed, with the rate-determining step being CO dissociation from the $[M(CO)_4bipy]^-$ radical anions.[139] The order of reactivity is Cr > Mo > W, which can be compared to the order found (*vide supra*) for *associative* substitution at 17-electron complexes of this triad: Mo, W ≫ Cr. Once again, however, the $[M(CO)_4bipy]^-$ complexes are not 19-electron, but rather "18 + δ" species in which almost all of the odd electron is situated on the bipy ligand.

There have been only three quantitative kinetic studies of truly 19-electron complexes. In the first, arene displacement from $[CpFe(arene)]$ by phosphines and phosphites was found to be associative.[140,141] Thus, the rate depends on the nature of the P-donor and the arene, and the activation entropy is large and negative. However, it is likely that successive arene ring slippage occurs to avoid the formation of 21-electron intermediates, and in this sense the mechanism is dissociative because bond breakage occurs prior to or within the transition state region. The authors propose a rapid preequilibrium slippage $[CpFe(\eta^6\text{-arene})] \leftrightarrow [CpFe(\eta^4\text{-}$ arene)], although the results are also consistent with slow $\eta^6 \to \eta^4$ slippage, provided the $\eta^4 \to \eta^6$ back reaction is rapid compared to the rate with which the η^4 intermediate is trapped by nucleophile.

In the second study, it was found that $[(MeCp)Mn(CO)_2NO]^+$ (**27⁺**) undergoes efficient ETC substitution of CO by a range of P-donors.[142,143]

27⁺

In this case it was possible to measure the rate constant (k) for the 19-electron substitution depicted in Scheme 8. The voltammetric data given in Fig. 7 nicely illustrate the essential features of the ETC process. As seen in Fig. 7B, the reduction wave for **27⁺** is completely suppressed at 25°C when the nucleophile $P(OEt)_3$ is present, even though there is no reaction in the bulk solution. This occurs because reduction of M—CO⁺ to M—CO at the electrode surface is rapidly followed by conversion to M—L. Now, M—L is more easily oxidized than is M—CO ($E_1^0 > E_2^0$),

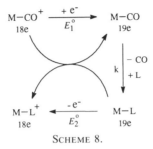

SCHEME 8.

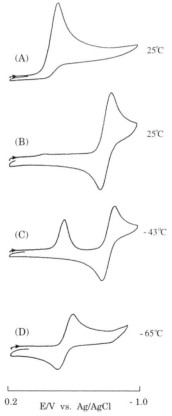

FIG. 7.   Cyclic voltammograms of 1.0 mM [(MeCp)Mn(CO)$_2$NO]$^+$ in CH$_2$Cl$_2$ at the indicated temperatures with P(OEt)$_3$ present at a concentration of (A) none and (B)–(D) 10 mM. The scan rate was 0.5 V/s, and the working electrode was a 1-mm-diameter glassy carbon disk.

and this means that M—L is spontaneously oxidized at a potential in the vicinity of E_1^0. The oxidation of M—L can occur heterogeneously at the electrode or homogeneously by cross-reaction with M—CO$^+$. Hence, the net current for the reduction of M—CO$^+$ is zero. The reversible wave in Fig. 7B is due to M—L$^+$/M—L. At $-65°C$ (Fig. 7D) the substitution step M—CO $\rightarrow$ M—L is too slow on the CV time scale for a significant amount of M—L to form, and the only electrochemical wave is that due to the reduction of M—CO$^+$. At $-43°C$ (Fig. 7C), the CV time scale and the reaction rate are competitive, and a reduction wave is observed for *both* M—CO$^+$ and M—L$^+$. It was found that variation of the nucleophile concentration or external CO pressure has no effect on the CV under these conditions, indicating a rate independent of [L] and [CO]. In addition, CVs obtained at $-43°C$ with a variety of P-donors were very similar, implying the same rate for all L. In other words, the mechanism is strictly dissociative, with the rate-determining step being CO loss to give the 17-electron intermediate [(MeCp)Mn(CO)NO], which is trapped rapidly and completely by nucleophile L. The activation parameters for CO dissociation from the 19-electron [(MeCp)Mn(CO)$_2$NO] (**27**) are: $\Delta H^‡ = 60 \pm 6$ kJ; $\Delta S^‡ = +37 \pm 15$ J K^{-1}.

Since the radical **27** contains a nitrosyl ligand, it would seem possible that the odd electron is largely localized on the NO group, so that **27** is really an $18 + \delta$ complex. However, MO calculations[143] as well as an ESR study[144] of the closely related complexes [CpM(CO)$_2$NO]$^-$ (M = Cr, Mo) implies that a large amount (>50%) of the unpaired spin density is on the metal in **27**. The simple observation that **27** reacts by a dissociative mechanism also argues against the possibility that **27** is a 17-electron complex (with the NO ligand being a one-electron donor) because 17-electron complexes almost always react by an associative mechanism (*vide supra*). The indenyl analogue of **27**$^+$, [(η^5-C$_9$H$_7$)Mn(CO)$_2$NO]$^+$ (**28**$^+$),

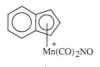

28$^+$

was also examined in order to test for the presence of any "indenyl effect." It was found that the CO substitutions in **27** and **28** are both dissociative and occur at very similar rates. This suggests that ring slippage does not occur when **27**$^+$ and **28**$^+$ are reduced. It is our general belief that slippage rarely occurs in 19-electron complexes. However, this does not mean that an indenyl effect cannot operate in organometallic radicals; its absence in **28** is due to the presence of the nitrosyl ligand, as explained later. In order to explore the dependence of the substitution rate on the metal, $[CpRe(CO)_2NO]^+$ was studied and found to be reduced at about the same potential as **27**$^+$, but unlike **27**, $[CpRe(CO)_2NO]$ is completely unreactive on the CV time scale toward nucleophiles. This indicates that the reactivity with respect to CO dissociation in these 19-electron complexes follows the order Mn $\geqslant$ Re, which likely reflects the stronger M—CO bond in the heavier transition metal. The same reactivity order was found for the 19-electron complexes $[(arene)M(CO)_3]$ (M = Mn, Re) (*vide supra*).

It has been known for many years that normally slow CO substitution in **27**$^+$ becomes rapid in the presence of a catalytic amount of reducing agent NEt$_3$.[145] This result and others suggest the possibility that many substitution reactions of 18-electron complexes thought to occur by conventional dissociative or associative pathways may in fact take place by an ETC catalyzed mechanism initiated by trace amounts of adventitious reductants in solution.

The third quantitative study of 19-electron centers involved an electrochemical investigation of reductively activated CO substitution in [CpFe (CO)$_3$]$^+$ (**29**$^+$) and $[(indenyl)Fe(CO)_3]^+$ (**30**$^+$).[146] It was shown that, upon

$$29^+ \qquad\qquad\qquad 30^+$$

reduction, **29**$^+$ very rapidly dissociates CO and dimerizes to [CpFe(CO)$_2$]$_2$. In the presence of P-donor nucleophiles, the dimerization is quenched, and an efficient ETC substitution occurs to afford $[CpFe(CO)_2L]^+$ by the mechanism shown in Scheme 8. The indenyl complex **30**$^+$ also follows an ETC mechanism for CO substitution and was chosen to determine if an indenyl effect is present in this case.

A cyclic voltammetric study of **29**$^+$ in the absence of nucleophiles showed its reduction to be chemically irreversible even at 100 V/s and −112°C in BuCN under CO. The chemistry involved is given in Eqs. (32)–(34). It was possible to show that the rate constant (k_{-CO}) for CO

$$[CpFe(CO)_3]^+ \quad \xrightleftharpoons{+ e^-} \quad [CpFe(CO)_3] \qquad\qquad (32)$$

$$[CpFe(CO)_3] \quad \xrightarrow{k_{-CO}} \quad [CpFe(CO)_2] \ + \ CO \qquad\qquad (33)$$

$$2\,[CpFe(CO)_2] \quad \xrightarrow{\qquad} \quad [CpFe(CO)_2]_2 \qquad\qquad (34)$$

dissociation from the 19-electron $[CpFe(CO)_3]$ must be greater than 10^3 s^{-1} at $-112°C$. With P-donor nucleophile (L) present, the 17-electron $[CpFe(CO)_2]$ is trapped according to Eq. (35) with the following large bimolecular rate constants $(M^{-1}s^{-1})$ at $25°C$: $P(OEt)_3$ ($> 10^9$), PPh_3 (2×10^8), $P(OPh)_3$ (5×10^7), $P(C_2H_4CN)_3$ (3×10^7), and $AsPh_3$ (8×10^5). Once formed, $[CpFe(CO)_2L]$ transfers an electron to the electrode or to $\mathbf{29^+}$, and the chain reaction is propagated.

$$[CpFe(CO)_2] + L \xrightarrow{k_L} [CpFe(CO)_2L] \tag{35}$$

The behavior of the indenyl complex $\mathbf{30^+}$ contrasts sharply with that of the Cp analogue, as illustrated in Fig. 8. Although reactions analogous to

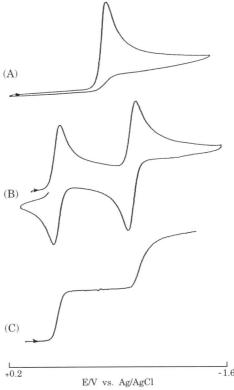

(A)

(B)

(C)

+0.2   -1.6
E/V vs. Ag/AgCl

FIG. 8.   Cyclic voltammograms of (A) $[CpFe(CO)_3]^+$ and (B), (C) $[(indenyl)Fe(CO)_3]^+$ in CH_2Cl_2 under CO. For (A) and (B) the scan rate was 0.5 V/s and the working electrode was a 1-mm-diameter glassy carbon disk. Curve (C) was obtained under steady-state conditions at 20 mV/s with a 10-μm-diameter gold disk. The temperature was 23°C for (A) and (B), and 0°C for (C).

Eqs. (32)–(34) take place, the CO dissociation is readily inhibited and dimer formation quenched by an atmosphere of CO, even at room temperature, so that the reduction becomes chemically reversible (Fig. 8B). The reduction product **30** undergoes further reversible reduction at a more negative potential to give **30**$^-$, which is known[147] from a structural study to have a slipped indenyl ring, i.e., $[(\eta^3\text{-}C_9H_7)Fe(CO)_3]^-$. Since **30**$^+$ is known to have an 18-electron η^5-indenyl structure and **30**$^-$ is known to be an 18-electron η^3-indenyl complex, the point of interest is the bonding in **30**. Although it has been suggested otherwise,[148] it is highly likely that **30** has η^5-bonding and is a true 19-electron complex. This is so because (i) CO substitution in **30** is strictly dissociative (*vide infra*) and (ii) the rate of heterogeneous charge transfer, as determined from steady-state voltammograms with microelectrodes (Fig. 8C), is fast for **30**$^+$ → **30** but slow for **30** → **30**$^-$. The latter is consistent with slippage during the second charge transfer and not during the first.

With P-donor nucleophiles, efficient ETC substitution occurs to give $[(\eta^5\text{-indenyl})Fe(CO)_2L]^+$ (Scheme 8). In spite of the rather striking differences in the voltammograms of the Cp and indenyl systems, the mechanism of CO substitution in the two 19-electron species **29** and **30** is the same, namely, dissociation of CO followed by nucleophile addition. There is, however, one big difference in that **29** is at least 10^6 times more reactive than is **30**. In other words, there is an *inverse* indenyl effect for CO substitution in **29** and **30** that applies to 19-electron species and does not involve ring slippage. A simple MO analysis[146] provides an explanation for the much reduced reactivity of **30** in comparison to **29**, without the need to invoke ring slippage. The LUMO for **29**$^+$ is approximately equally localized on the metal and the CO ligands, with only a small contribution from the Cp atomic orbitals. Furthermore, the Fe–CO interaction in the LUMO is antibonding, thus accounting for the ready dissociation of CO upon reduction. In **30**$^+$, the LUMO is calculated to contain about the same amount of metal character as in **29**$^+$. However, there is a marked difference in that this LUMO is significantly localized on the benzene ring of the indenyl ligand, with a concomitant reduction in the contribution from the COs. Thus, the LUMO on **30**$^+$ is less Fe–CO antibonding than the LUMO on **29**$^+$, in agreement with the observed reactivities. In effect, the indenyl complex, when reduced, is able to accommodate some of the extra electron density in the uncoordinated but conjugated benzene ring. This is electron density that would otherwise reside in the CO ligands in an Fe–CO π^*-derived orbital. In a sense, the indenyl group acts as an "electron sink," but slippage is not involved and there is no net decrease of electron density on the metal. The lack of an inverse indenyl effect in the reactions of **27** and **28** can be easily understood on the basis of MO calculations, which indicate that the electron sink nature of the indenyl benzene ring is largely overshadowed by the accepting power of the NO ligand.

The conclusion from this study is that extending the conjugation of a dienyl or other ligand can markedly influence the stability and reactivity of 19-electron complexes. The extended conjugation results in reduced metal–CO antibonding interactions, which results in increased stability and reduced rate of dissociation. Ironically, this inverse indenyl effect has the same origin as the *positive* effect seen in 18-electron systems. In 18-electron systems, associative nucleophilic attack is facilitated by the electron sink nature of the indenyl ligand, which may in the extreme case slip to η^3-bonding. In this way the energy of the activated complex is lowered and the rate increases. With 19-electron complexes, the indenyl ligand performs its electron sink role *in the ground state* without ring slippage, thereby stabilizing it with respect to CO dissociation and *increasing* the activation energy for CO substitution. The generalized conclusion is that indenyl ligands will accelerate substitutions at 18-electron centers (A or D mechanism) and 17-electron centers (A mechanism) but inhibit substitutions at 19-electron centers (D mechanism). Obviously, a similar analysis applies to the replacement of benzene by naphthalene or pyrrole by indole, etc.

IV

MIGRATORY INSERTION AND ISOMERIZATION REACTIONS

The equilibrium constant for the CO insertion shown in Eq. (36) (L = CO, PR_3) is greatly increased when the metal is oxidized from Fe(II) to Fe(III).[149] Electrochemical studies show that K_{eq} is ca. 10^{11} times greater for the Fe(III) analogues, and that this is largely due to an increase in the forward rate constant. This oxidatively induced insertion is not catalytic because the acyl product is more easily oxidized than is the reactant. The carbonylation reaction in Eq. (37), however, is subject to redox catalysis.[1,150] Thus, K_{eq} for Eq. (37) is quite large, but the reaction does not progress detectably after 5 days under CO at 0°C. The addition of a few mole percent of an oxidizing agent such as Ag^+ or $[Cp_2Fe]^+$ causes complete conversion to product within 2 minutes. The overall insertion rate

$$[CpFe(CO)(L)Me] + MeCN \longrightarrow [CpFe(L)(MeCN)(COMe)] \qquad (36)$$

$$[CpFe(CO)(PPh_3)Me] + CO \longrightarrow [CpFe(CO)(PPh_3)(COMe)] \qquad (37)$$

increases by a factor of ca. 10^7 upon oxidation. The use of resolved [CpFe(CO)(PPh$_3$)Me] leads to a racemic acyl product, a fact attributed to configurational instability of the 17-electron [CpFe(CO)(PPh$_3$)Me]$^+$ intermediate. The use of trityl cation as the oxidizing agent to initiate the catalytic carbonylation of [CpFe(CO)$_2$R] has also been described.[151] The alkyl to acyl migration in the 17-electron [CpFe(CO)(L)R]$^+$ occurs in two steps, and after some debate, there seems to be general agreement[150,152,153] that the mechanism consists of nucleophilic attack followed by migration and not vice versa. This is not surprising, because the latter mechanism requires a 15-electron intermediate. Reductive activation can also be used to induce alkyl to acyl migrations, in which case there is good evidence that migration occurs directly from the initially formed 19-electron intermediate.[154,155]

An interesting case of reductively induced hydride migration is summarized in Scheme 9.[156] Chemical oxidation of [Cp*Fe(dppe)H] affords the 17-electron monocation, which reversibly binds CO at −80°C. Spectroscopic studies established that [Cp*Fe(dppe)(CO)H]$^+$ is a genuine 19-electron complex, which undergoes hydride migration to an endo site of the C$_5$Me$_5$ ring when reduced with cobaltocene.

The reductive elimination of ethane from [Cp*Rh(PPh$_3$)Me$_2$] was found to be increased by a factor of at least 3×10^9 upon oxidation to the 17-electron radical cation.[157] The absence of solvent effects on the rate was interpreted to indicate direct elimination from the cation to give ethane and a 15-electron intermediate, which is rapidly trapped by solvent.

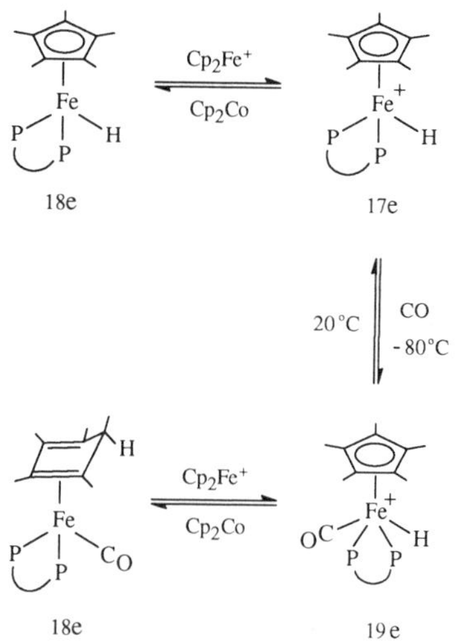

SCHEME 9.

There are many examples of redox-induced structural rearrangements in organometallic complexes; the reader is referred to a review by Connelly.[2] The simplest rearrangement is cis/trans or fac/mer isomerization, usually induced by oxidation. Thus, cis-[Mn(CO)$_2$(dppe)$_2$]$^+$ slowly converts to the trans isomer with a rate constant of 10^{-5} s^{-1} at room temperature. Upon oxidation, the cis → trans isomerization increases in rate by 7 powers of 10.[158] The process is not catalytic, so that stoichiometric oxidation followed by reduction is required to synthetically utilize the increased reactivity of the radicals in the conversion of 18-electron cis-[Mn(CO)$_2$(dppe)$_2$]$^+$ to trans-[Mn(CO)$_2$(dppe)$_2$]$^+$. An example of oxidatively induced fac → mer isomerization is given in Scheme 10.[159] The fac^0 → mer^0 reaction for the neutral 18-electron isomer is slow, with $k_2 = 2 \times 10^{-4}$ s^{-1}, $K_2 = 4$. The reaction fac$^+$ → mer$^+$ is much faster and also has a larger equilibrium constant: $k_1 = 1 \times 10^{-1}$ s^{-1}, $K_1 = 640$. In this case E^0 for the mer$^+$/mer couple is about 0.15 V negative of that for fac$^+$/fac, which means that the cross reaction is slightly unfavored ($K_3 = 1/160$). Nevertheless, the fac$^+$ → mer$^+$ conversion is rapid enough so that the cross reaction provides an effective pathway for the catalytic conversion fac^0 → mer^0. An analogous study has been reported concerning fac/mer isomerization in [M(CO)$_3$(η^3-P$_2$P$'$)] [M = Cr, Mo, W; P$_2$P$'$ = Ph$_2$PCH$_2$CH$_2$P(Ph)CH$_2$CH$_2$PPh$_2$].[160]

Upon oxidation, the vinylidene complex [(C$_6$Me$_6$)Cr(CO)$_2$\{C=C(Si Me$_3$)$_2$\}] undergoes rapid isomerization to the alkyne analogue, shown as **V$^+$** → **A$^+$** in Scheme 11.[161,162] While isomerizations in the direction alkyne → vinylidene are not rare when hydrogen migration is involved, the chemistry depicted in Scheme 11 is rare because the migrating group is SiMe$_3$ and unique because it is redox promoted in the direction vinylidene → alkyne. The conversion **A** → **V** is favored in the neutral complexes, and this isomerization is catalyzed by **V$^+$**.

Another interesting oxidatively promoted transformation is illustrated in Fig. 9. The syn-facial bimetallic complex (η^4,η^6-dimethylnaphthalene)Mn$_2$(CO)$_5$ (**31**) contains a fairly strong metal–metal bond.[163] The addition

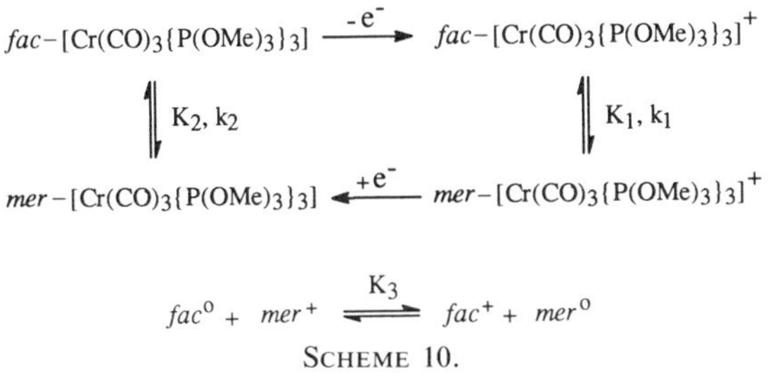

$$fac-[\text{Cr(CO)}_3\{\text{P(OMe)}_3\}_3] \xrightarrow{-e^-} fac-[\text{Cr(CO)}_3\{\text{P(OMe)}_3\}_3]^+$$

$$\Big\updownarrow K_2, k_2 \qquad\qquad \Big\updownarrow K_1, k_1$$

$$mer-[\text{Cr(CO)}_3\{\text{P(OMe)}_3\}_3] \xleftarrow{+e^-} mer-[\text{Cr(CO)}_3\{\text{P(OMe)}_3\}_3]^+$$

$$fac^0 + mer^+ \underset{}{\overset{K_3}{\rightleftharpoons}} fac^+ + mer^0$$

SCHEME 10.

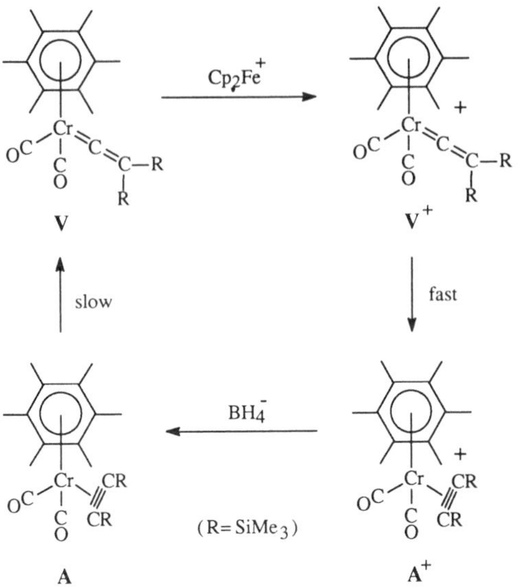

SCHEME 11.

of a trace of oxidant $[Cp_2Fe]^+$ to **31** under an atmosphere of CO results in the *catalytic* conversion to the zwitterionic **32**. Complex **32** retains the *syn*-facial structure, which is made possible by a large bending of the diene (η^4) plane from the η^6-ring.[164] Significantly, **32** can be converted back to **31** by the stoichiometric addition of the reducing agent cobaltocene or by addition of trimethylamine oxide.

Most redox-induced isomerizations are initiated by oxidation. Reductions that lead to isomerization are much less common. Two examples are Eqs. (38) and (39), both of which are accomplished by initial reduction with sodium amalgam followed by stoichiometric oxidation.[165,166] The neu-

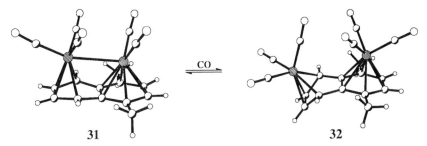

FIG. 9.   Redox promoted interconversion between *syn*-facial (η^4,η^6-dimethylnaphthalene) dimanganese carbonyl complexes.

$$fac\text{-}[Mn(CO)(CNBu^t)_3(bipy)]^+ \longrightarrow mer\text{-}[Mn(CO)(CNBu^t)_3(bipy)]^+ \quad (38)$$

$$cis,cis\text{-}[Mn(CO)_2(CNBu^t)_2(bipy)]^+ \longrightarrow cis,trans\text{-}[Mn(CO)_2(CNBu^t)_2(bipy)]^+ \quad (39)$$

tral radicals involved in this process have been shown to be of the 18 + δ variety. The isomerization of $[CpCo(1,5\text{-}COT)]^z$ to $[CpCo(1,3\text{-}COT)]^z$ is fairly slow in the 18-electron $z = 0$ complex, but is fast in the 19-electron anion $(z = -1)$.[167] Reductively initiated ETC catalysis has been used for the synthesis of the isomers of $[(CF_3)_2C_6Co_2\{P(OMe)_3\}_2]$ and other "flyover" organometallic clusters.[168]

V

REDOX SWITCHES

Organometallic molecules that undergo rapid chemically reversible oxidation or reduction may be able to function as effective redox switches. The goal is to modulate the chemical reactivity of a molecular system by attaching a redox active switch that can be turned on or off by electron transfer. A well-known example of this is the control of cation binding to crown ethers and other macrocyclic receptor molecules by covalently attaching a ferrocenyl group.[169,170] In this case, oxidation of the ferrocenyl moiety switches off the receptor site, in large part because of simple electrostatic effects. In a similar vein, a cobaltocene group can be used to modulate anion binding.[169,171] An example of this is provided by complex 33, in which reduction of the cobalt centers diminishes anion binding and recognition. The Lewis acidity of the boron center in the redox active ferrocenylboronic acid complex $[CpFe\{C_5H_4B(OH)_2\}]$ is easily modulated electrochemically, resulting in selective recognition of fluoride in the presence of other halides and common anions.[172] Similarly, a chiral ferrocenylboronic acid derivative has been used to detect saccharides.[173]

The chelating ferrocenyl ligand $FcOCH_2CH_2PPh_2$ in complex 34 provides electrochemical control over the electronic and steric environment

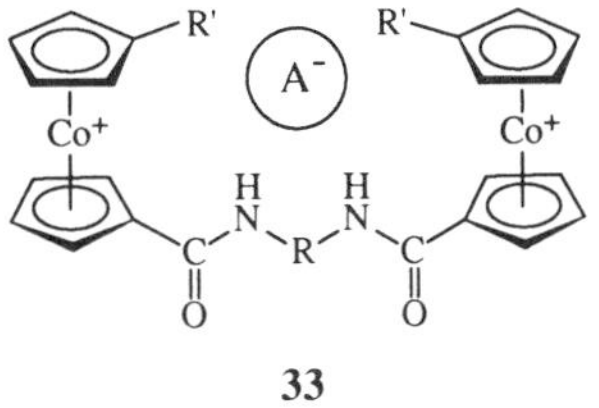

33

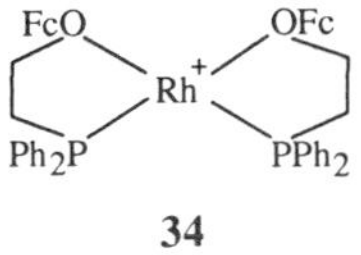

34

of the Rh(I) center.[174] Oxidation of **34** results in cleavage of the Rh–O links and subsequent dimerization via coordination of Rh to the Ph rings. The "hemilabile" ligand $FcOCH_2CH_2PPh_2$ is unusual in that the weak bond between the ether oxygen and a metal is easily broken upon oxidation of the ferrocene center, thereby providing some control over the coordination environment (and reactivity) of the metal. In another study, a ferrocenyl group was used as a redox switch to facilitate the reductive elimination from a Pt(II) complex, as shown in Eq. (40).[175] Oxidation of the

$$ Fc-C\equiv C-\underset{\underset{\bigcirc}{|}}{Pt}-PPh_2 \quad \xrightarrow{AgBF_4} \quad Fc-C\equiv C-\bigcirc-X \qquad (40) $$

analogous Pt(II) acetylide complex with the Fc moiety replaced by a phenyl group affords only very low yields of the reductive elimination product $PhCC(C_6H_4X)$. The conclusion is that the positive charge density localized on the Fc^+ center after oxidation is felt to a small extent by the platinum, thus inducing the reaction.

A novel self-switching process is summarized by the chemistry in Scheme 12.[176] The complex $[(C_6H_6)Cr(CO)_3]$ undergoes oxidation at the metal center to give an extremely reactive radical that under most conditions decomposes in an unproductive manner with liberation of benzene. In complex **35**, however, oxidation in CH_2Cl_2 occurs first at the ferrocenyl center rather than at chromium, with the result that the radical is by comparison much more stable. Nevertheless, the chromium is sufficiently activated in **35**$^+$ to undergo rapid CO substitution by $P(OEt)_3$. Concomitant with this substitution, the center of oxidation shifts from Fe to Cr, so that the ferrocenyl switch is shut off in the product, **36**$^+$, which has a 17-electron chromium atom. Oxidation of **36**$^+$ to **36**$^{2+}$ reopens the switch and the chromium is then attacked by $P(OEt)_3$ to give the *addition* product **37**$^{2+}$. Once again, the nucleophilic attack is accompanied by internal electron transfer and switch closing, so that the product has only 18-electron centers. The circled complexes **35**, **36**, and **37**$^{2+}$ were isolated and investigated separately, allowing the chemistry in Scheme 12 to be accurately defined. The synthesis of the addition complex **37**$^{2+}$ was only possible because of the internal self-switching ferrocenyl group; it is a thermally stable compound with a deep purple color due to a strongly allowed charge transfer transition.

Redox switches based on 1,1-bis(diphenylphosphino)cobaltocene have been used in complexes **38** and **39**.[177,178] In the former, the electrophilicity

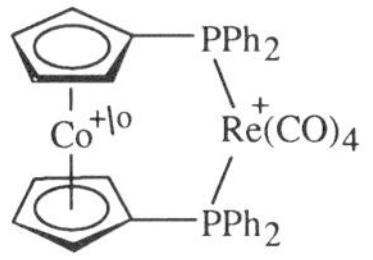

38 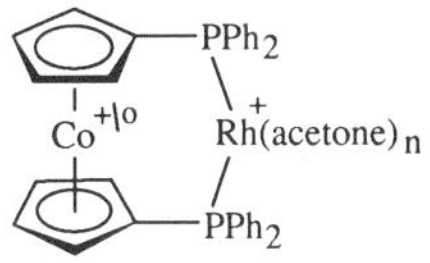   **39**

of the carbonyl carbons is strongly dependent on the cobalt oxidation state. Thus, the rate constant for attack by N_3^- on **38** is larger by a factor of 5400 for the oxidized form (Co^+) compared to the reduced form (Co^0). Complex **39** functions as a catalyst for the hydrogenation and hydrosilation of alkenes and ketones. For hydrogenation, the reaction rate is faster in the reduced state (Co^0), while for hydrosilation the rate is faster in the oxidized state (Co^+).

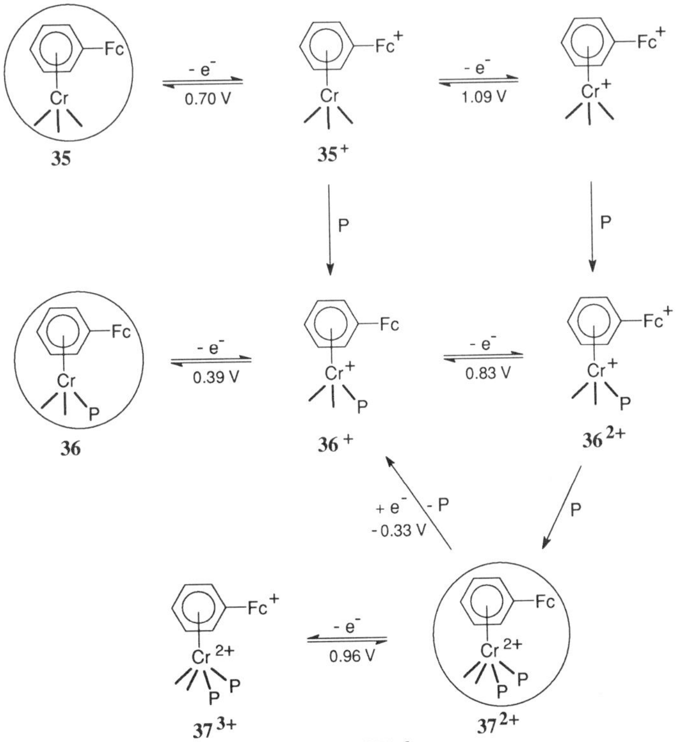

$[P = P(OEt)_3]$

SCHEME 12.

ACKNOWLEDGMENTS

It is a pleasure to acknowledge support from the National Science Foundation (CHE-9400800 and INT-9312709), as well as long-standing fruitful collaborations with Professors Gene Carpenter, Young Chung, John Edwards, and Philip Rieger.

REFERENCES

(1) Magnuson, R. H.; Meirowitz, R.; Zulu, S. J.; Giering, W. P. *Organometallics* **1983,** *2,* 460.
(2) Connelly, N. G. *Chem. Soc. Rev.* **1989,** *18,* 153.
(3) Tyler, D. R. *Prog. Inorg. Chem.* **1988,** *36,* 125.
(4) Astruc, D. *Chem. Rev.* **1988,** *88,* 1189.
(5) Baird, M. C. *Chem. Rev.* **1988,** *88,* 1217.
(6) Tyler, D. R.; Mao, F. *Coord. Chem. Rev.* **1990,** *97,* 119.
(7) Tyler, D. R. *Acc. Chem. Res.* **1991,** *24,* 325.
(8) *Organometallic Radical Processes;* Trogler, W. C., Ed.; Elsevier: Amsterdam, 1990.
(9) Ercoli, R.; Calderazzo, F.; Alberola, A. *J. Am. Chem. Soc.* **1960,** *82,* 2966.
(10) Kowaleski, R. M.; Basolo, F.; Trogler, W. C.; Gedridge, R. W.; Newbound, T. D.; Ernst, R. D. *J. Am. Chem. Soc.* **1987,** *109,* 4860.
(11) Bagchi, R. N.; Bond, A. M.; Brain, G.; Colton, R.; Henderson, T. L. E.; Kevekordes, J. E. *Organometallics* **1984,** *3,* 4.
(12) Lang, R. F.; Ju, T. D.; Kiss, G.; Hoff, C. D.; Bryan, J. C.; Kubas, G. J. *J. Am. Chem. Soc.* **1994,** *116,* 7917.
(13) Connelly, N. G.; Demidowicz, Z.; Kelly, R. L. *J. Chem. Soc., Dalton Trans.* **1975,** 2335.
(14) Meng, Q.; Huang, Y.; Ryan, W. J.; Sweigart, D. A. *Inorg. Chem.* **1992,** *31,* 4051.
(15) Cooley, N. A.; MacConnachie, P. T. F.; Baird, M. C. *Polyhedron* **1988,** *7,* 1965.
(16) Watkins, W. C.; Hensel, K.; Fortier, S.; Macartney, D. H.; Baird, M. C.; McLain, S. J. *Organometallics* **1992,** *11,* 2418.
(17) Scott, S. L.; Espenson, J. H.; Zhu, Z. *J. Am. Chem. Soc.* **1993,** *115,* 1789.
(18) Fortier, S.; Baird, M. C.; Preston, K. F.; Morton, J. R.; Ziegler, T.; Jaeger, T. J.; Watkins, W. C.; MacNeil, J. H.; Watson, K. A.; Hensel, K.; Le Page, Y.; Charland, J.-P.; Williams, A. J. *J. Am. Chem. Soc.* **1991,** *113,* 542.
(19) Legzdins, P.; McNeil, W. S.; Batchelor, R. J.; Einstein, F. W. B. *J. Am. Chem. Soc.* **1994,** *116,* 6021.
(20) Legzdins, P.; McNeil, W. S.; Shaw, M. J. *Organometallics* **1994,** *13,* 562.
(21) Poli, R.; Owens, B. E.; Linck, R. G. *Inorg. Chem.* **1992,** *31,* 662.
(22) Herrick, R. S.; Herrinton, T. R.; Walker, H. W.; Brown, T. L. *Organometallics* **1985,** *4,* 42.
(23) Hanckel, J. M.; Lee, K.-W.; Rushman, P.; Brown, T. L. *Inorg. Chem.* **1986,** *25,* 1852.
(24) Connelly, N. G.; Kitchen, M. D. *J. Chem. Soc., Dalton Trans.* **1977,** 931.
(25) Winter, A.; Huttner, G.; Gottlieb, M.; Jibril, I. *J. Organomet. Chem.* **1985,** *286,* 317.
(26) Connelly, N. G.; Freeman, M. J.; Orpen, A. G.; Sheehan, A. R.; Sheridan, J. B.; Sweigart, D. A. *J. Chem. Soc., Dalton Trans.* **1985,** 1019.
(27) Detty, M. R.; Jones, W. D. *J. Am. Chem. Soc.* **1987,** *109,* 5666.
(28) Baker, P. K.; Connelly, N. G.; Jones, B. M. R.; Maher, J. P.; Somers, K. R. *J. Chem. Soc., Dalton Trans.* **1980,** 579.
(29) Connelly, N. G.; Kelly, R. L.; Whiteley, M. W. *J. Chem. Soc., Dalton Trans.* **1981,** 34.
(30) Rosenblum, M. *Chemistry of the Iron Group Metallocenes;* Wiley: New York; 1965.

(31) Dixon, A. J.; George, M. W.; Hughes, C.; Poliakoff, M.; Turner, J. J. *J. Am. Chem. Soc.* **1992**, *114*, 1719.

(32) Connelly, N. G.; Gamasa, M. P.; Gimeno, J.; Lapinte, C.; Lastra, E.; Maher, J. P.; Le Narvor, N.; Rieger, A. L.; Rieger, P. H. *J. Chem. Soc., Dalton Trans.* **1993**, 2575.

(33) McKinney, R. J. *Inorg. Chem.* **1982**, *21*, 2051.

(34) Bard, A. J.; Garcia, E.; Kukharenko, S.; Strelets, V. V. *Inorg. Chem.* **1993**, *32*, 3528.

(35) Chin, T. T.; Sharp, L. I.; Geiger, W. E.; Rieger, P. H. *Organometallics* **1995**, *14*, 1322.

(36) Milukov, V. A.; Sinyashin, O. G.; Ginzburg, A. G.; Kondratenko, M. A.; Loim, N. M.; Gubskaya, V. P.; Musin, R. Z.; Morozov, V. I.; Batyeva, E. S.; Sokolov, V. I. *J. Organomet. Chem.* **1995**, *493*, 221.

(37) McCullen, S. B.; Brown, T. L. *J. Am. Chem. Soc.* **1982**, *104*, 7496.

(38) Kuksis, I.; Baird, M. C. *Organometallics* **1994**, *13*, 1551.

(39) Poli, R.; Owens, B. E.; Linck, R. G. *J. Am. Chem. Soc.* **1992**, *114*, 1302.

(40) Yao, Q.; Bakac, A.; Espenson, J. H. *Organometallics* **1993**, *12*, 2010.

(41) Caspar, J. V.; Meyer, T. J. *J. Am. Chem. Soc.* **1980**, *102*, 7795.

(42) Zhang, S.; Brown, T. L. *J. Am. Chem. Soc.* **1993**, *115*, 1779.

(43) Wegman, R. W.; Olsen, R. J.; Gard, D. R.; Faulkner, L. R.; Brown, T. L. *J. Am. Chem. Soc.* **1981**, *103*, 6089.

(44) Walker, H. W.; Herrick, R. S.; Olsen, R. J.; Brown, T. L. *Inorg. Chem.* **1984**, *23*, 3748.

(45) MacConnachie, C. A.; Nelson, J. M.; Baird, M. C. *Organometallics* **1992**, *11*, 2521.

(46) Huber, T. A.; Macartney, D. H.; Baird, M. C. *Organometallics* **1995**, *14*, 592.

(47) Turaki, N. N.; Huggins, J. M. *Organometallics* **1986**, *5*, 1703.

(48) Stone, N. J.; Sweigart, D. A.; Bond, A. M. *Organometallics* **1986**, *5*, 2553.

(49) Zoski, C. G.; Sweigart, D. A.; Stone, N. J.; Rieger, P. H.; Mocellin, E.; Mann, T. F.; Mann, D. R.; Gosser, D. K.; Doeff, M. M.; Bond, A. M. *J. Am. Chem. Soc.* **1988**, *110*, 2109.

(50) Lee, K. Y.; Kochi, J. K. *Inorg. Chem.* **1989**, *28*, 567.

(51) Zhang, Y.; Gosser, D. K.; Rieger, P. H.; Sweigart, D. A. *J. Am. Chem. Soc.* **1991**, *113*, 4062.

(52) Skagestad, V.; Tilset, M. *Organometallics* **1992**, *11*, 3293.

(53) Tilset, M. *Inorg. Chem.* **1994**, *33*, 3121.

(54) Kuchynka, D. J.; Amatore, C.; Kochi, J. K. *Inorg. Chem.* **1986**, *25*, 4087.

(55) Ryan, O. B.; Tilset, M.; Parker, V. D. *J. Am. Chem. Soc.* **1990**, *112*, 2618.

(56) Ammeter, J. H.; Swalen, J. D. *J. Chem. Phys.* **1972**, *57*, 678.

(57) Fischer, E. O.; Wawersik, H. *J. Organomet. Chem.* **1966**, *5*, 559.

(58) Gusev, O. V.; Denisovich, L. I.; Peterleitner, M. G.; Rubezhov, A. Z.; Ustynyuk, N. A. *J. Organomet. Chem.* **1993**, *452*, 219.

(59) Astruc, D. *Acc. Chem. Res.* **1991**, *24*, 36.

(60) Ruiz, J.; Guerchais, V.; Astruc, D. *J. Chem. Soc., Chem. Commun.* **1989**, 812.

(61) Trujillo, H. A.; Casado, C. M.; Astruc, D. *J. Chem. Soc., Chem. Commun.* **1995**, 7.

(62) Desbois, M. H.; Astruc, D.; Guillin, J.; Maroit, J. P.; Varret, F. *J. Am. Chem. Soc.* **1985**, *107*, 5280.

(63) Professor T. L. Brown first coined the term ''18 + δ'' complexes.

(64) Geiger, W. E. *J. Am. Chem. Soc.* **1974**, *96*, 2632.

(65) Lloyd, M. K.; McCleverty, J. A.; Orchard, D. G.; Connor, J. A.; Hall, M. B.; Hillier, I. H.; Jones, E. M.; McEven, G. K.; *J. Chem. Soc., Dalton Trans.* **1973**, 1743.

(66) Mao, F.; Philbin, C. E.; Weakley, T. J. R.; Tyler, D. R. *Organometallics* **1990**, *9*, 1510.

(67) Mao, F.; Tyler, D. R.; Rieger, A. L.; Rieger, P. H. *J. Chem. Soc., Faraday Trans.* **1991**, *87*, 3113.

(68) Mao, F.; Tyler, D. R.; Bruce, M. R. M.; Bruce, A. E.; Rieger, A. L.; Rieger, P. H. *J. Am. Chem. Soc.* **1992,** *114,* 6418.
(69) Yang, K.; Bott, S. G.; Richmond, M. G. *Organometallics* **1995,** *14,* 2387.
(70) Schut, D. M.; Keana, K. J.; Tyler, D. R.; Rieger, P. H. *J. Am. Chem. Soc.* **1995,** *117,* 8939.
(71) Bowyer, W. J.; Geiger, W. E. *J. Am. Chem. Soc.* **1985,** *107,* 5657.
(72) Nieldon, R. M.; Weaver, M. J. *Organometallics* **1989,** *8,* 1636.
(73) Merkert, J.; Nielson, R. M.; Weaver, M. J.; Geiger, W. E. *J. Am. Chem. Soc.* **1989,** *111,* 7084.
(74) Geiger, W. E. *Acc. Chem. Res.* **1995,** *28,* 351.
(75) Jonas, K.; Deffense, E.; Habermann, D. *Angew. Chem. Int. Ed. Engl.* **1983,** *22,* 716.
(76) Pierce, D. T.; Geiger, W. E. *J. Am. Chem. Soc.* **1992,** *114,* 6063.
(77) Scott, S. L.; Espenson, J. H.; Chen, W.-J. *Organometallics* **1993,** *12,* 4077.
(78) Stiegman, A. E.; Stieglitz, M.; Tyler, D. R. *J. Am. Chem. Soc.* **1983,** *105,* 6032.
(79) Avery, A.; Tyler, D. R. *Organometallics* **1992,** *11,* 3856.
(80) Won, T.-J.; Espenson, J. H. *Organometallics* **1995,** *14,* 4275.
(81) Stiegman, A. E.; Tyler, D. R. *Coord. Chem. Rev.* **1985,** *63,* 217.
(82) Avery, A.; Tenhaeff, S. C.; Weakley, T. J. R.; Tyler, D. R. *Organometallics* **1991,** *10,* 3607.
(83) Zhu, Z.; Espenson, J. H. *Organometallics* **1994,** *13,* 1893.
(84) Silavwe, N. D.; Goldman, A. S.; Ritter, R.; Tyler, D. R. *Inorg. Chem.* **1989,** *28,* 1231.
(85) Stiegman, A. E.; Tyler, D. R. *Inorg. Chem.* **1984,** *23,* 527.
(86) Goldman, A. S.; Tyler, D. R. *Inorg. Chem.* **1987,** *26,* 253.
(87) Castellani, M. P.; Tyler, D. R. *Organometallics* **1989,** *8,* 2113.
(88) Brown, D. S.; Delville-Desbois, M.-H.; Boese, R.; Vollhardt, K. P. C.; Astruc, D. *Angew. Chem., Int. Ed. Engl.* **1994,** *33,* 661.
(89) Dessy, R. E.; Stary, F. E.; King, R. B.; Waldrop, M. *J. Am. Chem. Soc.* **1966,** *83,* 471.
(90) Doxsee, K. M.; Grubbs, R. H.; Anson, F. C. *J. Am. Chem. Soc.* **1984,** *106,* 7819.
(91) Kuchynka, D. J.; Kochi, J. K. *Inorg. Chem.* **1988,** *27,* 2574.
(92) Kuchynka, D. J.; Kochi, J. K. *Inorg. Chem.* **1989,** *28,* 855.
(93) Kidd, D. R.; Brown, T. L. *J. Am. Chem. Soc.* **1978,** *100,* 4095.
(94) Fawcett, J. P.; Pöe, A. J.; Sharma, K. R. *J. Am. Chem. Soc.* **1976,** *98,* 1401.
(95) Boothe, B. L.; Haszeldine, R. N.; Reynolds, D. M. *J. Chem. Soc., Dalton Trans.* **1980,** 407.
(96) Fox, A.; Malito, J.; Pöe, A. *J. Chem. Soc., Chem. Commun.* **1981,** 1052.
(97) Pöe, A. *Trans. Met. Chem.* **1982,** *7,* 65.
(98) Herrinton, T. R.; Brown, T. L. *J. Am. Chem. Soc.* **1985,** *107,* 5700.
(99) McCullen, S. B.; Walker, H. W.; Brown, T. L. *J. Am. Chem. Soc.* **1982,** *104,* 4007.
(100) Shi, Q. Z.; Richmond, T. G.; Trogler, W. C.; Basolo, F. *J. Am. Chem. Soc.* **1984,** *106,* 71.
(101) Basolo, F. *Polyhedron* **1990,** *9,* 1503.
(102) Dixon, A. J.; Gravelle, S. J.; van de Burgt, L. J.; Poliakoff, M.; Turner, J. J.; Weitz, E. *J. Chem. Soc., Chem. Commun.* **1987,** 1023.
(103) Therien, M. J.; Ni, C.-L.; Anson, F. C.; Osteryoung, J. G.; Trogler, W. C. *J. Am. Chem. Soc.* **1986,** *108,* 4037.
(104) Song, L.; Trogler, W. C. *J. Am. Chem. Soc.* **1992,** *114,* 3355.
(105) Jensen, B. S.; Parker, V. D. *Electrochim. Acta* **1973,** *18,* 665.
(106) Parker, V. D.; Tilset, M. *J. Am. Chem. Soc.* **1987,** *109,* 2521.
(107) Zhang, K.; Gonzalez, A. A.; Mukerjee, S. L.; Chou, S.-J.; Hoff, C. D.; Kubat-Martin, K. A.; Barnhart, D.; Kubas, G. J. *J. Am. Chem. Soc.* **1991,** *113,* 9170.
(108) Howell, J. A. S.; Burkinshaw, P. M. *Chem. Rev.* **1983,** *83,* 557.

(109) Zhang, S.; Shen, J. K.; Basolo, F.; Ju, T. D.; Lang, R. F.; Kiss, G.; Hoff, C. D. *Organometallics* **1994**, *13*, 3692.
(110) Kowaleski, R. M.; Rheingold, A. L.; Trogler, W. C.; Basolo, F. *J. Am. Chem. Soc.* **1986**, *108*, 2460.
(111) Hershberger, J. W.; Klingler, R. J.; Kochi, J. K. *J. Am. Chem. Soc.* **1983**, *105*, 61.
(112) Zizelman, P. M.; Amatore, C.; Kochi, J. K. *J. Am. Chem. Soc.* **1984**, *106*, 3771.
(113) Hershberger, J. W.; Klingler, R. J.; Kochi, J. K. *J. Am. Chem. Soc.* **1982**, *104*, 3034.
(114) Angelici, R. J.; Loewen, W. *Inorg. Chem.* **1967**, *6*, 682.
(115) Huang, Y.; Carpenter, G. B.; Sweigart, D. A.; Chung, Y. K.; Lee, B. Y. *Organometallics* **1995**, *14*, 1423.
(116) Byers, B. H.; Brown, T. L. *J. Am. Chem. Soc.* **1975**, *97*, 947.
(117) Byers, B. H.; Brown, T. L. *J. Am. Chem. Soc.* **1977**, *99*, 2527.
(118) Hoffman, N. W.; Brown, T. L. *Inorg. Chem.* **1978**, *17*, 613.
(119) Alway, D. G.; Barnett, K. W. *Inorg. Chem.* **1980**, *19*, 1533.
(120) Coville, N. J.; Albers, M. O.; Singleton, E. *J. Chem. Soc., Dalton Trans.* **1983**, 947.
(121) Fabian, B. D.; Labinger, J. A. *Organometallics* **1983**, *2*, 659.
(122) Butts, S. B.; Schriver, D. F. *J. Organomet. Chem.* **1979**, *169*, 191.
(123) Keeley, D. F.; Johnson, R. E. *J. Inorg. Nucl. Chem.* **1959**, *11*, 33.
(124) Bezems, G. J.; Rieger, P. H.; Visco, S. *J. Chem. Soc., Chem. Commun.* **1981**, 265.
(125) Arewgoda, M.; Rieger, P. H.; Robinson, B. H.; Simpson, J.; Visco, S. J. *J. Am. Chem. Soc.* **1982**, *104*, 5633.
(126) Arewgoda, M.; Robinson, B. H.; Simpson, J. *J. Am. Chem. Soc.* **1983**, *105*, 1893.
(127) Sun, S.; Meng, Q.; Zhu, D.; Yao, Y.; Zhu, H.; You, X. *Huaxue Xuebao* **1992**, *50*, 444.
(128) Bruce, M. I.; Kehoe, D. C.; Matisons, J. G.; Nicholson, B. K.; Rieger, P. H.; Williams, M. L. *J. Chem. Soc., Chem. Commun.* **1982**, 442.
(129) Bruce, M. I.; Matisons, J. G.; Nicholson, B. K. *J. Organomet. Chem.* **1983**, *247*, 321.
(130) Osella, D.; Nervi, C.; Ravera, M.; Fiedler, J.; Strelets, V. V. *Organometallics* **1995**, *14*, 2501.
(131) Darchen, A.; Mahé, C.; Patin, H. *J. Chem. Soc., Chem. Commun.* **1982**, 243.
(132) Darchen, A. *J. Chem. Soc., Chem. Commun.* **1983**, 768.
(133) Narayanan, B. A.; Amatore, C.; Kochi, J. K. *Organometallics* **1987**, *6*, 129.
(134) Neto, C. A.; Baer, C. D.; Chung, Y. K.; Sweigart, D. A. *J. Chem. Soc., Chem. Commun.* **1993**, 816.
(135) Richmond, M. G.; Kochi, J. K. *Inorg. Chem.* **1986**, *25*, 656.
(136) Hinkelmann, K.; Mahlendorf, F.; Heinze, J.; Schacht, H.-T.; Field, J. S.; Vahrenkamp, H. *Angew. Chem., Int. Ed. Engl.* **1987**, *26*, 352.
(137) Donovan, B. T.; Geiger, W. E. *J. Am. Chem. Soc.* **1988**, *110*, 2335.
(138) Mao, F.; Tyler, D. R.; Keszler, D. *J. Am. Chem. Soc.* **1989**, *111*, 130.
(139) Miholova, D.; Vlcek, A. A. *J. Organomet. Chem.* **1985**, *279*, 317.
(140) Ruiz, J.; Astruc, D. *J. Chem. Soc., Chem. Commun.* **1989**, 815.
(141) Ruiz, J.; Lacoste, M.; Astruc, D. *J. Am. Chem. Soc.* **1990**, *112*, 5471.
(142) Neto, C. C.; Kim, S.; Meng, Q.; Sweigart, D. A.; Chung, Y. K. *J. Am. Chem. Soc.* **1993**, *115*, 2078.
(143) Huang, Y.; Neto, C. C.; Pevear, K. A.; Banaszak Holl, M. M.; Sweigart, D. A.; Chung, Y. K. *Inorg. Chim. Acta* **1994**, *226*, 53.
(144) Geiger, W. E.; Reiger, P. H.; Tulyathan, B.; Rausch, M. D. *J. Am. Chem. Soc.* **1984**, *106*, 7000.
(145) James, T. A.; McCleverty, J. A. *J. Chem. Soc., Dalton Trans.* **1970**, 850.
(146) Pevear, K. A.; Babaszak Holl, M. M.; Carpenter, G. B.; Rieger, A. L.; Rieger, P. H.; Sweigart, D. A. *Organometallics* **1995**, *14*, 512.
(147) Forschner, T. C.; Cutler, A. R.; Kullnig, R. K. *Organometallics* **1987**, *6*, 889.

(148) Wuu, Y.-M.; Zou, C.; Wrighton, M. S. *J. Am. Chem. Soc.* **1987,** *109,* 5861.
(149) Magnuson, R. H.; Meirowitz, R.; Zulu, S.; Giering, W. P. *J. Am. Chem. Soc.* **1982,** *104,* 5790.
(150) Prock, A.; Giering, W. P.; Greene, J. E.; Meirowitz, R. E.; Hoffman, S. L.; Woska, D. C.; Wilson, M.; Chang, R.; Chen, J.; Magnuson, R. H.; Eriks, K. *Organometallics* **1991,** *10,* 3479.
(151) Bly, R. S.; Silverman, G. S.; Hossain, M. M.; Bly, R. K. *Organometallics* **1984,** *3,* 642.
(152) Golovin, M. N.; Meirowitz, R.; Rahman, M. M.; Liu, H. Y.; Prock, A.; Giering, W. P. *Organometallics* **1987,** *6,* 2285.
(153) Therien, M. J.; Trogler, W. C. *J. Am. Chem. Soc.* **1987,** *109,* 5127.
(154) Miholova, D.; Vlcek, A. A. *J. Organomet. Chem.* **1982,** *240,* 413.
(155) Donovan, B. T.; Geiger, W. E. *Organometallics* **1990,** *9,* 865.
(156) Hamon, P.; Hamon, J.-R.; Lapinte, C. *J. Chem. Soc., Chem. Commun.* **1992,** 1602.
(157) Pedersen, A.; Tilset, M. *Organometallics* **1993,** *12,* 56.
(158) Kochi, J. In *Organometallic Radical Processes;* Trogler, W. C., Ed.; Elsevier: Amsterdam, 1990, pp. 2215–2218.
(159) Bond, A. M.; Colton, R.; Kevekordes, J. E. *Inorg. Chem.* **1986,** *25,* 749.
(160) Bond, A. M.; Colton, R.; Feldberg, S. W.; Mahon, P. J.; Whyte, T. *Organometallics* **1991,** *10,* 3320.
(161) Connelly, N. G.; Orpen, A. G.; Rieger, A. L.; Rieger, P. H.; Scott, C. J.; Rosair, G. M. *J. Chem. Soc., Chem. Commun.* **1992,** 1293.
(162) Connelly, N. G.; Geiger, W. E.; Lagunas, C.; Metz, B.; Rieger, A. L.; Rieger, P. H.; Shaw, M. J. *J. Am. Chem. Soc.* **1995,** *117,* 12202.
(163) Sun, S.; Dullaghan, C. A.; Carpenter, G. B.; Rieger, A. L.; Rieger, P. H.; Sweigart, D. A. *Angew. Chem., Int. Ed. Engl.* **1995,** *34,* 2540.
(164) Sun, S.; Carpenter, G. B.; Sweigart, D. A. To be published.
(165) Alonso, F. J. G.; Riera, V.; Valin, M. L.; Moreiras, D.; Vivanco, M.; Solans, X. *J. Organomet. Chem.* **1987,** *326,* C71.
(166) Brown, N. C.; Carriedo, G. A.; Connelly, N. G.; Alonso, F. J. G.; Quarmby, I. C.; Rieger, A. L.; Rieger, P. H.; Riera, V.; Vivanco, M. *J. Chem. Soc., Dalton Trans.* **1994,** 3745.
(167) Richards, T. C.; Geiger, W. E. *J. Am. Chem. Soc.* **1994,** *116,* 2028.
(168) Arewgoda, C. M.; Robinson, B. H.; Simpson, J. *J. Chem. Soc., Chem. Commun.* **1982,** 284.
(169) Beer, P. D. *Chem. Soc. Rev.* **1989,** *18,* 409.
(170) Plenio, H.; Yang, J.; Diodone, R.; Heinze, J. *Inorg. Chem.* **1994,** *33,* 4098.
(171) Beer, P. D.; Hesek, D.; Kingston, J. E.; Smith, D. K.; Stokes, S. E.; Drew, M. G. B. *Organometallics* **1995,** *14,* 3288.
(172) Dusemund, C.; Sandanayake, K. R. A. S.; Shinkai, S. *J. Chem. Soc., Chem. Commun.* **1995,** 333.
(173) Orin, A.; Shinkai, S. *J. Chem. Soc., Chem. Commun.* **1995,** 1771.
(174) Singewald, E. T.; Markin, C. A.; Stern, C. L. *Angew. Chem., Int. Ed. Engl.* **1995,** *34,* 1624.
(175) Sato, M.; Mogi, E.; Kumakura, S. *Organometallics* **1995,** *14,* 3157.
(176) Yeung, L. K.; Kim, J. E.; Chung, Y. K.; Rieger, P. H.; Sweigart, D. A. *Organometallics* **1996,** *15.* In press.
(177) Lorkovic, I. M.; Wrighton, M. S.; Davis, W. M. *J. Am. Chem. Soc.* **1994,** *116,* 6220.
(178) Lorkovic, I. M.; Duff, R. R.; Wrighton, M. S. *J. Am. Chem. Soc.* **1995,** *117,* 3617.

ADVANCES IN ORGANOMETALLIC CHEMISTRY, VOL. 40

A Review of Group 2 (Ca, Sr, Ba) Metal-Organic Compounds as Precursors for Chemical Vapor Deposition

WILLIAM A. WOJTCZAK, PATRICK F. FLEIG, and MARK J. HAMPDEN-SMITH

Department of Chemistry and
The Center for Micro-Engineered Materials
University of New Mexico, Albuquerque, New Mexico

I

INTRODUCTION

There is considerable interest in the deposition of films containing the Group 2 elements because these films exhibit a variety of technologically important properties. These materials include perovskite phase ferroelectric ceramics such as $BaTiO_3$ and $Ba_{1-x}Sr_xTiO_3$,[1] dielectric materials such as Sr_2AlTaO_6, ceramic superconductors such as $YBa_2Cu_3O_{7-x}$,[2] magnetic

215

materials such as $BaFe_{12}O_{19}$,[3,4] and display materials such as Ce-doped MS, where M = Ca, Sr, Ba, and Ce-doped $CaGa_2S_4$.[5] Films of these materials have been deposited by a variety of vapor-phase chemical and physical methods, each of which has various advantages and disadvantages.[6,7] Chemical vapor deposition (CVD) is a method that relies on the transport of a volatile metal-containing compound to a heated substrate where it undergoes a chemical reaction to form a film of the final desired phase.[7-13] A schematic representation of the steps involved in film formation by chemical vapor deposition is presented in Fig. 1.[6,7,12]

The key processes can be generalized as follows.[7,12,14,15] The precursor is delivered to the CVD reactor. Gas-phase reactions can occur under certain conditions that are generally (although not always) detrimental to the deposition process. In the absence of gas-phase reactions, the precursor is transported to the heated substrate where it physisorbs, followed by chemisorption and chemical reactions that are often very complex and in which it is generally desired that the organic supporting ligands be removed completely from the surface to avoid impurity incorporation from the reaction by-products. For films described here, it is generally necessary that more than one precursor react on the heated substrate to form a crystalline phase that contains many elements. As a result, the system is very complex. Any of the preceding steps may be rate limiting, which can have significant consequences on the control of the composition, homogeneity, crystallinity, and microstructure of the film, especially in a process that involves more than one reagent. For example, if a chemi-

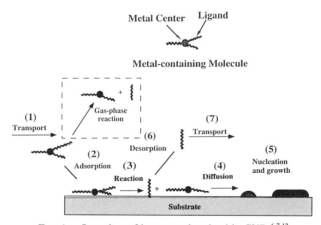

FIG. 1.   Overview of key steps involved in CVD.[6,7,12]

cal step (such as precursor evaporation or a surface reaction) is rate limiting, the film composition is likely to be very sensitive to the deposition conditions compared to a situation where a physical process (such as diffusion or the precursor feed) is rate limiting.[14]

The potential benefits of CVD over other film deposition techniques are that CVD-derived films can be deposited under conditions that give conformal coverage, they can be deposited at low temperatures, there can be a high level of compositional control, thin layers can be deposited, the technique can be scaled to coat large areas uniformly, and there is also the possibility for area-selective deposition[13] as a result of the *chemical* nature of this process. The details of CVD and related chemical deposition processes such as atomic layer epitaxy (ALE), organometallic vapor-phase epitaxy (OMVPE), and others have been described elsewhere.[6]

Chemical vapor deposition is a process that critically depends on the chemistry of molecular species and therefore provides an opportunity for chemists to make a major contribution to an important area of materials science. This is exacerbated in the case of deposition of Group 2 metal-containing films because it is particularly difficult to prepare molecular compounds of the heavier elements as precursors with all the appropriate characteristics.

The major difficulty in depositing the materials just described has been in the availability of suitable precursors for the transport and deposition of the Group 2 elements. The origin of this problem lies in the tendency of the Group 2 elements to form oligomeric species as a result of their high coordination numbers (typically, Ca, Sr, and Ba prefer coordination numbers of 8–12) and the resulting small charge-to-size ratio, which is not conducive to the formation of high-vapor-pressure species.

The goal of this review is to describe and review the molecular design considerations that should be incorporated into the formation of Group 2 compounds that can be used as precursors for CVD; describe the appropriate synthetic and structural chemistry of Ca, Sr, and Ba compounds that fulfill some or all of these design requirements; describe the results of the vaporization and transport of Group 2 precursors used in the CVD of Group 2 metal-containing films; and finally, draw some conclusions for future directions of this field. A number of reviews cover various aspects of the deposition of Group 2 metal-containing films including ceramic superconductors[2,16–22] and ferroelectric materials,[1,23–27] and reference to these reviews is made where appropriate to avoid duplication. The figures have been redrawn based on the figures that appeared in the original literature. For detailed information see the original reference. For definitions of ligand abbreviations, see footnotes to tables.

II

LIGAND DESIGN FOR GROUP 2 COMPOUNDS AS PRECURSORS FOR CVD

There are a number of chemical and physical properties that should be considered when designing or selecting metal-containing compounds as precursors for CVD of metal-containing materials. These properties can be broadly categorized into precursor volatility, reactivity, and solubility (for reasons that are described later).

A. *Volatility*

For a metal-containing compound to be volatile, it is generally desirable that the molecular weight and lattice energy (if it is a crystalline solid) be minimized, and hence that it be preferably monomeric with low molecular weight. Strategies to reduce the degree of oligomerization can be classified into two classes, the use of sterically demanding ligands and the use of polydentate ligands. Both methods have advantages and disadvantages. The use of sterically demanding ligands results in a reduction of the tendency for oligomerization by rendering some of the coordination sites on the central metal ion sterically inaccessible to other ligands in the system. As a result, low-coordination-number species are produced, such as the three-coordinate compound $[Ba(N(SiMe_3)_2)_2]_2$.[28] There are several potential disadvantages of this approach. One is that the metal center is rendered kinetically inert by the large ligands that prevent oligomerization. However, it is likely that smaller molecules can penetrate the protective organic ligand shell and access the metal center. Consequently, these species are likely to be highly reactive toward small reagents such as H_2O or CO_2, which could result in handling difficulties, short shelf lives, short-term stability at elevated temperatures, and the potential for gas-phase reactions during CVD through reactions with other reagents or by-products. Another possible drawback is that increasing the size of the ligands also increases the molecular weight of the compound and may reduce the volatility compared to a lower-molecular-weight species. Group 2 metal compounds described in Section III that fit into this category are the metal amides, metal phosphides, metal cyclopentadienides, metal poly(pyrazolyl)borates, and most examples of the metal alkoxide compounds.

An alternative strategy involves the use of polydentate ligands to prevent oligomerization by satisfying the coordination number of the metal center. To date, this strategy has proved the more successful of the two in that polydentate ligand adducts of Group 2 compounds are more frequently used for the CVD of Group 2 element containing materials. Polydentate ligand adducts of the Group 2 elements have the advantages of low molecular weights, they are frequently monomeric, and they have relatively low reactivity toward other reagent such as H_2O because of the lack of available coordination sites. The disadvantage of the polydentate ligands is that they can bridge between two metal centers, which can lead to oligomerization and, at worst, polymerization, resulting in reduced volatility (and probably reduced solubility). Group 2 metal compounds described in Section III that fit into this category are some examples of the metal alkoxides, particularly the metal β-diketonate and carboxylate compounds.

The β-diketonate ligands, especially the fluorinated derivatives, have been used extensively in the deposition of metal-containing films.[29] These ligands have the benefits that they impart volatility, presumably through reduction in cohesive forces between molecules, and also that they are bidentate, occupying two coordination sites at the metal center for each ligand. In addition, the fluorinated derivatives are well known to impart the highest volatility among this family, but with the drawback in the case of Group 2 elements that thermal decomposition can result in formation of metal fluorides that can be severely detrimental to film properties. These aspects are described in more detail later.

B. Reactivity

Another aspect of precursor design that requires consideration is reactivity. Chemical vapor deposition presents a dichotomy in the sense that it is often desirable to deposit films at relatively low temperatures and high deposition rates. However, the precursor should have a long shelf life and be stable at temperatures where it exhibits a reasonable vapor pressure if traditional delivery methods are used. In the case of films containing Group 2 elements, it is often not necessary to deposit films below substrate tempertures of 500°C for many applications. In many cases, deposition of high-quality films at or below this temperature would be considered a significant breakthrough. Therefore, less reactive molecular precursors can be tolerated, which is likely to result in longer shelf lives. The sensitivity of the precursor to other environmental factors such as H_2O and air is another important practical consideration. Molecular precursors that are unreactive toward air and water offer a considerable practical advantage over those that are not.

The other aspect of reactivity that is crucial to the deposition of films by CVD is the nature of the surface reaction. This has been studied in more detail in some simpler systems, especially the deposition of elementary materials such as metals.[7] It is desirable that the precursors undergo surface reactions that result in the complete elimination of the organic supporting ligands to give a crystalline film with the desired stoichiometry, free of contamination. These are all key issues for Group 2 metal-containing films because the properties of these materials depend critically on the crystallinity of the films (e.g., ferroelectricity requires crystalline material with a polar space group), a high level of control over stoichiometry (e.g., solid solutions such as ternary $Ba_{1-x}Sr_xTiO_3$ often have improved properties relative to either individual binary counterparts such as $BaTiO_3$ or $SrTiO_3$), and a high level of purity, especially where the films will be incorporated into a Si-based device structure or the property relies on electrical characteristics where small amounts of charge carriers such as F are intolerable. Surface reactions have not been studied in detail for the deposition of Group 2 metal-containing films mainly because the highest priority has been to create a ligand set that is capable of transporting the precursor to the substrate. Other reagents such as H_2O, O_2, or N_2O have been used to help remove contaminants, which is reasonably successful because of high deposition temperatures (e.g., in the presence of O_2 this can lead to oxidation of organic fragments), for films where low levels (~1 at.%) of contaminants can be tolerated.

C. *Transport Processes: Precursor Delivery*

It is valuable to consider the method by which molecular species are delivered to CVD reactors before describing the chemistry and use of Group 2 compounds, because this can affect the design criteria for these compounds.

The traditional method of molecular precursor delivery involves heating a precursor in a container, often a bubbler, under either reduced pressure or a flow of carrier gas. This transports the precursor into the CVD reactor through evaporation of the precursor into the vapor phase. An example of a typical apparatus that has been used to deposit $BaTiO_3$ films is shown in Fig. 2.[30] This method has a number of limitations. Metal-organic compounds in general do not exhibit particularly high vapor pressures, and Group 2 compounds generally exhibit low vapor pressures. Mass flow controllers normally require ~1 torr gas pressure to operate correctly, and so a carrier gas is often used. In order to establish a sufficient vapor pressure to obtain a reasonable film deposition rate, it is generally necessary to heat the precursors to a temperature that, in the case of Group 2 compounds, is often as high as 200°C. This can lead to thermal decomposition, which is clearly detrimental to the deposition process. In addition,

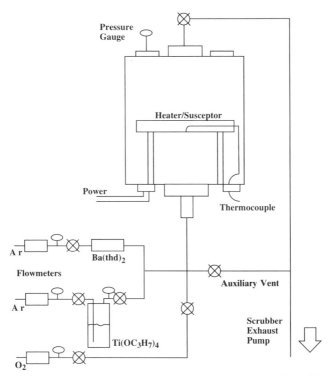

FIG. 2.   Schematic diagram of the MOCVD apparatus used to deposit $BaTiO_3$ thin films. (Redrawn from Ref. 30.)

the use of precursors that are solids at their evaporation temperature can lead to irreproducible transport rates because the surface area of the precursor is reduced with time. To avoid irreproducible transport of solid precursors, there has been significant interest in developing molecular species that are liquids at their evaporation temperature.

Because of the low volatility of Group 2 compounds, even more extreme methods to ensure that the precursor can reach the substrate have been used. In many cases described in Section IV, a container with the Group 2 element precursor was placed directly inside the CVD reactor adjacent to the substrate. A schematic diagram of a typical reactor setup is shown in Fig. 3.[31] This has often led to successful transport of the Group 2 precursor to the heated substrate surface, but it is unlikely that these conditions can be scaled up for practical applications.

One way to solve these problems is to employ an alternative method of precursor transport.[14] A number of different methods have been employed,

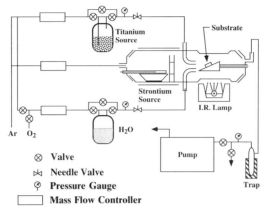

FIG. 3.   Schematic of a low-pressure CVD reactor used to deposit $SrTiO_3$ films emphasizing the proximity of the Sr source, $Sr(dpm)_2$, to the substrate. (Redrawn from Ref. 31.)

and they generally have the common feature that they rely on introduction of the precursor into the deposition system as a liquid. These methods include aerosol delivery, liquid delivery with flash evaporation, and supercritical fluid delivery. It should be noted that there is often some ambiguity in the literature considering this type of precursor delivery. It is crucial that the precursors evaporate into the gas phase before reaching the heated substrate; otherwise, the benefits of CVD are lost. In many cases involatile precursors have been delivered and films have been deposited by droplet or particle deposition, which would not be considered a CVD process. Aerosol delivery [also known by a variety of other names such as aerosol-assisted (AA)CVD or spray metal-organic (MO)CVD] involves the dissolution of the precursor(s) in a solvent, the generation of an aerosol from this liquid, and the delivery of the aerosol in a carrier gas into the deposition system.[32-39] The aerosol droplets evaporate and provide a method to deliver precursors that have low volatility and thermal stability. For example, it has been demonstrated that precursors that cannot be transported by conventional methods can be used to deposit high-quality films by AACVD.[33,40] This method also has the potential advantage that it is simple to deliver complex mixtures of precursors to substrates and exhibit a high level of control over film composition, provided that the precursors do not react with each other, or the solvent, at ambient temperatures.[33,41]

Liquid delivery with flash evaporation is a method in which a liquid containing the precursor(s) dissolved in a solvent is sprayed onto a heated porous ceramic frit where evaporation occurs, and the resulting gas-phase species are delivered into a CVD reactor.[42] This method has the potential advantages of aerosol delivery, but has the disadvantage that thermal decomposition can occur on the heated mesh, which may deplete the gas stream with one or more reagents. However, this method has been used successfully to deposit Group 2 metal-containing films, as described later in this review.

Supercritical fluid transport involves dissolving the precursor(s) in a supercritical fluid and spraying this liquid into a CVD reactor.[43-45] The rapid expansion of the supercritical fluid enhances evaporation of the solute and so aids transport of the precursor to the substrate. This method has the potential advantages of the other methods, but is more limited in the choice of solvent, e.g., CO_2. It should be noted that this method may not require the evaporation of the precursors and under these conditions would not be classified as a CVD method.

These precursor delivery methods have been described in more detail in other texts and are not discussed in further detail here.[14] The consequence of the availability of these methods is that the problems of thermal decomposition, low volatility, and prolonged heating of a precursor in a bubbler can be circumvented to some extent. However, this imposes another design criterion on the precursor, which is that it must have significant solubility in a solvent that does not play a detrimental role in film deposition.

III

SYNTHESIS AND CHARACTERIZATION OF GROUP 2 COMPOUNDS

In this section, the chemistry of Group 2 compounds, focusing on Ca, Sr, and Ba, is reviewed. The classes of complexes included in this review have either demonstrated ability or, in our opinion, show promise as precursors for Group 2 element containing films. Within these classes, we have not reviewed the chemistry of all the Group 2 element containing compounds that are known, only the chemistry that is relevant to the CVD of Group 2 element containing films. For example, the chemistry of Group 2 metal alkoxide compounds is much more widely studied than is apparent from this discussion,[46] because we have only included those compounds that have, or are likely to have, properties that are necessary for these compounds to be used as precursors for CVD. Since precursor volatility is a primary issue, we have excluded large molecular clusters (oligomeric species) except those in which volatility has been documented.

We have tried to identify and quantify trends in precursor properties, especially those required for CVD precursors. However, it is very difficult to provide a quantitative comparison of the key property, namely volatility, when there are often few quantitative data available in the literature. Most reports that focus on the synthesis or the use of Group 2 compounds in CVD experiments report the sublimation temperatures at a given pressure. There are very few reports of equilibrium vapor pressures, which are required to establish a quantitative comparison of the volatility between these species. The vapor pressure data that are available are discussed in Section IV. There is also some ambiguity in the literature concerning the sublimation data. In some reports, sublimation temperatures are reported as vapor pressures.[47] In other reports, it is not clear that the species being sublimed did so intact, which needs to be established by characterizing the sublimate.[48,49]

The paucity of vapor pressure data is not surprising because these compounds generally have low volatility compared to other species that have been used as precursors for CVD (e.g., WF_6 exhibits a vapor pressure of 1000 torr at room temperature!)[50] and are frequently plagued with problems such as lack of long-term thermal stability (ligand dissociation). Unfortunately, it is not possible to make an unambiguous direct comparison of the volatility of different species from comparison of their sublimation temperature, because no information is provided on the *rate of sublimation*. In this review, we report the sublimation temperatures and use this as an approximate guide to make qualitative comparisons of *apparent volatility* between precursors where we feel it is reasonable to do so.

The following discussion is organized according to a description of the synthesis of the Group 2 compounds, followed by descriptions of their chemical and physical properties, with particular reference to those properties most important to CVD processes. In some cases the solid-state structures of the precursors are presented because this provides an unambiguous verification of the degree of oligomerization of these species in the solid state. The gas-phase structures of most of these compounds, which are more relevant to this discussion, are not available. However, it is generally true that the degree of oligomerization of a compound is likely to be no larger in the gas phase than in the solid state.

A. *Group 2 Amides and Phosphides*

The only well-characterized amides and phosphides of the Group 2 metals, with the exception of the carbazoles, are those containing the bis(trimethylsilyl)amido(phosphido) ligand. This ligand has been successfully used to produce soluble, low-nuclearity, and low-coordination-number complexes of the heavier Group 2 elements. One feature of this steri-cally encumbering ligand is its ability to accommodate changes in ionic radius of the metal, from Mg (0.72 Å) to Ba (1.35 Å),[51] without imposing changes in the gross structural features of the compounds (*vide infra*). Although base-free compounds with this ligand are usually dimeric, except those of Be, monomeric species can be isolated by addition of Lewis bases such as THF and DME.

1. *Group 2 Amides and Phosphides: Syntheses*

The three general methods that have been used to prepare Group 2 amide complexes are transamination (transmetallation), metathesis, and redox. A listing of the Ca, Sr, and Ba amide and phosphide compounds that have been synthesized, along with some of their physical properties, is located in Table I.[28,52–67] By taking advantage of the high reduction potentials of the Group 2 metals, Bradley *et al.*, have employed transamination reactions to prepare Group 2 amides with Lewis bases [Eq. (1)].[54] The transamination–redox reaction uses mercuric bis(trimethylsilyl)amide, $Hg[N(SiMe_3)_2]_2$, as the amide transfer agent, with the corresponding Group 2 metal in THF.

$$M^0 + Hg^{II}[N(SiMe_3)_2]_2 \longrightarrow M^{II}[N(SiMe_3)_2]_2(THF)_2 + Hg^0$$

$$(M = Mg, Ca, Sr, and Ba) \tag{1}$$

The reaction takes place slowly in refluxing THF. The driving force is the oxidation of the Group 2 metal and concomitant reduction to form elemental mercury. Yields range from 50% for M = Ba up to 80% for M = Ca, Sr. Separation of the amide compounds from the elemental mercury is facilitated by their good solubility in THF. An analogous trans-amination reaction reported by Westerhausen and Schwartz involves the use of tin bis(trimethylsilyl)amide, $Sn[N(SiMe_3)_2]_2$, and the Group 2 metal for the preparation of the Lewis-base-free Mg [Eq. (2)], Ca and Sr, and Lewis base Ca and Sr [Eq. (3)] amide adducts.[52,53,56,68]

$$M^0 + Sn^{II}[N(SiMe_3)_2]_2 \longrightarrow \{M^{II}[N(SiMe_3)_2]_2\}_2 + Sn^0$$

$$(M = Mg, Ca, and Sr, solvent = toluene) \tag{2}$$

$$M^0 + Sn^{II}[N(SiMe_3)_2]_2 \longrightarrow M^{II}[N(SiMe_3)_2]_2(DME)_n + Sn^0$$

$$(M = Ca \ (n = 1) \ and \ Sr \ (n = 2), solvent = 1,2\text{-dimethoxymethane (DME)}) \tag{3}$$

Yields range from 40% for $Mg[N(SiMe_3)_2]_2(DME)_2$ up to 80–90% for $Sr[N(SiMe_3)_2]_2(DME)_2$ and $[Ca(N(SiMe_3)_2)_2]_2$. The advantage of this reaction over that of the mercury reaction is in the ability to isolate the

TABLE I

Ca, Sr, and Ba Amide and Phosphide Compounds and Their Physical Properties[a]

Compound	Color/state	m.p. (°C)	Sublim. temp. (°C)	Solubility[b]	Ref.
[Ca(N(SiMe$_3$)$_2$)$_2$]$_2$	light orange-brown solid	>120, dec.	*	s = b; ss = a	52, 53
Ca(N(SiMe$_3$)$_2$)$_2$(THF)$_2$	colorless solid	151–153[c]	*	s = a, b, c	54
		181[c]			53
Ca(N(SiMe$_3$)$_2$)$_2$(pyridine)$_2$	pale yellow solid	*	*	s = a, b	54
Ca(N(SiMe$_3$)$_2$)$_2$(DME)$_2$	colorless solid	>110, dec.	volatile[d]	s = a, b, d	53
Ca(N(SiMe$_3$)$_2$)$_2$(DME)	white solid	144, dec.	*	s = a, b, d	52, 53
[Ca(N(SiMe$_3$)$_2$)(μ_2-N(SiMe$_3$)$_2$)(THF)]$_2$	colorless solid	*	*	s = a, b, c	55
[Sr(N(SiMe$_3$)$_2$)$_2$]$_2$	colorless solid	164, dec.	*	s = b; ss = a	53, 56
Sr(N(SiMe$_3$)$_2$)$_2$(DME)$_2$	colorless solid	114, dec.	*	s = b, d	53, 56
Sr(N(SiMe$_3$)$_2$)$_2$(THF)$_2$	colorless solid	128–130	*	s = a, b, c	54, 57
[trans-Sr(N(SiMe$_3$)$_2$)$_2$(μ-1,4-dioxane]$_x$	colorless solid	*	*	s = b	57
[Ba(N(SiMe$_3$)$_2$)$_2$]$_x$	fine white powder[e]	186–188[e]	50 mtorr[f]	s = b, e; ss = a	58
[Ba(N(SiMe$_3$)$_2$)$_2$]$_2$	orange solid[e]	>150, dec.[e]	*	s = b; ss = a	28
					53
Ba(N(SiMe$_3$)$_2$)$_2$(THF)$_2$	colorless solid	88–90	*	s = a, b, c	28, 54
[(SiMe$_3$)$_2$N]$_2$BaL$_2$ L = 2,2'-bipyridyl	white solid		*	s = b; ss = a	58
[(SiMe$_3$)$_2$N]$_2$BaL$_2$ L = 4-t-butylpyridine	white solid		*	s = b; ss = a	58
[(SiMe$_3$)$_2$N]$_2$BaL$_2$ L = 4-(dimethylamino)pyridine	white solid		*	s = b; ss = a	58
[Ba(N(SiMe$_3$)$_2$)$_2$(THF)]$_2$	white powder	*	*	s = a, b	28
Ba(N(SiMe$_3$)$_2$)$_2$(DME)$_{2.2}$	green solid	76, dec.	*	s = b, d	53
Ca(N(SiMe$_3$)$_2$)(clox)(THF)$_3$[g]	colorless solid	*	*	s = b, c	59
Ca(N(SiMe$_3$)$_2$)(C$_5$Me$_5$)(THF)$_3$	solid	*	120, 10^{-6} torr[h]	s = b, c	60
Ca(N(SiMe$_3$)$_2$)(1,2,3,4-(i-Prop)$_4$C$_5$H)(THF)	white solid	160–165		s = b, c	61

[(C$_5$Me$_4$Et)Ca (μ_2-NSiMe$_2$CH$_2$CH$_2$SiMe$_2$)]$_2$	off-white solid	*	*	s = a, b, c	60
Li[Ba(C$_5$Me$_5$)$_2$N(SiMe$_3$)$_2$] · 2THF	solid	*	*	s = c; is = a	60
[(η^1-N-carbazolyl)$_2$Ca(pyridine)$_4$]· 3(pyridine)	yellow solid	*	*	s = f; is = a	62
(η^1-N-carbazolyl)$_2$Sr(NH$_3$)(DME)$_2$	colorless solid	*	*	s = f; is = a	62
(η^1-N-carbazolyl)$_2$Ba(DME)$_3$	colorless solid	*	*	s = f; is = a	62
{Ba[μ_2-OSi(t-Bu)$_2$((CH$_2$)$_3$NMe$_2$) [N(SiMe$_3$)$_2$]}$_2$	colorless solid	143–145	200, 10^{-4} torr, dec.	s = a, b	63
{Ba[μ_2-OSi(t-Bu)$_2$((CH$_2$)$_3$NMe$_2$) [N(SiMe$_3$)$_2$]][THF]}$_2$	colorless solid	94–95	200, 10^{-4} torr, dec.	s = a, b	63
[Ba(N(SiMe$_3$)$_2$)(P(SiMe$_3$)$_2$)(DME)]$_2$	colorless solid	143, dec.	*	s = b	64
Ca(P(SiMe$_3$)$_2$)$_2$(PMDETA)^i	colorless solid	102	*	s = b, c	65
Ca(P(SiMe$_3$)$_2$)$_2$(TMTA) · THF^i	colorless solid	89	*	s = b, c	65
Ca(P(SiMe$_3$)$_2$)$_2$(THF)$_3$	colorless solid	117	*	s = c	65
Sr(P(SiMe$_3$)$_2$)$_2$(THF)$_3$	colorless solid	*	*	s = b, c	66
[Sr(P(SiMe$_3$)$_2$)$_2$(THF)$_{1.5}$]$_2$ · (toluene)	colorless solid	87–89	*	s = b, c	66
Sr(P(SiMe$_3$)$_2$)$_2$(THF)$_3$	colorless solid	119	*	s = c	65
Ba(P(SiMe$_3$)$_2$)$_2$(THF)$_4$	colorless solid	70, dec.^{j}	*	s = c	65
[Ba(P(SiMe$_3$)$_2$)$_2$(THF)$_2$]$_n$	colorless solid	*	*	s = c	67

a * = information not available.

b s = soluble, ss = slightly soluble, is = insoluble; a = aliphatic hydrocarbons. b = aromatic hydrocarbons. c = tetrahydrofuran. d = dimethoxyethane. e = diethylether. f = pyridine.

c Different melting points reported for this compound.

d Sublimate is Ca(N(SiMe$_3$)$_2$)$_2$(DME).

e Different properties reported for this compound.

f Infrared heat source.

g clox is the alkoxide ligand: OC(Ph)$_2$CH$_2$C$_6$H$_4$-4-Cl.

h Sublimate is a (3:3:1) mixture of Ca(N(SiMe$_3$)$_2$)$_2$(1,2,3,4-(i-Prop)$_4$C$_5$H)(THF), Ca(1,2,3,4-(i-Prop)$_4$C$_5$H$_2$)$_2$, and Ca(N(SiMe$_3$)$_2$)$_2$(THF)$_n$.

i PMDETA and TMTA are 2,5,8-trimethyl1-2,5,8-triazanonane and 1,3,5-trimethyl-1,3,5-triazinane, respectively.

j Decomposes by solvent loss.

corresponding base-free compounds. The same authors have also prepared the Mg bis(trimethylsilyl)amide adducts and base-free compound by an acid–base reaction between a magnesium alkyl and amine, eliminating the alkane [Eq. (4)].[69,70]

$$MgX_2 + 2 (Me_3Si)_2NH \longrightarrow Mg[N(SiMe_3)_2]_2L_n + 2 XH$$

$(n = 0, X = butyl, solvent = hexane; n = 1, X = ethyl, solvent = Et_2O, L = Et_2O)$ (4)

Boncella et al. reported the synthesis of the base-free adduct, $[Ba[N(SiMe_3)_2]_2]_x$, by a halide metathesis reaction between anhydrous $BaCl_2$ and 2 equivalents of $NaN(SiMe_3)_2$ in diethyl ether according to Eq. (5).[58]

$$BaCl_2 + 2 NaN(SiMe_3)_2 \longrightarrow \{Ba[N(SiMe_3)_2]_2\}_x + 2 NaCl(s)$$ (5)

The degree of aggregation of this compound could not be clarified, and it is not known if it is monomeric or a "highly fluxional associated structure" by ^1H and ^{13}C NMR. A compound of similar composition has been synthesized by Caulton et al., using a different method; see the later discussion. And, the metal triflate $M(OSO_2CF_3)_2$ has also been used along with sodium bis(trimethylsilyl)amide to give the calcium and strontium bis-THF amide adducts and the heterobimetallic compound $[Ba(N(SiMe_3)_2)(\mu\text{-}N(SiMe_3)_2)Na(THF)_2]$.[71]

The compound $[Ba[N(SiMe_3)_2]_2]_x$ undergoes reaction with Lewis bases such as 2,2'-bipyridyl, 4-t-butylpyridine, and 4-(dimethylamino)pyridine according to Eq. (6).[58]

$$\{Ba[N(SiMe_3)_2]_2\}_x + nL \longrightarrow Ba[N(SiMe_3)_2]_2L_n$$

$(n = 1, L = 2,2'$ bipyridyl; $n = 2, L = 4$-t-butylpyridine and 4-(dimethylamino)pyridine)   (6)

Metathesis reactions between Group 2 metal alkoxides and lithium amides in hexane yield the metal amide and lithium alkoxide according to Eq. (7).[60] The lithium alkoxide is insoluble in hexane and is easily separated from the hexane-soluble amide.

$$M(OC_6H_2\text{-}t\text{-}Bu_2\text{-}2, 6\text{-}Me\text{-}4)_2(THF)_n + \tfrac{2}{3}[Li(\mu\text{-}N(SiMe_3)_2)]_3 \longrightarrow$$

$$\tfrac{1}{2}[M(N(SiMe_3)_2)(\mu\text{-}N(SiMe_3)_2)(THF)]_2 + [Li(\mu\text{-}OC_6H_2\text{-}t\text{-}Bu_2\text{-}2, 6\text{-}Me\text{-}4)(THF)]_2 (s)$$

$(M = Ca, n = 3; M = Ba, n = 4)$ (7)

The complexes $Ca(N(SiMe_3)_2)(clox)(THF)_3$,[59] $Ca(N(SiMe_3)_2)(C_5Me_5)(THF)_3$,[60] $Ca(N(SiMe_3)_2)(1,2,3,4\text{-}(i\text{-}Prop)_4C_5H)(THF)$,[61] and $[(C_5Me_4Et)Ca(\mu_2\text{-}NSiMe_2CH_2CH_2SiMe_2)]$,[60] have also been prepared by metathesis reactions. The mixed amide-cyclopentadienyl compound $Ca(N(SiMe_3)_2)(C_5Me_5)(THF)_3$ was obtained in 56% yield by a metathesis reaction between $(C_5Me_5)_2Ca(THF)_2$ and $LiN(SiMe_3)_2$ in THF with elimination of LiC_5Me_5. The analogous reaction with barium produces the adduct $Li[Ba(C_5Me_5)_2N(SiMe_3)_2] \cdot 2THF$.[60] The stability difference between barium–lithium aggregates and calcium–lithium aggregates appears to be the main determinant in the isolation of different products from these otherwise identical reactions. The other three compounds were obtained by metathesis reactions between the heteroleptic iodides $Cp(^{*,4i})MI(THF)_n$, $CaI(clox)(THF)_3$, and $KN(SiMe_3)_2$. Other mixed amide–alkoxide compounds have been prepared by partial alcoholysis of the amide–THF adducts $(M[N(SiMe_3)_2]_2(THF)_2)$, e.g.: $\{Ba[\mu_2\text{-}OSi(t\text{-}Bu)_2((CH_2)_3NMe_2)] [N(SiMe_3)_2]\}_2$ and its THF adduct from $Ba[N(SiMe_3)_2]_2(THF)_2$, and one equivalent of $HOSi(t\text{-}Bu)_2(CH_2)_3NMe_2$.[63] The mixed amide–phosphide compound $[Ba(N(SiMe_3)_2)(P(SiMe_3)_2)(DME)]_2$ was prepared in a similar manner, but with $HP(SiMe_3)_2$ and the metal amide in DME.[64]

Another route to Group 2 amides, as well as to other metal-organic compounds,[72] has been reported by Caulton et al. and involves the direct reaction of elemental Ba with the silylamine, $HN(SiMe_3)_2$, in THF to yield the Lewis base adduct $Ba[N(SiMe_3)_2]_2(THF)_2$ [Eq. (8)].[28]

$$Ba^0 + 2\ HN(SiMe_3)_2 \longrightarrow Ba^{II}[N(SiMe_3)_2]_2(THF)_2 + H_2 \qquad (8)$$

Ammonia is initially bubbled through the reaction mixture for 5 minutes and then bubbled several times again throughout the reaction time. The NH_3 reacts very quickly with the Ba metal, and it is presumed that the resulting species ($[Ba(NH_2)_2]_x$ or $Ba(NH_3)_x$) is responsible for further reaction with the silylamine. It should be noted that direct reaction between the metal and amine without the ammonia present resulted in poor yields, irreproducible results, and long (on the order of days) reaction times. If $Ba[N(SiMe_3)_2]_2(THF)_2$ is dissolved in toluene and stirred for 15 minutes with subsequent removal of the solvent in vacuo, the dimeric complex $\{Ba[N(SiMe_3)_2]_2(THF)_1\}_2$ is obtained. The species $Ba[N(SiMe_3)_2]_2(THF)_2$ may be completely desolvated by sublimation at an unspecified temperature at approximately 50 mtorr to give the dimeric Lewis-base-free complex $\{Ba[N(SiMe_3)_2]_2\}_2$.[28] The strontium compound $Sr[N(SiMe_3)_2]_2(THF)_2$ has been prepared by a similar redox reaction between the metal, activated by co-condensation of metal vapor and toluene, and $HN(SiMe_3)_2$ in THF.[57]

This ammonia route has also been employed by Schleyer et al. to prepare alkaline-earth metal carbazoles of the formula $M(\eta^1\text{-}N\text{-}carbazolyl)_2ML_n$ (M = Ca, Sr and Ba; L = py, DME, NH_3; n = 2–4).[62]

By taking advantage of the basicity differences between amines and phosphines, Westerhausen et al., prepared homoleptic Group 2 phosphide compounds with Lewis bases such as THF, PMDETA (2,5,8-trimethyl-2,5,8-triazanonane), and TMTA (1,3,5-trimethyl-1,3,5-triazinane) by reaction of the metal amides with two equivalents of $HP(SiMe_3)_2$.[65,66] Most of the compounds were obtained in yields of 70% or better. Drake *et al.* prepared the barium bis(bistrimethylsilyl)phosphide in 43% yield by an ammonia method similar to that described for the amide compounds.[67] The preparation of heterometallic phosphide compounds (Ba–Sn and Ca–Sn) has also been described.[64,73]

2. Group 2 Amides and Phosphides: Chemical and Physical Properties

All but one of the compounds in Table I are believed to be either monomeric or dimeric in solution. Coordination numbers for the amides range from a low of 3 in $[Ba(N(SiMe_3)_2)_2]_2$[28] to a high of 6 in $Sr[N(SiMe_3)_2]_2(DME)_2$,[56] for those with the same type of anionic ligands. Little information is available regarding volatility of these compounds, although many have low melting points in the range of 90–150°C. Most compounds show good solubility in ethers and aliphatic and aromatic hydrocarbons. This good solubility in hydrocarbon solvents is probably derived from the bis(trimethylsilyl)amido ligand and its ability to encapsulate the ionic metal centers in an organic sheath of trimethylsilyl groups.

The structures of a monomeric Ba base adduct and a dimeric Sr base-free complex are shown in Figs. 4[28] and 5,[56] respectively. Removal of one THF molecule from $Ba[N(SiMe_3)_2]_2(THF)_2$ (Fig. 4) by dissolving it in

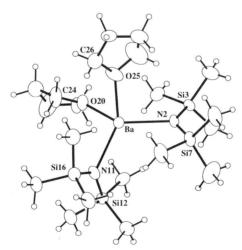

FIG. 4.   ORTEP drawing (50% probability level) of $Ba(N(SiMe_3)_2)_2(THF)_2$ showing the distorted tetrahedral ligand geometry around the barium atom. The Ba—N and Ba—O bond distances have average values of 2.592(6) and 2.731(20) Å, respectively, and the O–Ba–O, O–Ba–N, and N–Ba–N bond angles have a range of 91.44(2)–128.54(21)°, respectively. (Redrawn from Ref. 28.)

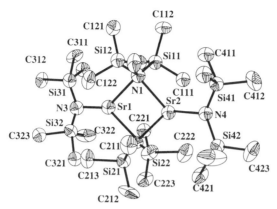

FIG. 5.   ORTEP drawing (30% probability level) of $[Sr(N(SiMe_3)_2)_2]_2$ showing the roughly trigonal disposition of nitrogen atoms around the metal center. (Redrawn from Ref. 56.)

toluene and stripping the solution to dryness gives the dimeric species $\{Ba[N(SiMe_3)_2]_2(THF)\}_2$, shown in Fig. 6.[28] This compound achieves the same metal coordination number as in $Ba[N(SiMe_3)_2]_2(THF)_2$ through bridging of two bis(trimethylsilyl)amide ligands. Upon removal of two THF molecules from $Ba[N(SiMe_3)_2]_2(THF)_2$ by sublimation, the dimeric species $\{Ba[N(SiMe_3)_2]_2\}_2$ is formed. It has a structure quite similar to that of the strontium base-free compound (Fig. 5) and $\{Ba[N(SiMe_3)_2]_2(THF)\}_2$, but with the THF ligands removed.[28] Most of the monomeric complexes discussed here are bidentate Lewis base adducts (Fig. 7), whereas the dimeric species are either the monosubstituted Lewis base or Lewis-base-free

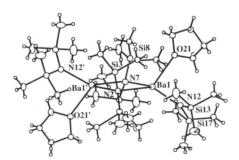

FIG. 6.   ORTEP drawing of $\{Ba[N(SiMe_3)_2]_2(THF)\}_2$. The molecule resides on a crystallographic two-fold axis that intersects the bridging nitrogen atoms. The geometry around each barium atom is crudely tetrahedral with internal angles in the range of 83.3(3)–136.1(3)°. (Redrawn from Ref. 28.)

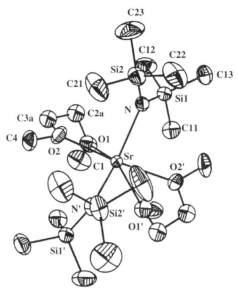

FIG. 7.   ORTEP drawing (50% probability level) of $Sr(N(SiMe_3)_2)_2(DME)_2$ showing the trans arrangement of the amide ligands and the octahedral geometry around the strontium atom. X–M–X' angles range from 61.5(2)° for the chelating cis DME oxygen atoms up to 160.9(3)° for the trans DME oxygen atoms. (Redrawn from Ref. 56.)

adducts (Fig. 6). This appears to be a general trend in aggregation for the heavier Group 2 amide compounds. However, greater aggregation can be achieved by employing suitably disposed ligands; e.g., addition of 1,4-dioxane to monomeric $Sr[N(SiMe_3)_2]_2(THF)_2$ gives the novel chain polymer $[trans\text{-}Sr(N(SiMe_3)_2)_2(\mu\text{-}1,4\text{-}dioxane)]_\infty$, where the strontium atoms are in a square planar geometry and the amide groups are trans to one another (Fig. 8).[57]

The carbazole compounds, $(\eta^1\text{-}N\text{-}carbazolyl)_2Ca(pyridine)_4$, $(\eta^1\text{-}N\text{-}carbazolyl)_2Sr(NH_3)(DME)_2$, and $(\eta^1\text{-}N\text{-}carbazolyl)_2Ba(DME)_3$ (Fig. 9),[62] are colorless or yellow solids that have good solubility in donor solvents such as pyridine.[62] They are monomeric in the solid state with the carbazolyl ligands bonded only through the nitrogen atom (η^1) to the metals. The metal coordination spheres are completed by incorporation of the Lewis bases pyridine, DME, and ammonia. A smooth increase in coordination number with metal size, from 6 (Ca) to 7 (Sr) to 8 (Ba), is observed in these compounds. The reason for the $\eta^1\text{-}N$-bonding of the carbazolyl ligand instead of multihapto bonding as occurs in the alkali metal compounds of carbazole and the fluorenyl compounds of the alkaline-earth metals (see Section III,G,2) is presently unclear.[62]

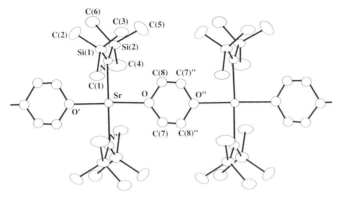

FIG. 8.   ORTEP drawing of [{*trans*-Sr(N(SiMe$_3$)$_2$)$_2$(μ-1,4-dioxane)}]$_x$ showing the trans arrangement of the amide groups and the repeating square planar unit. The O–Sr–N angles have values of 89.6(3)° and 90.4(3)°. (Redrawn from Ref. 57.)

Mixed anion compounds can be obtained by metathesis reactions or alcoholysis reactions as previously described. One of these, Ca(N (SiMe$_3$)$_2$)(clox)(THF)$_3$, is particularly interesting in that it has both an alkoxide and amide ligand in the same molecule and is also monomeric

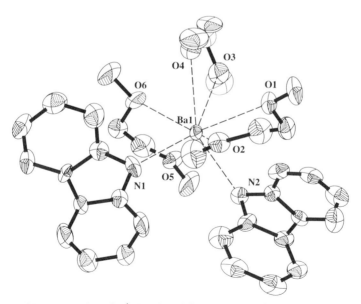

FIG. 9.   ORTEP drawing of (η^1-N-carbazolyl)$_2$Ba(DME)$_3$. The carbazole ligands are cis to one another, and the Ba—N bond distances are 2.749 and 2.752 Å, with an N–Ba–N angle of 98.1°. (Redrawn from Ref. 62.)

(Fig. 10).[59] The geometry is that of a distorted trigonal bipyramid with the large anionic ligands located in the equatorial plane. This heteroleptic compound is stable toward disproportionation to $Ca(clox)_2(THF)_3$ (see Section III,D) and $Ca[N(SiMe_3)_2]_2(THF)_2$. In aromatic solvents, NMR experiments show that only 5% of the homoleptic complexes exist, indicating that the equilibrium lies strongly in favor of the heteroleptic species.[59] However, the dimeric compounds $\{Ba[\mu_2\text{-}OSi(t\text{-}Bu)_2((CH_2)_3NMe_2)][N(SiMe_3)_2]\}_2$ and $\{Ba[\mu_2\text{-}OSi(t\text{-}Bu)_2((CH_2)_3NMe_2)][N(SiMe_3)_2][THF]\}_2$, containing terminal amide and bridging μ_2-siloxide groups, undergo ligand redistribution in THF to give the homoleptic species $Ba[N(SiMe_3)_2]_2$ $(THF)_2$ and $Ba[OSi(t\text{-}Bu)_2(CH_2)_3NMe_2]_2$.[63] The mono-ring amide $Ca[N(SiMe_3)_2](C_5Me_5)(THF)_3$ disproportionates with $K_d = 9.8 \times 10^{-3}$ in saturated benzene, while the more sterically encumbered $Ca(N(SiMe_3)_2)(1,2,3,4\text{-}(i\text{-}Prop)_4C_5H)(THF)$ (Fig. 11[61]) undergoes some disproportionation only after sublimation at 120°C and 10^{-6} torr (for more on disproportionation, see Section III,G,1).[61] $Ca(N(SiMe_3)_2)(1,2,3,4\text{-}(i\text{-}Prop)_4C_5H)(THF)$ is monomeric in the solid state with the calcium atom lying in the plane defined by the oxygen and nitrogen atoms and the ring centroid. Interestingly, an agostic interaction between the calcium center and one trimethylsilyl group

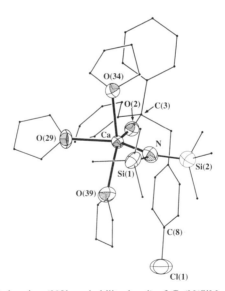

FIG. 10.   ORTEP drawing (30% probability level) of $Ca(N(SiMe_3)_2)(clox)(THF)_3$. The clox and amide groups are cis to one another in the equatorial plane. The Ca–N and Ca–O bond distances are 2.353(5) Å and 2.087(4) Å, respectively, and the N–Ca–O angle is 124.2(2)°. (Redrawn from Ref. 59.)

seems to be indicated by short $Ca \cdots Si$ and $Ca \cdots C_{Me}$ [Si(4) and C(48), Fig. 11] distances and a significantly more acute Ca–N(2)–Si(4) angle [105.9(6)°] as compared to the Ca–N(2)–Si(3) angle [128.9(6)°].

The phosphide compounds are colorless solids with low melting points and good solubility in ethers and aromatic solvents. To date, all Group 2 bis(trimethylsilyl)phosphide compounds have been isolated as base adducts and are either monomeric or dimeric, depending on the number and type of ligated base and the temperature at which they are isolated. The tridentate bases PMEDTA and TMTA, and THF, give monomeric compounds with calcium, while larger strontium forms a monomeric base adduct when four THF ligands are coordinated and a dimeric species when only three THF ligands are provided.[65,66] However, the species $Sr(P(SiMe_3)_2)_2(THF)_{4.5}$ and $[Sr(P(SiMe_3)_2)_2(THF)_{1.5}]_2$ are intimately related in solution through an equilibrium by which the dimer exists predominantly at high temperatures (+60°C) and the monomer at low temperatures (−60°C). The monomer $Sr(P(SiMe_3)_2)_2(THF)_4$ and its barium analogue (Fig. 12[65]) have distorted octahedral coordination geometry with the phosphide ligands trans and the four THF molecules bonded equatorially. The

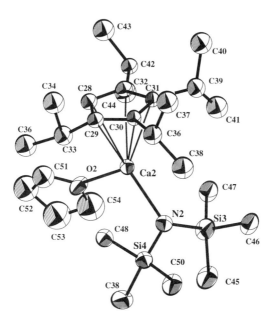

FIG. 11. ORTEP drawing (30% probability level) of $Ca(N(SiMe_3)_2)(1,2,3,4-(i\text{-}Prop)_4C_5H)$ (THF) (conformer B). The complex has a pseudotrigonal planar coordination geometry with Ca—C bond distances in the range 2.64(1)–2.70(1) Å and an average Ca—N bond distance of 2.295(7) Å. (Redrawn from Ref. 61.)

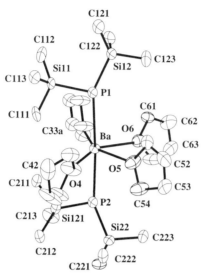

FIG. 12. ORTEP drawing (40% probability level) of $Ba(P(SiMe_3)_2)_2(THF)_4$. The complex has distorted octahedral geometry with average Ba—P and Ba—O bond distances of 3.174(22) and 2.719(26) Å, respectively. The P–Ba–P angle is 174.9(1)°. (Redrawn from Ref. 65.)

dimer exists in the solid state with one terminal and three μ_2-bridging phosphide ligands. The three THF molecules are located on one metal center. A more symmetrical structure is observed for the mixed amide–phosphide dimer, $[Ba(N(SiMe_3)_2)(P(SiMe_3)_2)(DME)]_2$, where two phosphide ligands bridge and one amide ligand is terminally bonded to each metal center. The coordination sphere around each metal is then completed by chelating DME ligands.[64]

B. Metallocenes

Metallocene complexes of the Group 2 elements have been known for many years. The first synthesized were those containing the $C_5H_5^-$ (Cp) ligand, reported by Ziegler in 1954. These compounds are thought to be polymeric (as is known for the case of $[CaCp_2]_n$) in the solid state. Apparently, the Cp ligand's steric demands (it is commonly considered to occupy three coordination sites when η^5-bonded) are not great enough to give low-nuclearity compounds for these large metals. As a result, a variety of derivatized cyclopentadiene systems have been used since in the preparation of metallocene compounds of Ca, Sr, and Ba, resulting in the synthesis of many monomeric complexes that show both good solubility in organic solvents and good volatility. In this regard, cyclopentadienide ligands are particularly well suited for the Group 2 metals, since they are capable of occupying up to three coordination sites and can be tailored, by employing sterically demanding substituents (i.e., $-Me$, $-SiMe_3$, $-i$-Pr, $-t$-Bu), to protect additional sites. In addition, the stability of the aromatic system ensures that the compounds will remain intact even at relatively high temperatures. For example, ferrocene, $Fe(\eta^5-C_5H_5)_2$, is thermally inert on heating to $>500°C$.[51] For recent reviews of Group 2 cyclopentadienyl chemistry, see Refs. 74 and 75.

1. Metallocenes: Syntheses

Group 2 metallocenes have the general formula $M(Cp)_2L_n$, where Cp is a derivatized or underivatized cyclopentadienyl ring (see Table II) and L is a two- or four-electron Lewis base donor such as THF, PR_3, Et_2O, or DME ($n = 0–2$). Several methods have been used to prepare alkaline-earth metallocene complexes.

One very common method of preparation is by reaction of MI_2 (M = Ca, Sr, or Ba) with the Na, K, or Li salt of the desired cyclopentadienyl ligand according to Eq. (9).[76–86]

$$MI_2 + 2\ M'Cp \longrightarrow M(Cp)_2 + 2\ M'I(s)$$

$$(M = Ca, Sr, or Ba, M' = Na, K, or Li) \qquad (9)$$

This reaction, when carried out in the absence of a coordinating solvent such as THF or Et_2O, yields the base-free form of the complex ($M(Cp)_2$). When a Lewis base such as THF or Et_2O is present in the reaction mixture, an adduct of the formula $M(Cp)_2L$ or $M(Cp)_2L_2$ is generally obtained [Eq. (10)].

$$MI_2 + 2\ M'Cp + L \longrightarrow M(Cp)_2(L)_n + 2\ M'I(s)$$

$$(M = Ca, Sr, or Ba, M' = Na, K, or Li)$$

$$(L = Lewis\ base,\ n = 1–2) \qquad (10)$$

Once Lewis base adducts of these metallocenes have formed, the coordinated bases can be removed by either repeated sublimation[76] or the "toluene reflux" method.[82,83,87] The toluene reflux method involves dissolving the compound in toluene, and then distilling off the toluene at its reflux temperature over several hours (alternatively, the toluene can be removed under reduced pressure). The procedure can be repeated as many times as is necessary to remove the base. However, for the method to work properly, the compound should be completely dissolved in the toluene or another refluxing solvent, and the solvent should have a higher boiling point than the coordinated Lewis base that is to be removed. The method works well for removal of all of the coordinated base from most compounds. For some compounds, only partial removal of the Lewis base can be achieved by successive sublimations; partially solvated base complexes of formula $M(Cp)_2L_n$, where n is between zero and one, are then usually obtained.[76] The general trend reported by Gardiner and Raston and others for the ease in removal of Lewis bases from Ca, Sr, and Ba metallocenes is Ca $\ll$ Sr $<$ Ba.[77,88–90] One explanation for this ordering is that the lower charge density of the larger metal ions results in weaker binding of the Lewis bases. The greater coordination needs of the larger metal ions are apparently of much less importance.

Reaction of cyclopentadiene directly with metal chips under reflux or in liquid ammonia has been used in the preparation of the $M(C_5H_5)_2$[91,92] and $Ca(C_5H_4Me)_2(DME)$[93] compounds, respectively. However, a more general reaction method is co-condensation of metal vapor, the vapor of the cyclopentadiene of choice, and the solvent, which is typically hexane or toluene, according to Eq. (11).[88,89,94]

$$M^0 + CpH \longrightarrow M^{II}(Cp)_2 + H_2$$

$$(M = Ca, Sr, or Ba; CpH = cyclopentadiene of choice) \qquad (11)$$

The co-condensed reactants and solvent form a frozen matrix on a liquid-nitrogen ($-195.8°C$) cooled reaction vessel and react on warming to room temperature to give the metallocene compound and hydrogen gas. One major advantage of this method is in the formation of base-free complexes. The base-free complexes, however, can then easily be converted to their respective Lewis base adducts by stirring them in solutions containing the Lewis base according to Eq. (12).[83,89] Reaction of the metal, activated with $HgCl_2$, and 6,6-dimethylfulvene in THF gives base and base-free *ansa*-metallocenes $(M(R_2R_2'(C_5H_4)_2)$ in moderate yield. These base-free compounds can be likewise converted to base adducts by stirring them in solutions containing a base.[95]

$$M(Cp)_2 + L \longrightarrow M(Cp)_2(L)_n$$

$$(M = Ca, Sr, or Ba; Cp = cyclopentadiene of choice; n = 1 or 2)$$

(L = THF, OEt_2, DME, bipy, PEt_3, 2,6-xylylisocyanide, pyrazine, or $Me_3SiCCCCSiMe_3$)

$$(12)$$

There have also been a few reports of the synthesis of alkaline-earth metallocenes by other methods. For example, $Ca(C_5Me_5)_2(THF)_2$, $Ca(1,3$-t-$Bu_2C_5H_3)_2(THF)$, $Ca((SiMe_3)C_5H_4)_2(THF)$, and $Ca(1,3$-$(SiMe_3)_2C_5H_3)_2$

(THF) were prepared by reaction of the corresponding cyclopentadiene ligand with calcium amide and THF in liquid ammonia.[96] The barium amide compound $Ba[N(SiMe_3)_2]_2(THF)_2$ has been used to deprotonate $C_5Ph_4H_2$ to give solvated octaphenylbarocene, $Ba(C_5Ph_4H)_2(THF)$.[97] And, for cyclopentadiene ligands that are strong acids such as $HC_5(CO_2Me)_5$, an acid–base reaction in water can be used to prepare $M(C_5(CO_2Me)_5)_2$ compounds as shown in Eq. (13).[98]

$$MCO_3 + 2\ HC_5(CO_2Me)_5 \longrightarrow M(C_5(CO_2Me)_5)_2 + H_2O + CO_2(g)$$

$$(M = Ca,\ Ba) \tag{13}$$

The reaction is driven to completion by evolution of $CO_2(g)$. For M = Sr, a halide metathesis reaction between $Tl(C_5(CO_2Me)_5)$ and $SrCl_2 \cdot 6H_2O$ in methanol yielded $Sr(C_5(CO_2Me)_5)_2 \cdot 3H_2O$.[98]

The preparation of $Be(C_5Me_5)Cl$ has been reported from the reaction of $Mg(C_5Me_5)_2$[81] and $BeCl_2(OEt_2)_2$.[83] The reaction was conducted in the hope that $Be(C_5Me_5)_2$ would be the product. However, it appears that two C_5Me_5 ligands, although unable to sterically saturate the heavier alkaline earth metals, are too large to fit around the much smaller beryllium atom without causing serious repulsive interactions.

2. Metallocenes: Chemical and Physical Properties

The metallocene complexes of the Group 2 elements have been well characterized by spectroscopic and physical techniques to determine their structures and degree of aggregation. The Ca, Sr, and Ba metallocene complexes that have been prepared and characterized are listed in Table II.[47,60,76–86,88,89,91–104] From Table II, it is apparent that many different substituted cyclopentadiene ring systems have been used in the preparation of a wide variety of metallocene complexes. Some general trends in reactivity, volatility, and solubility can be extracted from Table II. First, with the exception of $(Cp^{4i})_2Ca$, all base-free metallocenes will readily accept a Lewis base donor, suggesting that even with large substituents on the Cp rings the compounds are not sterically or coordinatively saturated. Second, lower degrees of molecular aggregation in the solid state can be associated with greater volatility and solubility. For example, Gardiner and Raston have compared the volatility and solubility of $Ba(C_5H_5)_2$ with that of $Ba(C_5Me_5)_2$.[88] In terms of volatility, $Ba(C_5H_5)_2$, which is known to be polymeric in the solid state, sublimes at 420–460°C under high vacuum,[92] whereas $Ba(C_5Me_5)_2$, with only relatively weak intermolecular interactions in the solid state, sublimes under much more mild conditions, 130–140°C and 10^{-3} torr.[88] In terms of solubility, $Ba(C_5H_5)_2$ is insoluble in all noncoordinating and many coordinating solvents, whereas

TABLE II

Ca, Sr, and Ba Metallocene Compounds and Their Physical Properties[a]

Compound	Color/state	m.p. (°C)	Sublim. temp. (°C)	Solubility[b]	Ref.
$[Ca(C_5H_5)_2]_n$	colorless solid	*	265, high vacuum	s = a, b; is = c, d, e	91, 92, 94, 99, 100
$[Sr(C_5H_5)_2]_n$	colorless solid	*	360–440, high vacuum	s = a; is = c, d, e	92, 94
$[Ba(C_5H_5)_2]_n$	colorless solid	*	420–460, high vacuum	s = a; is = c, d, e	92, 94
$Ca(C_5H_5)_2(THF)_2$	white solid	>100, dec.	*	s = a, b	84, 96
$Sr(C_5H_5)_2(THF)$	white solid	>400	*	s = a; is = b	84
$Ba(C_5H_5)_2(THF)_{0.25}$	pale pink solid	>400	*	s = a; is = b	84
$Ca(C_5Me_5)_2$	colorless solid	207–210	75, 10^{-3} torr	s = c, d, e	81, 82, 101
$Sr(C_5Me_5)_2$	white solid	216–218	100–110, 10^{-3} torr	s = c; e; ss = d	77, 80, 83
$Ba(C_5Me_5)_2$	white solid	265–268	130–140, 10^{-3} torr	s = c; e; ss = d	77–80, 83
$Ca(C_5Me_5)_2(THF)_2$	white solid	208–210[c,d] 168	*	s = b, c; is = d	84, 96
$Sr(C_5Me_5)_2(THF)_2$	white solid	191–192[e]	100, 4–5 torr	s = b, c; is = d	84, 102
$Ba(C_5Me_5)_2(THF)_2$	white solid	185[c,d] 187–191	130, 4–5 torr	s = b, c, f, g; is = d	83, 84, 102
$Ca(C_5Me_5)_2(OEt_2)$	white solid	186–191	*	s = e, f	81
$Sr(C_5Me_5)_2(OEt_2)$	white solid	210–215	*	s = e, f	83
$Ca(C_5Me_5)_2(DME)$	colorless solid	260	*	s = e; ss = f	96
$Ca(C_5Me_4Et)_2(THF)$	white solid	*	*	s = b, c, f; ss = d	60
$Ca(C_5Me_5)_2(bipy)$	orange-red solid	>300	*	is = c, h	83
$Sr(C_5Me_5)_2(bipy)$	orange-red solid	*	*	is = c, d	83
$Ba(C_5Me_5)_2(bipy)$	orange-red solid	*	*	is = c, d	83
$Ca(C_5Me_5)_2(2,6\text{-xylylisocyanide})_2$	yellow solid	213–215 dec.	*	s = c, f, g	83
$Sr(C_5Me_5)_2(2,6\text{-xylylisocyanide})_2$	solid	230–233 dec.	*	s = c, f, g	83
$Ba(C_5Me_5)_2(2,6\text{-xylylisocyanide})_2$	solid	211–212 dec.	*	s = c, f, g	83
$[Ba(C_5Me_5)_2]_2(\mu\text{-pyrazine})$	red solid	226–227	100, 10^{-3} torr	s = c, g	103
$[Ca(C_5Me_5)_2]_2(\mu\text{-pyrazine})$	purple solid	>300	*	s = c, g	103

Compound	Form	mp (°C)	Sublimation	Spectra	Ref.
[Ca(C$_5$Me$_5$)$_2$]$_2$(μ-2,3,5,6-tetramethypyrazine)	red solid	320–325 dec.	*	s = c, g	103
[Ba(C$_5$Me$_5$)$_2$]$_2$(μ-2,3,5,6-tetramethypyrazine)	orange solid	>325	*	s = c, g	103
Ca(C$_5$Me$_5$)$_2$(PEt$_3$)	white solid	111–113	*	s = c, f, g, h	83
Sr(C$_5$Me$_5$)$_2$(PEt$_3$)	white solid	140–142	*	s = c, f, g, h	83
Ba(C$_5$Me$_5$)$_2$(PEt$_3$)	white solid	250–255	*	s = c, f, g, h	83
Ca(C$_5$Me$_5$)$_2$(Me$_3$SiCCCSiMe$_3$)	orange-yellow solid	*	*	s = g, i	104
Ca(C$_5$H$_4$Me)$_2$(DME)	colorless solid	*	*	s = e	93
Ca(1,2,4-(i-Prop)$_3$C$_5$H$_2$)$_2$	sticky yellow solid	*	100, 10^{-6} torr	s = b, d, e, i	76
Ca(1,2,4-(i-Prop)$_3$C$_5$H$_2$)$_2$(THF)	viscous yellow oil	*	*	s = g	76
Sr(1,2,4-(i-Prop)$_3$C$_5$H$_2$)$_2$	yellow-brown viscous oil	*	120, 10^{-6} torr	s = g	76
Sr(1,2,4-(i-Prop)$_3$C$_5$H$_2$)$_2$(THF)	off-white solid	*	120, 10^{-6} torr	s = b, g, i	76
Sr(1,2,4-(i-Prop)$_3$C$_5$H$_2$)$_2$(THF)$_{0.6}$	sub. prod. of above off-white solid yellow oil	*	*	*	76
Sr(1,2,4-(i-Prop)$_3$C$_5$H$_3$)$_2$(THF)$_2$	cream-colored solid	*	120, 10^{-6} torr	ss = b, i	76
Ba(1,2,4-(i-Prop)$_3$C$_5$H$_3$)$_2$	off-white waxy solid	92–94	120, 10^{-6} torr	s = c, d, e	76
Ba(1,2,4-(i-Prop)$_3$C$_5$H$_2$)$_2$(THF)	pale yellow solid	*	110, 10^{-3} torr	s = (1:1, b:i), g	76
Ba(1,2,4-(i-Prop)$_3$C$_5$H$_2$)$_2$(THF)$_2$	solid	107–120 dec.	*	s = (1:1, b:i)	76
Ca(1,2,3,4-(i-Prop)$_4$C$_5$H$_2$)	white solid	196–200	190, 10^{-6} torr	s = b, c, e, g, i	86
Ba(1,2,3,4-(i-Prop)$_4$C$_5$H$_2$)	white solid	149–150	90, 10^{-2} torr (start) 160, 10^{-3} torr (rapid)	s = b, c, e, g, i	86
Ca((SiMe$_3$)C$_5$H$_4$)$_2$(THF)	colorless solid	125–132	*	s = b	96
Ca((SiMe$_3$)C$_5$H$_4$)$_2$(DME)	colorless solid	*	*	s = f	96
Ca(1,3-(SiMe$_3$)$_2$C$_5$H$_3$)$_2$	solid	189	180, 10^{-3} torr	s = c, e, f, g	89
Sr(1,3-(SiMe$_3$)$_2$C$_5$H$_3$)$_2$	solid	150	200, 10^{-3} torr	s = c, e, f, g	89
Ba(1,3-(SiMe$_3$)$_2$C$_5$H$_3$)$_2$	solid	222	340, 10^{-3} torr	s = c, e, f, g	89
Ca(1,3-(SiMe$_3$)$_2$C$_5$H$_3$)$_2$(THF)	solid	186^c 189–190	*	s = b, f	89, 96
Ca(1,3-(SiMe$_3$)$_2$C$_5$H$_3$)$_2$(DME)	colorless solid	*	*	s = f	96
Sr(1,3-(SiMe$_3$)$_2$C$_5$H$_3$)$_2$(THF)	solid	145	*	s = f	89
Ba(1,3-(SiMe$_3$)$_2$C$_5$H$_3$)$_2$(THF)	solid	>260	*	ss = b, f	89

(continued)

TABLE II (continued)

Compound	Color/state	m.p. (°C)	Sublim. temp. (°C)	Solubility[b]	Ref.
Ca(t-BuC$_5$H$_4$)$_2$	colorless solid	301–304	180, 10^{-4} torr	ss = f, g, i; s = e	88
Sr(t-BuC$_5$H$_4$)$_2$	colorless solid	375–380	260, 10^{-4} torr	ss = f, g, i; s = e	88
Ba(t-BuC$_5$H$_4$)$_2$	glasslike solid	320, dec.	320, 10^{-4} torr	ss = f, g, i	88
Ca(t-BuC$_5$H$_4$)$_2$(THF)$_2$	colorless solid	325–330	180, 10^{-4} torr	s = b (5 g/L); ss = f	88
Sr(t-BuC$_5$H$_4$)$_2$(THF)$_2$	colorless solid	375–376	260, 10^{-4} torr^f	s = b (5 g/L); is = f	88
Ba(t-BuC$_5$H$_4$)$_2$(THF)$_2$	glasslike solid	*	320, 10^{-4} torr^f	is = b, d	88
Ca(1,3-t-Bu$_2$C$_5$H$_3$)$_2$(THF)	colorless solid	125–130	*	s = b; ss = f	96
Ca(C$_5$(CO$_2$Me)$_5$)$_2$	white solid	295–298	*	s = j, k	98
Sr(C$_5$(CO$_2$Me)$_5$)$_2$ · 3H$_2$O	white solid	218, dec.	*	s = j, k	98
Ba(C$_5$(CO$_2$Me)$_5$)$_2$	white solid	>300	*	s = j, k	98
BaC$_5$H$_4$(CH$_2$CH$_2$O)$_2$CH$_3$)$_2$	brown liquid	—	100, 10^{-3} torr (reported as vapor pressure)	s = c, d	47, 85
Ba(C$_5$Ph$_4$H)$_2$(THF)	yellow solid	156–184, dec.	110–120, 10^{-6} torr, dec.	s = b, g	97
ansa-Ca(Me$_4$C$_2$(C$_5$H$_4$)$_2$)	colorless solid	>163, dec.	*	s = b, g; is = i	95
ansa-Ca(Me$_4$C$_2$(C$_5$H$_4$)$_2$)(t-BuNCHCHN-t-Bu)	colorless solid	157	*	s = b, g	95
ansa-Ca(Me$_2$Et$_2$C$_2$(C$_5$H$_4$)$_2$)	colorless solid^g	170, dec.	*	s = b	95
ansa-Ca(Me$_2$Et$_2$C$_2$(C$_5$H$_4$)$_2$)(t-BuNCHCHN-t-Bu)	colorless solid^g	166	*	s = b, g	95
ansa-Sr(Me$_4$C$_2$(C$_5$H$_4$)$_2$) · 2THF	colorless solid	>181, dec.	*	s = b	95

^{a} * = information not available.

^{b} s = soluble, ss = slightly soluble, is = insoluble; a = dimethyl sulfoxide, b = tetrahydrofuran, c = aromatic hydrocarbons, d = alkanes, e = ethers. f = benzene, g = toluene, h = pentane, i = hexane, j = alcohols, k = water.

^{c} Two melting points reported.

^{d} Solvent loss at 100°C.

^{e} Decomposes with solvent loss.

^{f} Sublimate is M(t-BuC$_5$H$_4$)$_2$.

^{g} Mixture of isomers.

$Ba(C_5Me_5)_2$ has good solubility in aromatic hydrocarbons and ethers and is even slightly soluble in alkanes. Although decreases in sublimation temperatures have been achieved by introduction of sterically demanding substituents on the Cp ring and would portend widespread use of these compounds as precursors in MOCVD (see Section IV,B), it is difficult to quantitatively assess these data in that respect because data on rates of sublimation are lacking in the literature.

Single-crystal X-ray diffraction data confirm that most of the metallocenes in Table II are either monomeric or dimeric and have only weak intermolecular interactions in the solid state, which is consistent with their higher volatilities. One exception to the preceding statement, as noted earlier, is $[CaCp_2]$; the unit cell diagram is presented in Fig. 13.[75,91] In this compound, each calcium atom interacts with four Cp rings in an η^1, η^3, η^5, η^5 fashion; discrete $CaCp_2$ units do not exist in the crystal. However, by addition of five methyl groups to each Cp ring (Cp*), the monomeric metallocene $CaCp_2^*$ can be obtained. The structure is shown in Fig. 14.[78] The barium analogue exists as a one-dimensional chain polymer linked together by $Ba-Me_{Cp}$ interactions.[78,79] This greater degree of aggregation is no doubt a consequence of the larger size of barium. $CaCp_2^*$ readily accepts Lewis bases to give compounds such as $Ca(C_5Me_5)_2(THF)_2$,[84,96] $Ca(C_5Me_5)_2(bipy)$,[83] $Ca(C_5Me_5)_2(2,6-xylylisocyanide)_2$,[83] $Ca(C_5Me_5)_2(PEt_3)$,[83] etc. The barium and strontium base-free de-

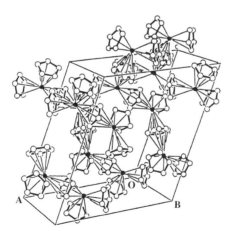

FIG. 13. Unit cell of $[CaCp_2]_n$. Each calcium atom interacts with four Cp rings in an η^1, η^3, η^5, η^5 fashion; three rings are trigonally disposed around the calcium atom. One of the η^5-C_5H_5 ligands is rotationally disordered (right center of cell). The Ca–C distances are 2.75(1) Å (η^5), 2.85(1) Å (η^5), 2.701(1), 2.789(2), and 2.951(2) Å (η^3), and 3.100(2) Å (η^1). (Redrawn from Refs. 75, 91.)

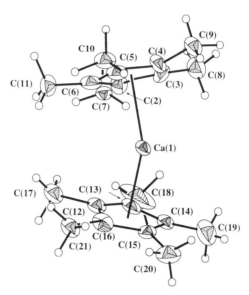

FIG. 14.   ORTEP view of CaCp$_2^*$ showing the bent structure found in the Ca, Sr, and Ba metallocene compounds. The molecule has an average Ca—C bond distance of 2.64(1) Å and a Cp$_{centroid}^*$–Ca–Cp$_{centroid}^*$ bond angle of 147.7°. (Redrawn from Ref. 78.)

camethylmetallocenes behave similarly. By introduction of three or four i-Pr groups to the Cp ring, in an attempt to form more "encapsulated" metallocenes, the compounds M(Cp^{3i})$_2$ and M(Cp^{4i})$_2$ were obtained.[76,86,105] Interestingly, the greater steric requirements of six i-Pr groups in Ca(Cp^{3i})$_2$, as compared to 10 methyl groups in Ca(C$_5$Me$_5$)$_2$, occludes the calcium atom to such a degree that only monosolvated complexes such as Ca(Cp^{3i})$_2$(THF)[76] can be isolated. When four i-Pr groups are placed on each ring, as in Ca(Cp^{4i})$_2$, it becomes impossible even to isolate a monosolvated species.[75,86] The molecular structure of Ca(Cp^{4i})$_2$ is presented in Fig. 15,[86] and a space-filling model comparing Ca(Cp^{4i})$_2$ and CaCp$_2^*$ is shown in Fig. 16.[86] From Fig. 16, one can clearly see why the degree of THF solvation in these metallocenes is CaCp$_2^*$(THF)$_2$ > Ca(Cp^{3i})$_2$ (THF) > Ca(Cp^{4i})$_2$(THF)$_0$ and related to the steric bulk of the rings (Cp^{4i} > Cp^{3i} > Cp*).[76] The Cp^{3i} compounds of barium still accept two THF molecules to give Ba(Cp^{3i})$_2$(THF)$_2$,[76] and even Ba(Cp^{4i})$_2$ interacts with small bases.[86] It is not surprising that the barium metallocene compound with one of the highest apparent volatilities, Ba(Cp^{4i})$_2$ (90, 10^{-2} torr), also has the lowest degree of aggregation of any base-free barium metallocene.[86] Its apparent volatility based on sublimation data compares favorably with that of many barium β-diketonate compounds.

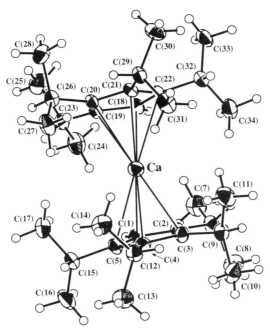

FIG. 15.   ORTEP view of Ca(Cp^{4i})$_2$. Although the average Ca—C bond distance is identical to that of CaCp$_2^*$, the Cp$_{centroid}^{4i}$–Ca–Cp$_{centroid}^{4i}$ angle of 162.3° is 14.5° more obtuse than the corresponding angle of CaCp$_2^*$. (Redrawn from Ref. 86.)

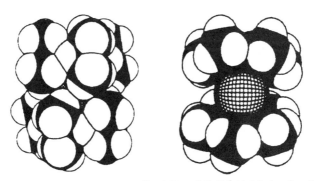

FIG. 16.   Space-filling models of Ca(Cp^{4i})$_2$ (left) and CaCp$_2^*$ (right) showing the degree of encapsulation of the metal centers in each compound (cross-hatched area). (Redrawn from Ref. 86.)

Mass spectral data have shown that base-free metallocenes, with the possible exception of the $[MCp_2]_n$ compounds, exist as true monomers in the gas phase (e.g., $Ba(C_5Me_5)_2$) with the most prominent peak corresponding to the molecular ion.[81–83] The mass spectrum of $[CaCp_2]_n$ has not only the molecular ion (M$^+$) and $CaCp^+$ fragment peaks, but also more complex fragments such as $[Ca_2Cp_2]^+$ and $[Ca_2Cp_3]^+$.[91] One feature of base-free metallocenes that does not appear to change on going from the solid state to the gas phase, or with derivatization of the Cp ring, is their bent structures. In the solid, their deviation from linearity tends to increase with an increase in the metal radius. This is demonstrated by the $Cp^*_{centroid}$–M–$Cp^*_{centroid}$ angles of $Ca(C_5Me_5)_2$ and $Ba(C_5Me_5)_2$, which have values of 147.7° and 130.9°, respectively.[78] The Ca compound has an angle that deviates 32.3° from 180°, while the angle in the Ba analogue deviates 49.1°. Molecular structure determinations of $Ca(C_5Me_5)_2$,[81] $Sr(C_5Me_5)_2$,[77,80] and $Ba(C_5Me_5)_2$[77,80] by gas-phase electron diffraction indicates that all three compounds have thermal averaged bent structures with $Cp^*_{centroid}$–M–$Cp^*_{centroid}$ angles near 150°. Bent structures in the solid have previously been attributed either to intermolecular interactions, e.g., Ba–Me$_{Cp}$ in $Ba(C_5Me_5)_2$, or to crystal packing forces. However, such factors should be of little importance in the gas phase. So the origin of the nonlinearity in these alkaline-earth metallocenes in the gas phase is unclear. Many calculations have been conducted, and a number of explanations have been put forth.[77,106–109] One recent calculation has suggested that the van der Waals attractive interactions between the methyl groups of the Cp* rings may cause the bending.[80,110] The calculated energy difference between the bent and linear orientations was placed at <2.1 kcal.[110] This calculation, using a molecular mechanics force field approach, was able to predict accurately the bend angle in $Ba(Cp^{4i})_2$. The assertion by the authors that $Ba(Cp^{5i})_2$ should be linear because of interligand methyl–methyl repulsions will have to await this compound's synthesis and structural characterization to be verified.[110]

As mentioned earlier, many Lewis base metallocenes have been synthesized and structurally characterized. An examination of their structures shows that they are all predictably bent, more so than their base-free congeners, and have one two-electron, two two-electron or one four-electron Lewis base donor. $Ca[1,3-(SiMe_3)_2C_5H_3]_2(THF)$,[89] shown in Fig. 17, is an example of a Lewis base monosubstituted Ca metallocene complex, whereas $Sr(t-BuC_5H_4)_2(THF)_2$,[88] shown in Fig. 18, is an example of a disubstituted metallocene. Even weak donor base such as alkynes, $Me_3SiCCCCSiMe_3$, will bind to the alkaline-earth metal centers (Fig. 19).[104] In this case, the weak donor is easily displaced by stronger bases.

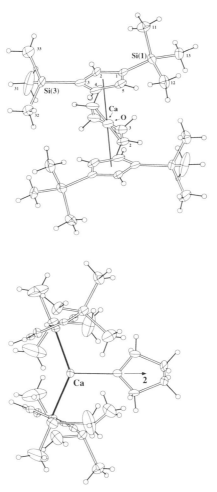

FIG. 17.   ORTEP diagrams (20% probability level for heavy atoms, arbitrary for H-atoms) of Ca(1,3-(SiMe$_3$)$_2$C$_5$H$_3$)$_2$(THF) from two different side perspectives, showing the trigonal planar geometry around the metal. (Redrawn from Ref. 89.)

The lone set of dimers in Table II, [M(C$_5$Me$_5$)$_2$]$_2$(μ-pyrazine) (Fig. 20; M = Ba), were synthesized from the base-free metallocenes by addition of pyrazine to MCp$_2^*$ (1 : 2), analogous to the dioxane reaction for the amide compounds (Fig. 8).[103] The pyrazine compounds are intensely colored owing to both charge transfer and intraligand electronic transitions. The pyrazine ligand has been shown to be labile by uv-vis spectroscopy.

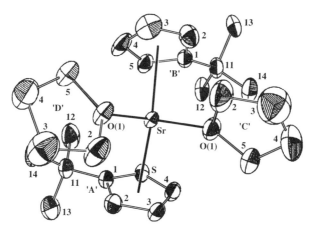

Fig. 18. ORTEP diagram (20% probability level) of Sr(t-BuC$_5$H$_4$)$_2$(THF)$_2$ showing the pseudotetrahedral coordination environment of the metal center. (Redrawn from Ref. 88.)

The *ansa*-metallocenes, M(R$_2$R$_2'$C$_2$(C$_5$H$_4$)$_2$, accept both mondentate Lewis bases such as THF and bidentate bases such as N,N'-di(t-Bu)glyoxaldial-dimine. The structure of *ansa*-CaC$_2$(Me$_4$(C$_5$H$_4$)$_2$) with a bidentate base is similar in geometry to that of Sr(t-BuC$_5$H$_4$)$_2$(THF)$_2$ (Fig. 18).[95]

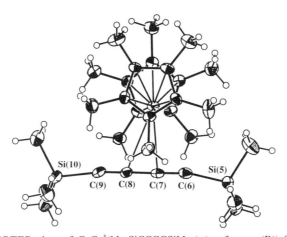

Fig. 19. ORTEP view of CaCp$_2^*$(Me$_3$SiCCCCSiMe$_3$) (conformer (B)) from an above perspective. The –SiMe$_3$ groups are bent back from the metal center in both conformers. In conformer (B), three carbon atoms (C(6)–C(8)) are within 3.2 Å of the metal center, whereas in conformer (A) a more symmetrical structure with the calcium bonding to two carbon atoms (C(7) and C(8)) is observed. (Redrawn from Ref. 104.)

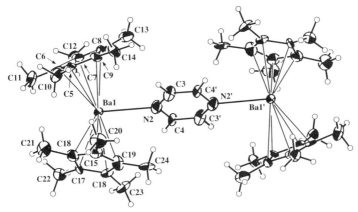

FIG. 20.   ORTEP drawing of $[BaCp_2^*]_2(\mu\text{-}1,4\text{-pyrazine})$ showing the pyramidal geometry around each barium atom. Each barium atom lies 0.65 Å out of the plane defined by the two ring centroids and the nitrogen atom of the pyrazine ligand. (Redrawn from Ref. 103.)

C. Poly(pyrazolyl)borates

As the interest in finding alkaline-earth compounds suitable for MOCVD applications has increased, investigation of ligands that are both thermally robust and able to encapsulate large metal centers has intensified. Some of the previously discussed cyclopentadienides meet these criteria, and the related poly(pyrazolyl)borates have recently received a great deal more attention.

1. Poly(pyrazolyl)borates: Syntheses

The main method of preparation of the alkaline-earth poly(pyrazolyl) borates is by halide metathesis. Belderrain et al. have reported the synthesis of the Ca, Sr, and Ba poly(pyrazolyl)borates MTp_2^* by reaction of the metal iodides, MI_2, with KTp^* ($Tp^* = HB(3,5\text{-}Me_2pz)_3$) in THF, similar to the method described for the metallocene compounds.[111]

$$MI_2 + 2\ KTp^* \longrightarrow MTp_2^* + 2\ KI$$

$$(M = Ca, Sr, and Ba) \tag{14}$$

They have also demonstrated that reaction of BaI_2 with only one molar equivalent of KTp^* in THF yields the heteroleptic compound [Tp^*BaI $(THF)_n$], which upon drying under vacuum loses crystallinity and solvent to form the monosolvated species ($n = 1$). The mixed Tp^*–Bp^* ($Bp^* = H_2B(3,5\text{-}Me_2pz)_2$) complex of Ba, $BaTp^*Bp^*$, may be prepared from [$Tp^*BaI(THF)_n$] by addition of KBp^* to this species in THF. Caulton and Chisholm *et al.* have synthesized and structurally characterized the barium BaTp$_2^*$ complex.[112]

Sohrin *et al.* have reported the synthesis of Ca[HB(pz)$_3$]$_2$ and M[HB(3,5-Me$_2$pz)$_3$]$_2$ (M = Ca, Sr) by addition of an aqueous solution of K[HB(pz)$_3$] or K[HB(3,5-Me$_2$pz)$_3$] to an acidic solution of MCl$_2 \cdot$ 2H$_2$O.[113,114] To this mixture, 1 M NaOH was then added until white precipitate ceased to form (pH = 9).

2. *Poly(pyrazolyl)borates: Chemical and Physical Properties*

Table III[111–114] contains the physical properties of several poly(pyrazolyl)borate complexes found in the literature. Group 2 poly(pyrazolyl)borates are white monomeric complexes that exhibit some volatility. Ca [HB(pz)$_3$]$_2$, which appears to be one of the most volatile of the poly(pyrazolyl)borates, is reported to sublime at 160°C and 10^{-2} atm.[113] The Ba[HB(3,5-Me$_2$pz)$_3$]$_2$ compound is soluble in THF and hot toluene, and the M[HB(pz)$_3$]$_2$ compounds have good solubility in aromatic and chlorinated hydrocarbons. Ca[HB(3,5-Me$_2$pz)$_3$]$_2$ and Sr[HB(3,5-Me$_2$pz)$_3$]$_2$ are also soluble in aromatic and chlorinated hydrocarbons, but less so than their nonmethylated counterparts.[114]

The solid-state structure of the heteroleptic compound [Tp*BaI (HMPA)$_2$] is shown in Fig. 21.[111] The steric bulk of the poly(pyrazolyl)borate ligand is apparent in this figure, effectively blocking oligomerization from occurring. This is even more apparent in the bis(3,5-dimethylpoly (pyrazolyl)borate) complex Ba[HB(3,5-Me$_2$pz)$_3$]$_2$ shown in Fig. 22, where two poly(3,5-dimethylpyrazolylborate) ligands have sterically saturated the coordination sphere of the barium center. The shortest intermolecular distance is 5.23 Å.[112]

Although the Group 2 element analogues have not yet been prepared, there are a number of derivatives of *t*-Bu-substituted pyrazolylborates that provide good steric protection of coordination sites because of the large size of the *t*-Bu substituent. For example, In[HB(3-*t*-Bupz)$_3$] (Fig. 23) was recently prepared and shown to be a rare example of an inert, monomeric In(I) compound.[115] Pyridyl-substituted pyrazolylborates have also been employed in the synthesis of homoleptic transition metal (Ti) and Group 13 pyrazolylborate compounds (Tl).[116] The pyridine substituent acts in the dual capacity of both sterically and coordinatively saturating the metal centers.

D. *Alkoxides*

Many alkoxide compounds of the heavier Group 2 elements are known, but most are either not volatile enough for use in CVD or are not thermally

TABLE III

Ca, Sr, and Ba Poly(pyrazolyl)borate Compounds and Their Physical Properties[a]

Compound	Color/state	m.p. (°C)	Sublim. temp. (°C)	Solubility[b]	Ref.
Ca[HB(pz)$_3$]$_2$	white solid	275–276	160, 10^{-2} atm	s = a	113
Ca(HB(3,5-Me$_2$pz)$_3$)$_2$	white solid	*	200, 10^{-3} torr	ss = a; is = b	111, 114
Sr(HB(3,5-Me$_2$pz)$_3$)$_2$	white solid	*	200, 10^{-3} torr	ss = a; is = b	111, 114
[Ba(HB(3,5-Me$_2$pz)$_3$)$_2$]	white solid	308–314	200, 10^{-3} torr 143, 3×10^{-5} torr	s = c; ss = a; is = b	111 112
(HB(3,5-Me$_2$pz)$_3$)BaI(THF)$_n$	white solid	*	*	s = c	111
(HB(3,5-Me$_2$pz)$_3$)Ba(H$_2$B(3,5-Me$_2$pz)$_2$) · THF	white solid	*	*	*	111
BaI(HB(3,5-Me$_2$pz)$_3$)(HMPA)$_2$	white solid	*	*	*	111

[a] * = information not available.

[b] s = soluble, ss = slightly soluble, is = insoluble; a = common organic solvents, b = water, c = tetrahydrofuran.

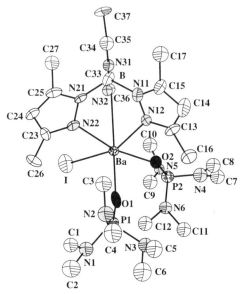

Fig. 21.   ORTEP drawing (20% probability level) (Tp*BaI(HMPA)$_2$ showing the distorted octahedral geometry of the six-coordinate bariu   atom. (Redrawn from Ref. 111.)

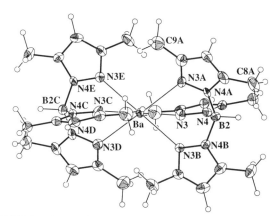

Fig. 22.   ORTEP drawing (20% probability level) of Ba[HB(3,5-Me$_2$pz)$_3$]$_2$. The molecule
has S_6 symmetry with the S_6 axis coincident with a line drawn through the two boron atoms.
(Redrawn from Ref. 112.)

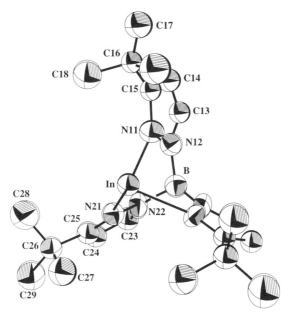

FIG. 23.   ORTEP drawing of the structure of In[HB(3-*t*-Bupz)₃]. (Redrawn from Ref. 115.)

robust. The amount of oligomerization in these species has been shown to decrease with increasing steric hindrance of the alkoxide ligand or presence of "pendent donor" functionalities on the ligand. Polyether ligands have also been shown to lower aggregation by occupying multiple coordination sites. Application of these principles has recently resulted in the synthesis of more volatile alkoxide complexes. Heterometallic alkoxide compounds are not reviewed here. This area and Group 2 alkoxide chemistry in general has been reviewed elsewhere.[26,27,46,75,117–120]

1. *Alkoxides: Syntheses*

Only a few methods have been used to prepare Group 2 alkoxide complexes. One of the most common methods is reaction between the metal or metal hydride and alcohol in either coordinating or noncoordinating solvents under refluxing conditions.[55,121–130] However, this method yields compounds of varying stoichiometry. In many cases, oxo-alkoxide compounds are formed through unspecified reactions of the alkoxide ligand (e.g., $H_2Ba_8(\mu_5\text{-}O)(OPh)_{14}(HMPA)_6$,[125] $H_4Ba_6(\mu_6\text{-}O)(OCH_2CH_2OCH_3)_{14}$,[126] $HBa_5(O)(OPh)_9 (THF)_8$[127]). A better understanding of these reactions may lead to better control over metal oxide film formation. For example, ether elimination is a clean method for removing two alkoxide ligands and replacing them with one oxo ligand. The preceding reaction has also been conducted with HMPA (HMPA = $OP(NMe_2)_3$) and TMEDA (TMEDA = $Me_2NC_2H_4NMe_2$) as ligands to produce both mixed metal alkoxide (HMPA or TMEDA) complexes and large aggregates, for example, $H_2Sr_6Ba_2(\mu_5\text{-}O)(OPh)_{14}(HMPA)_6$,[125] $H_2Ba_8(\mu_5\text{-}O)(OPh)_{14}(HMPA)_6$,[125] and $Ba_6(OPh)_{12}$ (TMEDA)$_4$.[122] When HMPA or TMEDA is used, the synthesis is similar to that described earlier, but with addition of the base into the reaction mixture.

Another method of producing Group 2 alkoxides that has proven to be very fruitful is the use of ammonia-gas-saturated solutions as solvent for the reaction of the metal (M^0) and the alcohol.[48,49,72,131–137] This procedure involves bubbling ammonia gas through a suspension of the metal granules in the alcohol and the solvent (toluene or THF) for an initial period of up to 10 min.[49] The ammonia is thought to react with the metal to form the more reactive metal amide or $M(NH_3)_x$ or both. The resulting species can then undergo an anion metathesis reaction via proton transfer with the alcohol (amide), or redox reaction ($M(NH_3)_x$), to form the alkoxide. This reaction method has produced a wide variety of products (see Table IV): compounds of type $[M(Ar^m)_2]_p(L)_n$ (M = Ca, Sr, Ba; m = 1,4,5; L = THF, HMPA, DME, 18-crown-6, etc.; n = 0,1,2,3; p = 1, 2) and $[M(OXR_3)_2]_p(L)_n$ (M = Ca, Sr, Ba; X = C, Si; R = Ph, t-Bu); L = THF, NH_3. These compounds are of varying stoichiometries, depending on the nature and relative amounts of the metal, alcohol, and ancillary ligands used. The mild conditions of the reactions generally prevent ligand degradation from occurring, and reaction stoichiometry is observed in the products. Yields are generally moderate to good by this method (50–80%).

A few compounds have been synthesized by halide metathesis reactions between MI_2 or $MI(R)(THF)_n$ (M = Ca, Sr or Ba; R = alkoxide, aryl oxide, or Cp) and M(OR) (M = Na or K). These include $[M(Ar^m)_2]_p(L)_n$ (M = Ca, Ba; m = 2–3; L = THF; n = 1,2,3; p = 1,2),[138] $[MI(Ar^m)]_p(L)_n$ (M = Ca, Ba; m = 2–3; L = THF; n = 3,4; p = 1,2),[59] $Ba_4(\mu_4\text{-}O)(\mu_2\text{-}Ar^8)_6$,[144] $Ca(C_5(Me)_4Et)(Ar^2)$,[60] $Ca(C_5(Me)_4Et)(Ar^8)$,[60] and $Ca(1,2,3,4\text{-}(i\text{-}Prop)_4C_5H)$ $(Ar^2)(THF)$.[61] Good yields are generally obtained; however, problems can be encountered in the separation of the alkali metal halide or incomplete substitution of halide for alkoxide ligands.

A series of very volatile alkaline-earth compounds with pendent ether groups, $[M(OC\text{-}t\text{-}Bu(CH_2OR)_2)_2]_2$ (M = Ca, Sr, Ba; R = i-Pr or Et), have been synthesized by the halide metathesis and metal/ammonia methods as well as by reaction between the metal amides $M(N(SiMe_3)_2)_2(THF)_n$ and the alcohols $HOC\text{-}t\text{-}Bu(CH_2OR)_2$.[48] Alcoholysis of metal amides has also been used to prepare $[Ba(OCMe_3)_2(HOCMe_3)_2]_4$ and $Ba(OSi(t\text{-}Bu)_2(CH_2)_3NMe_2)_2$.[63,147]

2. Alkoxides: Chemical and Physical Properties

Group 2 alkoxides tend to be either colorless or white solids. In fact, all but two of the compounds in Table IV are solids; the exceptions are $Ba(O(CH_2CH_2O)_2CH_3)_2$ and $Ba(O(CH_2CH_2O)_3CH_3)_2$.[124] Many alkaline-earth metal alkoxides show good solubility in a variety of solvents. For most compounds solubility increases with the donor ability of the solvent. The compounds display varying degrees of oligomerization from one metal up to nine. Sublimation temperatures range from 80 to 280°C at reduced pressures. Some general trends can be extracted from Table IV concerning apparent volatility based on the sublimation data available, with the caveat mentioned earlier. The best apparent volatilities are associated with compounds that are either perfluorinated, $M[OC(CF_3)_3]_2$ (140–280°C, $<10^{-5}$ torr),[121] or contain alkoxides with pendent donors, $[M(OCt\text{-}Bu(CH_2OR)_2)_2]_2$ (150–185°C, 10^{-2} mbar),[48] $Ba(O(CH_2CH_2O)_2CH_3)_2$ (200°C, 10^{-3} torr, vapor pressure),[124] and $H_4Ba_6(\mu_6\text{-}O)(OCH_2CH_2OCH_3)_{14}$ (160°C, 10^{-1} torr).[126] These compounds are among the few that appear to sublime intact. Clearly, ligands that can act in the dual capacity of encapsulating metal centers (i.e., isolating them from one another) and coordinatively saturating them, while also imparting thermal stability to the resulting compound, should give the most volatile alkaline-earth compounds. Although some perfluorinated alkoxide ligands have molecular weights more than two times those of their nonfluorinated congeners, the volatilities of the metal–fluoroalkoxide compounds are usually greater (e.g., compare $Ba[OCH(CMe_3)_2]_2$ with $Ba[OCH(CF_3)_2]_2$, Table IV). Apparently, the fluorine atoms act to disrupt intermolecular interactions.

In the past 5 years, the number of single-crystal X-ray structures of Group 2 alkoxide compounds reported in the literature has increased dramatically. This is in large part due to growing interest in them as precursors to materials. The phenoxide ligand and its derivatives have produced by far the majority of low-nuclearity Group 2 "alkoxides." However, it is apparent from Fig. 24[122] and Table IV that the phenoxide ligand itself, $^-O(C_6H_5)$, is unable to completely prevent oligomerization and that more sterically hindered aryloxides, such as $^-OC_6H_2\text{-}t\text{-}Bu_3$ found in the complex $Sr(OC_6H_2\text{-}t\text{-}Bu_3)_2(THF)_3 \cdot 0.5THF$ (Fig. 25),[132] are required to obtain monomeric species. The cationic cluster $\{Ca_3(OPh)_5[OP(NMe_2)_3]_6\}^+$ is more accurately described by the formula $\{Ca_3(\mu_3\text{-}OPh)_2(\mu_2\text{-}OPh)_3[OP(NMe_2)_3]_6\}^+$.[122] The three-metal unit is bridged on each of three edges and capped on its two faces by phenoxide ligands. Each metal has

TABLE IV

Ca, Sr, and Ba Aryl Oxide, Alkoxide, and Siloxide Compounds and Their Physical Properties[a]

Compound[b]	State/color	m.p. (°C)	Sublim. temp. (°C)	Solubility[c]	Ref.
[Ca(Ar¹)₂]₂	solid colorless	110–112	<250, 10^{-4} torr[d]	is = b; s = c; vs = a	132
Ca(Ar¹)₂(THF)₃(toluene)	solid colorless	133–136	*	vs = a; is = b; s = c	132
[Sr(Ar¹)₂]₂	solid colorless	142–145	<250, 10^{-4} torr[d]	vs = a; is = b; s = c	132
Sr(Ar¹)₂(THF)₃ · 0.5THF	solid colorless	>210, dec.	*	vs = a; is = b; s = c	132
[Ba(Ar¹)₂]₂	solid colorless	180–183	<250, 10^{-4} torr[d]	vs = a; is = b; s = c	132
Ba(Ar¹)₂(THF)₃ · THF	solid colorless	152–155	*	vs = a; is = b; s = c	132
Ba(Ar¹)₂(HMPA)₃(toluene)	solid colorless	72–75	*	vs = a; is = b,c	132
Ba(Ar¹)₂(DME)₂ · 3(toluene)	solid colorless	307–310	*	vs = a; is = b; s = c	132
Ca(Ar²)₂(THF)₃ · THF	solid colorless	*	80, 10^{-2} torr[e]	vs = a; s = d	55, 138
[Ca(Ar²)(μ₂-Ar²)]₂	solid colorless	*	*	vs = a; s = d	55
Ba(Ar²)₂(THF)₃ · THF	solid colorless	*	*	vs = a; s = d	55,138
Ca(Ar³)₂(THF)₃ · THF	solid colorless	*	*	vs = a; s = d	138
[Ca(Ar³)(μ₂-Ar³) · THF]₂ · 2(toluene)	solid colorless	*	*	vs = a; s = d	138
[Ba(Ar²)(THF)₃]₂	solid colorless	*	*	vs = a; is = d	59
[Ca(Ar³)(THF)₄	solid colorless	*	*	vs = a; s = d (dec.)	59
[Ba(Ar⁴)₂(HOCH₂CH₂NMe₂)₄] · 2(toluene)	solid colorless	94–97	*	s = a, c, d, e	135
Ba(Ar⁵)₂(18-crown-6) · (HAr⁵) · (18-crown-6)	solid colorless	*	10^{-1} torr,[f] heat, dec.	s = a, c	130
Ca(Ar⁶)(H₂O)₆ · (Ar⁶) · (H₂O)	solid yellow	*	*	s = f	139
Ba(Ar⁶)₂ · 3(1,10-phenanthroline)	solid yellow	*	*	s = j	140
Ca(Ar⁷)₂ · (HO(CH₂CH₂O)₃CH₂CH₂OH) · (H₂O)	solid	203	*	s = j	141
Ba(Ar⁷)₂ · 2(1,10-phenanthroline) · (acetone)	solid yellow	*	*	s = j	142
Ba(Ar⁷)₂ · (dibenzo-24-crown-8) · 2H₂O	solid yellow	104	*	s = k	143
Ca(C₅(Me)₄Et)(Ar⁸)	solid off-white	*	*	s = c	60
Ca(C₅(Me)₄Et)(Ar⁹)	solid	*	*	ss = d; s = a (dec.)	60

Compound	Color/state	mp (°C)	Sublimation/volatility	Solubility	Ref
Ca(1,2,3,4-(i-Prop)$_4$C$_5$H)(Ar^2)(THF)	white solid		*	s = c, g	61
[Ba(HOCH$_2$CH$_2$NH$_2$)(μ_2-OH)(μ_1, μ_2-HOCH$_2$CH$_2$NH$_2$)(HAr^5)]$_x$	solid colorless		involatile	ss = b; s = a, c, e	131
Ba$_3$(μ_5-OH)(μ_3-Ar^5)$_4$(μ_5-Ar^5)$_4$(Ar^5)(THF)$_5$	solid colorless	*	100, 10^{-1} torr,[f] dec.	ss = b, c; s = a, e	131
Ba$_4$(μ_4-O)(μ_2-Ar^8)$_6$ · 3(toluene)	solid pale yellow	*	135, 10^{-6} torr, dec.	s = a, c, d	144
Sr$_3$(OPh)$_6$(HMPA)$_5$ · (toluene) · C$_5$H$_{12}$	solid colorless	127–130	*	s = c	122
[Ca$_3$(OPh)$_5$(HMPA)$_6$]|[(OPh)(2PhOH)] · (toluene)	solid colorless	78–80	*	s = c	122
[Ba$_6$(OPh)$_{12}$(TMEDA)$_4$] · 4(toluene)	solid colorless	96–99	*	s = c	122
Ca$_9$(OCH$_2$CH$_2$OCH$_3$)$_{18}$ · (HOCH$_2$CH$_2$OCH$_3$)$_2$	solid colorless		*	s = g	129
Ca$_6$(μ_4-O)$_2$(μ_3-OEt)$_4$(OEt)$_4$ · 14EtOH	solid colorless		involatile	s = a, c, g	145
Ba$_3$(μ_2-OCPh$_3$)$_3$(OCPh$_3$)(THF)$_3$ · (toluene) · THF	solid colorless	>250, dec.	*	s = a, c	49
Sr$_4$(μ_3-OPh)$_2$(μ_2-OPh)$_4$(OPh)$_2$(THF)$_6$(PhOH)$_2$	solid colorless	96–99[d]	*	s = a	128
HBa$_5$(O)(OPh)$_9$(THF)$_8$	solid colorless	226–228	*	s = a, c	127
H$_3$Ba$_6$(O)(O-t-Bu)$_{11}$(OCEt$_2$CH$_2$O)(THF)$_3$	solid colorless	>340	*	s = a, c	127
H$_4$Ba$_6$(μ_6-O)(OCH$_2$CH$_2$OCH$_3$)$_{14}$	solid colorless	217–218	160, 10^{-1} torr	s = c	126
H$_3$Ba$_8$(μ_5-O)(OPh)$_{14}$(HMPA)$_6$	solid colorless	311–314	*	s = c, h	125
H$_5$Sr$_6$Ba$_2$(μ_5-O)(OPh)$_{14}$(HMPA)$_6$	solid colorless		*	s = c, h	125
Ba(O(CH$_2$CH$_2$O)$_2$CH$_3$)$_2$	liquid brown	—	200, 10^{-3} torr (reported as vapor pressure)	vs = a, c, d, i	124
Ba(O(CH$_2$CH$_2$O)$_3$CH$_3$)$_2$	liquid brown		—	vs = a, c, d, i	124
[Ca(OC-t-Bu(CH$_2$O-i-Pr)$_2$)$_2$]$_2$	solid colorless		150, 10^{-2} mbar	vs = a, b, c	48
[Sr(OC-t-Bu(CH$_2$O-i-Pr)$_2$)$_2$]$_2$	solid colorless		170, 10^{-2} mbar	vs = a, b, c	48
[Ba(OC-t-Bu(CH$_2$O-i-Pr)$_2$)$_2$]$_2$	solid colorless		185, 10^{-2} mbar	vs = a, b, c	48
[Ba(OC-t-Bu(CH$_2$OEt)$_2$)$_2$]$_2$	solid colorless		150, 10^{-2} mbar	vs = a, b, c	48
[Ca(OC(t-Bu)$_3$)$_2$]$_2$	solid colorless		185, 10^{-3} mbar dec.[h]	vs = c; s = b, i	48
[Ca(OCH(t-Bu)$_2$)$_2$]$_3$	solid colorless		*	s = a, c	48
Ca(OCEt$_3$)$_2$	white solid	>350, dec.	*	*	54

(continued)

TABLE IV (*continued*)

Compound[b]	State/color	m.p. (°C)	Sublim. temp. (°C)	Solubility[c]	Ref.
$Sr(OCEt_3)_2$	white solid	>300	*	*	54
$Ba(OCEt_3)_2$	white solid	*	265, <10^{-5} torr[i]	is = b, d; ss = a	54, 121
$Ba(OCMeEt\text{-}i\text{-}Pr)_2$	white solid	*	240, <10^{-5} torr[i]	is = b, d; ss = a	121
$Ba(OCMe_3)_2$	white solid	*	270, <10^{-5} torr[i]	is = b, d; s = a	121
$Ba[OCH(CMe_3)_2]_2$	white solid	*	260, <10^{-5} torr[i]	ss = b; s = d; vs = a	121
$Ba[OCH(CF_3)_2]_2$	white solid	*	230, <10^{-5} torr[i]	is = b, d; vs = a	121
$Ba_5(\mu_5\text{-}OH)(\mu_3\text{-}OCH(CF_3)_2)_4(\mu_2\text{-}OCH(CF_3)_2)_4(OCH(CF_3)_2)_2(THF)_4(H_2O)$	solid colorless	*	200, 10^{-3} torr	s = a	146
$Ca[OC(CF_3)_3]_2$	yellow-white powder	*	140, <10^{-5} torr[i]	is = b, d; vs = a	121
$Sr[OC(CF_3)_3]_2$	white solid	*	230, <10^{-5} torr[j]	is = b, d; vs = a	121
$Ba[OC(CF_3)_3]_2$	white solid	*	280, <10^{-5} torr[j]	is = b, d; vs = a	121
$Ba(OCH_2C_4H_7O)_2$	white solid	*	involatile	is = b, d; ss = a	121
$[Ba(OCMe_3)_2(HOCMe_3)_2]_4$	colorless solid	*	*	s = b, c	147
$Ba(N(C_2H_4O)(C_2H_4OH)_2)_2 \cdot 2EtOH$	colorless solid	*	*	s = j	123
$Ba(OCPh_3)_2(15\text{-}crown\text{-}5)(THF) \cdot THF$	solid colorless	*	200, 10^{-2} torr dec., THF, crown loss	vs = l; is = b; s = a: ss = c	136
$Ca_2(OSiPh_3)_4(NH_3)_4 \cdot 0.5(toluene)$	solid colorless	110–115, >165 dec.	*	s = c	137
$[Ca(OSiPh_3)_2(THF)]_n$	solid colorless	120–125, de-solvates	*	s = a	137

Compound	Appearance	mp/dec.	Sublimation	Solubility	Ref.
$Ba(OSiPh_3)_2(15\text{-crown-}5)(THF) \cdot THF$	solid colorless	141 dec.[g]	220, 10^{-5} torr dec., THF, crown loss	vs = a, l; is = b; s = c	136
$Sr(OSiPh_3)_2(15\text{-crown-}5)(THF) \cdot THF$	solid colorless	*	*	vs = l; is = b; s = a; ss = c	136
$Ba_3(OSiPh_2OSiPh_2O)_3(\text{tetraglyme})_2$	white powder	*	*	s = c, l	134
$Ba_3(OSiPh_2OSiPh_2O)_3(HMPA)_5(H_2O)$	white powder	*	*	s = c, l	134
$Ba_3(OSiPh_2OSiPh_2O)(H_2O)(NH_3)_{0.33}$	white powder	*	*	s = c, l	134
$Ba_3(OSiPh_3)_6(THF) \cdot 0.5THF$	solid colorless	*	200–220, 10^{-2} torr dec.[k]	s = l	133
$Ba_2(\mu_2\text{-}OSi(t\text{-}Bu)_3)_3(OSi(t\text{-}Bu)_3)(THF)$	solid colorless	*	80, 1 torr[l]	s = a, c, d	49
$Ba(OSi(t\text{-}Bu)_2(CH_2)_3NMe_2)_2$	colorless oil	*	200, 10^{-4} torr, dec.	s = c, g	63

[a] * = information not available.

[b] vs = very soluble, s = soluble, ss = slightly soluble, is = insoluble; a = tetrahydrofuran, b = n-pentane, c = toluene, d = benzene, e = dichloromethane, f = water, g = hexane, h = hexamethylphosphoramide, i = diethyl ether, j = alcohols, k = ethyl acetate, l = dimethyl sulfoxide.

[c] $Ar^1 = OC_6H_2\text{-}2,4,6\text{-}t\text{-}Bu_3$, $Ar^2 = OC_6H_3\text{-}2,6\text{-}t\text{-}Bu_2\text{-}4\text{-}Me$, $Ar^3 = OC(Ph_2)CH_2C_6H_4\text{-}4\text{-}Cl$, $Ar^4 = OC_6H_3\text{-}2,6\text{-}t\text{-}Bu_2$, $Ar^5 = OC_6H_3\text{-}3,5\text{-}t\text{-}Bu_2$, $Ar^6 = OC_6H_3\text{-}2,4\text{-}(NO_2)_2$, $Ar^7 = OC_6H_3\text{-}2,4,6\text{-}(NO_2)_3$, $Ar^8 = OC_6H_2(CH_2NMe_2)_3\text{-}2,4,6$.

[d] Sublimes with some decomposition.

[e] Sublimate is $[Ca(Ar^2)(\mu_2\text{-}Ar^2)]_2$.

[f] Decomposes with liberation of HAr^5.

[g] Melts with loss of THF.

[h] Decomposition product is $[Ca(OCH(t\text{-}Bu)_2)_2]_3$.

[i] For appreciable rates of sublimation, temperature should be increased by 20–30°C above sublimation point in Ref. 47.

[j] Rapid and almost complete sublimation.

[k] Forms $[Ba(OSiPh_3)_2]_n$.

[l] The volatility data on this compound have been challenged; see Ref. 48.

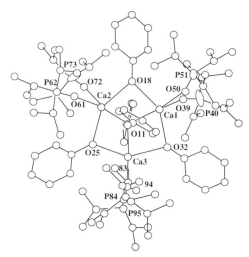

FIG. 24.   ORTEP drawing (50% probability level) of $\{Ca_3(OPh)_5[OP(NMe_2)_3]_6\}^+$ looking down the pseudo-threefold axis that intersects the μ_3-oxygen atoms of the phenoxide ligands. The compound has a distorted octahedral coordination geometry around each calcium atom, with Ca—O bond distances of 2.38(1)–2.44(1) Å (μ_3), 2.33(1)–2.36(1) Å (μ_2), and 2.28(1)–2.34(1) Å (HMPA). (Redrawn from Ref. 122.)

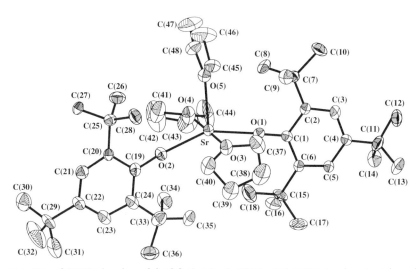

FIG. 25.   ORTEP drawing of $Sr(OC_6H_2\text{-}t\text{-}Bu_3)_2(THF)_3 \cdot 0.5$ THF showing the trigonal bipyramidal ligand arrangement around the metal center. The more spatially demanding aryl oxide ligands are located at the equatorial sites. (Redrawn from Ref. 132.)

a coordination number of six with the compound possessing pseudo-D_{3h} symmetry. The anion is a phenoxide ligand remote from the cation and hydrogen bonded to two phenol molecules [this unit is represented as $[H_2(OPh)_3]^-$ in Eq. (15)]. The following equilibrium process in toluene was determined by variable-temperature NMR spectroscopy:[122]

$$\{Ca_3(\mu_3\text{-OPh})_2(\mu_2\text{-OPh})_3[OP(NMe_2)_3]_6\}^+ + [H_2(OPh)_3]^- \rightleftharpoons$$

$$Ca_3(\mu_3\text{-OPh})_2(\mu_2\text{-OPh})_3(OPh)[OP(NMe_2)_3]_5 + OP(NMe_2)_3 + 2PhOH \quad (15)$$

The molecular geometry of $Sr(OC_6H_2\text{-}t\text{-Bu}_3)_2(THF)_3$ can best be described as distorted trigonal bipyramidal. The large aryl oxide groups occupy two of the equatorial sites, which allows the t-Bu groups to better extend into space without causing serious repulsive contacts. The metal center in this compound has an unusually low coordination number of five.[132]

The very large alkoxide ligand $^-OCPh_3$, and its silicon analogue $^-OSiPh_3$, have been used in conjunction with ethers and polyethers to give monomeric, dimeric, and trimeric alkoxide (siloxide) compounds of barium. The dimeric compound $Ba_2(OCPh_3)_4(THF)_3$ shown in Fig. 26 is obtained from a THF solution after an ammonia-catalyzed reaction between the metal and the alcohol.[49] The structure is close to that found for face-bridging bioctahedra, but with two terminal ligands removed. Three $^-OCPh_3$ ligands bridge the metal centers in a μ_2-fashion. Each metal is five-coordinate; however, inequality in ligand distribution, with one barium possessing two THF molecules and the other a terminal $^-OCPh_3$ ligand,

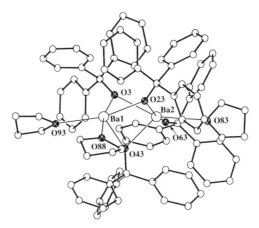

FIG. 26.   ORTEP drawing of $Ba_2(OCPh_3)_4(THF)_3$ showing the square pyramidal coordination geometry around each barium atom. The two square pyramids share a common triangular face comprising the three μ_2-oxygen atoms of the alkoxide ligands. (Redrawn from Ref. 49.)

gives the two barium centers distinct chemical environments. The structure is nonrigid in solution at $-40°C$ with only one type of alkoxide ligand observed. A similar structure to $Ba_2(OCPh_3)_4(THF)_3$ is found for both $Ba_2[\mu_2-OSi(t-Bu)_3]_3[OSi(t-Bu)_3](THF)$[49] and $Ca_2(OSiPh_3)_4(NH_3)_4$.[137] The less sterically encumbered $^-OSiPh_3$ ligand, under similar reaction conditions with barium metal, behaves somewhat differently in giving the trinuclear species $Ba_3(OSiPh_3)_6(THF)$.[133] This compound has one terminal, two μ_3-bridging, and three μ_2-bridging siloxides. ^{29}Si NMR experiments suggest that the terminal and μ_2-bridging siloxides exchange in solution at room temperature. The heterometallic compound $KBa_2(OSiPh_3)_5(DME)_2$, prepared by reaction of $KOSiPh_3$ with $Ba_3(OSiPh_3)_6(THF)$ (3 : 2) in THF and crystallization of the precipitate from DME, has a similar structure, but with a potassium atom occupying one vertex of the metal triangle with ligated DME molecules.[148]

If the reaction to form $Ba_2(OCPh_3)_4(THF)_3$ or $Ba_3(OSiPh_3)_6(THF)$ is carried out in the presence of 15-crown-5, the isostructural monomeric compounds $Ba(OCPh_3)_2(15\text{-crown-}5)(THF)$, $Sr(OSiPh_3)_2(15\text{-crown-}5)(THF)$, and $Ba(OSiPh_3)_2(15\text{-crown-}5)(THF)$ are obtained.[136] The structure of these is shown in Fig. 27. Clearly, the presence of the polyether, in these otherwise identical reactions, serves to reduce aggregation of the metal centers by occupying five metal coordination sites and blocking one side of the metal center. The size of the strontium ($r_{Sr} = 1.18$ Å) and barium (1.35 Å) ions, as compared to the hole size of the 15-crown-5 ether (1.7–2.2 Å), indicates that only side-on bonding of 15-crown-5 rings to these metal centers should be observed.[51] This is in contrast to the two types of bonding found for the 18-crown-6 (hole size 2.6–3.2 Å) compounds of these metals where the ring usually coordinates around the center of the metal but in some instances may sit to one side (see later discussion). The two alkoxides (siloxides) of $M(OXPh_3)_2(15\text{-crown-}5)(THF)$ are found cis to one another

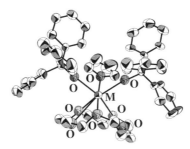

FIG. 27.   A generalized ORTEP drawing of the piano-stool type structure of the isostructural series $M(OXPh_3)_2(15\text{-crown-}5)(THF)$ (M = Sr, Ba; X = C or Si). (Redrawn from Ref. 136.)

below and to one side of the 15-crown-5 ring, with the coordination sphere completed by a THF molecule to give a roughly trigonal arrangement of these three ligands. The M—O_{ether} and M—$O_{alkoxide}$ bond distances are similar to those of $Ba_2(OCPh_3)_4(THF)_3$ and $Ba_3(OSiPh_3)_6(THF)$. The only other monomeric alkoxide compound with a polyether ligand is $Ba(OC_6H_3$-3,5-t-$Bu_2)_2$(18-crown-6) (Fig. 28).[130] In this compound the barium atom sits at the center of the 18-crown-6 ring with aryl oxide groups to each side. The molecular structure also includes a hydrogen-bonded alcohol molecule and a nonbonded 18-crown-6 molecule.

Although the ^{1}H and ^{13}C NMR spectra of $Ba(OCPh_3)_2$(15-crown-5)(THF) remain unchanged after reflux in toluene, both the strontium and barium siloxide compounds, under similar conditions, develop new peaks that can be unambiguously assigned to the formation of the silyl ether Ph_3SiO $SiPh_3$.[136] It appears that these siloxide compounds undergo ether elimination at elevated temperatures to form metal oxo-siloxide clusters. Similar reactivity has been proposed to account for the formation of $Pb_4(\mu_4$-$O)(SiPh_3)_6$ from $Pb(OSiPh_3)_2$.[149] There are also NMR and single-crystal X-ray diffraction data that support ether elimination between $Ba(OSiPh_3)_2$(15-crown-5)(THF) and $Ti(O$-i-$Pr)_4$ in toluene to give the Ph_3SiO-i-Pr ether and presumably a barium–titanium oxo-alkoxide compound.[136,150] Similar ether elimination processes may be occurring to give the oxo-alkoxide clusters mentioned at the beginning of this section. This type of reactivity, if tunable, may be useful with regard to CVD applications, because clean elimination of the supporting ligand framework is important in obtaining pure inorganic phases such as MO or $MTiO_3$ (M = Ca, Sr, Ba). Unfortu-

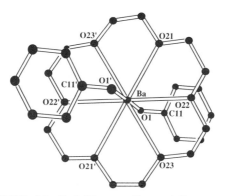

FIG. 28.   The $Ba(OC_6H_3$-3,5-t-$Bu_2)_2$(18-crown-6) core (t-Bu groups have been omitted for clarity) of $Ba(OC_6H_3$-3,5-t-$Bu_2)_2$(18-crown-6) · 2(ArOH) · (18-crown-6). The coordination geometry can be described as hexagonal bipyramidal with the oxygen atoms of the BaO_6O_2 core defining the vertices of the polyhedron. (Redrawn from Ref. 130.)

nately, on sublimation, the $M(OXPh_3)_2(15\text{-crown-}5)(THF)$ compounds lose both THF and 15-crown-5 at 220°C, 10^{-5} torr. Similar behavior has been observed for $Ba(OC_6H_3\text{-}3,5\text{-}t\text{-}Bu_2)_2(18\text{-crown-}6)$.[130] The dissociation of the ligands on attempted sublimation of these compounds may be remedied through further tailoring of the polyether and siloxide ligands.

To date, the most potentially useful alkaline-earth alkoxides for CVD applications based on their apparent volatilities are the $H_4Ba_6(\mu_6\text{-}O)$ $(OCH_2CH_2OCH_3)_{14}$[126] (Fig. 29) and the $[M(OC\text{-}t\text{-}Bu(CH_2OR)_2)_2]_2$ (M = Ca; R = i-Pr) (Fig. 30)[48] compounds, because of their facile sublimation and good solubility. The $[M(OC\text{-}t\text{-}Bu(CH_2OR)_2)_2]_2$ compounds were reported to sublime nearly quantitatively, while $H_4Ba_6(\mu_6\text{-}O)(OCH_2CH_2OCH_3)_{14}$ sublimes in only 30% yield. In $H_4Ba_6(\mu_6\text{-}O)(OCH_2CH_2OCH_3)_{14}$, the barium atoms are arranged in an octahedral array with an oxygen atom at the center equally bonded to each barium. Each barium atom is also coordinated to seven additional oxygen atoms; four oxygen atoms from the μ_3-$OCH_2CH_2OCH_3$ ligands, two ether oxygen atoms, and either a η^1-$OCH_2CH_2OCH_3$ (for apical barium atoms) or η^2-$OCH_2CH_2OCH_3$ (for equatorial barium atoms) ligand. Four of the 14 ligands are in the form of methoxyethanols, but all 14 ligands are equivalent in solution on the NMR

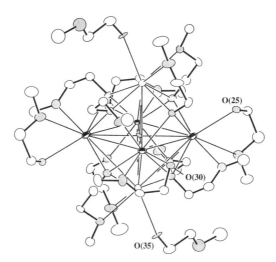

FIG. 29.   An ORTEP drawing of $H_4Ba_6(\mu_6\text{-}O)(OCH_2CH_2OCH_3)_{14}$. Three of the four alcohol hydrogen atoms are thought to be associated with O(25), O(30), and O(35), which is supported by the long Ba—O bond distances for these ligands (2.71–2.81 Å). The fourth hydrogen atom is thought to be disordered over the $Ba_6(\mu_6\text{-}O)(\mu_3\text{-}O)(O)_{18}$ moiety. (Redrawn from Ref. 126.)

time scale. The apparently high volatility of this cluster is rather surprising, considering it has a high molecular weight; the volatility may be due to extremely low packing forces.

[Ca(OC-t-Bu(CH$_2$O-i-Pr)$_2$)$_2$]$_2$ exists as a dimer in both the solid state and gas phase (by mass spectrometry). Three alkoxide groups μ_2-bridge the two metal units with the fourth alkoxide at a terminal site. The coordination spheres of the two metals are completed by bonding of five of eight ether appendages to give each metal a distorted octahedral geometry.

E. β-Diketonates

The chemical and physical properties of many of the β-diketonate and ketoiminate complexes found in the literature are presented in Table V, and a summary of the abbreviations of the β-diketones and ketoimines used in the literature and in this paper is given in Table VI. The reader should be aware that the original authors' nomenclature is used in Table V; e.g., Ca(dpm)$_2$, Ba$_4$(tmhd)$_8$, and Ba$_2$(thd)$_4$(NH$_3$)$_2$ are all compounds that contain the same β-diketonate (2,2,6,6-tetramethylheptane-3,5-dionate). This may seem confusing at first. However, since there is no consensus on the acronyms to use for these ligands, one is liable to come across all the permutations listed in Table VI in the literature, and it seems appropriate to introduce them here. Since many methods have been used to prepare the

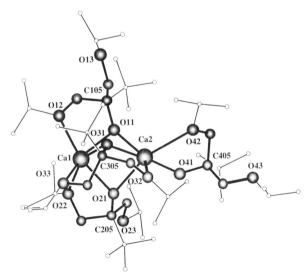

FIG. 30.   Molecular structure of [Ca(OC-t-Bu(CH$_2$O-i-Pr)$_2$)$_2$]$_2$. (Redrawn from Ref. 48.)

TABLE V

Ca, Sr, and Ba β-Diketonate and β-Ketoiminate Compounds and Their Physical Properties[a]

Compound	Color/state	m.p. (°C)	Sublim. temp. (°C)	Solubility[b]	Ref.
Ca(tpm)$_2$ · 2H$_2$O	white solid	*	190–195, 1 × 10^{-1} torr (rapid)	s = a	151
Sr(tpm)$_2$ · 2H$_2$O	white solid	*	190–220, 1 × 10^{-1} torr	s = a	151
Ba(tpm)$_2$ · 2H$_2$O	white solid	*	190–220, 1 × 10^{-1} torr	*	151
Ca(ppm)$_2$ · 2H$_2$O	white solid	*	190–195, 1 × 10^{-1} torr (rapid)	s = a	151
Sr(ppm)$_2$ · 2H$_2$O	white solid	*	190–220, 1 × 10^{-1} torr	s = a	151
Ba(ppm)$_2$ · 2H$_2$O	white solid	*	190–220, 1 × 10^{-1} torr	*	151
Ca(hpm)$_2$ · 2H$_2$O	white solid	*	190–195, 1 × 10^{-1} torr (rapid)	s = a	151
Sr(hpm)$_2$ · 2H$_2$O	white solid	*	190–220, 1 × 10^{-1} torr	s = a	151
Ba(hpm)$_2$ · 2H$_2$O	white solid	*	190–220, 1 × 10^{-1} torr	*	151
Ca(fod)$_2$	solid	90–93	volatile	s = b	152
Ba(fod)$_2$	solid	194–196	150, 1 × 10^1 torr	s = b	152
Ca(hfac)$_2$(H$_2$O)$_2$	solid	*	*	s = c	153
[Ba(hfac)$_2$(H$_2$O)]$_x$	solid	*	*	s = a, d	153
Ca(hfac)$_2$	pale yellow solid	170–230 dec.	130–150, <10^{-5} torr	s = a, e	154
Sr(hfac)$_2$	light yellow solid	220–275 dec.	180–190, <10^{-5} torr	s = a, e	154
Ba(hfac)$_2$	off-white solid	>210 dec.	205–220, <10^{-5} torr	s = a, e	154, 155
[Ba$_2$(hfac)$_4$(Et$_2$O)]$_x$	colorless solid	*	*	s = a:f (1:1)	156
(enH$_2$)$_{1.5}$[Ba(hfac)$_5$]C$_2$H$_5$OH	colorless solid	*	280, 7.6 × 10^2 torr	s = b	157, 158
Ca(tfac)$_2$	pale yellow solid	170–255 dec.	190–200, <10^{-5} torr	s = e	154
Sr(tfac)$_2$	off-white solid	160–230 dec.	200–225, <10^{-5} torr	s = e	154

Compound	Appearance	mp (°C)	Sublimation	s	Ref.
Ba(tfac)₂	solid	170–230 dec.	205–215, <10⁻⁵ torr	s = e	154
Ba(dfhd)₂	white solid	229.1	220, <1 torr 39 mg/min	s = a, b	159
Ba(ofhd)₂	white solid	233.8	220, <1 torr 3.0 mg/min	s = a, b	159
Ba(tdfnd)₂	white powder	186	220, 7.6 × 10² torr	s = b	160, 161
Ca(tdfnd)₂ · H₂O	white powder	*	160, 4 × 10¹ torr	s = b, g	160
Sr(tdfnd)₂ · H₂O	white powder	*	165, 4 × 10¹ torr	s = b, g	160
Ba(tdfnd)₂ · H₂O	white powder	187	195, 4 × 10¹ torr	s = b, g	160, 162
Ca(dfhd)₂ · H₂O	white powder	*	170, 4 × 10¹ torr	s = b, g	160
Sr(dfhd)₂ · H₂O	white powder	160	195, 4 × 10¹ torr	s = b, g	160
Ba(dfhd)₂ · H₂O	white powder	200	240, 4 × 10¹ torr	s = b, g	160
Ca(acac)₂(H₂O)₂ · H₂O	colorless solid	*	does not sublime	s = g	163
Ba(acac)₂	white solid	180 dec.	*	*	164
Ca₄(dpp)₈(EtOH)₂	yellow solid	*	*	s = h, i	165
Sr₄(dpp)₈(acetone)₂	pale yellow solid	*	*	s = b	166
Ba(dmmod)₂	colorless liquid	—	85, 1 × 10² torr (reported as vapor pressure)	s = h	47, 167
Ba(tclac)₂ · 0.5H₂O[c]	off-white solid	110–112 dec.	*	s = b, e	168
[Ba(thd)₂(Et₂O)]₂	colorless solid	*	*	s = a	169
Ba(thd)₂(tmeda)₂	solid	*	*	*	170
Ba(thd)₂(CH₃OH)₃ · 2CH₃OH	colorless solid	*	240, 5 torr, sublimes with loss of CH₃OH	s = b	171, 172
Ba(thd)₂(CH₃OH)₂(H₂O)₂	colorless solid	*	100, 10⁻¹ torr	s = b : g (8 : 1)	172, 173
Ba₂(thd)₄(NH₃)₄	white solid	153.3(170.3–173.1)[d]	213–224, 5 × 10⁻⁴ torr 2 h, Ba(thd)₂(43%)[e]	s = j	174
[Ba(μ₁, μ₂-dpm)(dpm)(HOC₆H₃-t-Bu₂-3,5)₂(THF)]₂	colorless solid	*	200–220, 10⁻² torr dec.[f]	s = h	131

TABLE V (continued)

Compound	Color/state	m.p. (°C)	Sublim. temp. (°C)	Solubility[b]	Ref.
$Ba_2(thd)_4(bipy)_2$	colorless solid	*	heat, 10^{-2} torr[g]	s = (a : f; 1 : 1)	175
$Ca(dpm)_2$	white solid	222–225	*	s = d	176
$Sr(dpm)_2$	white solid	199–201	*	s = d	176
$Sr_5(tmhd)_6(Htmhd) \cdot (toluene) \cdot (C_5H_{12})$	colorless solid	172–176	180–210, 10^{-3} torr	s = f, j	177
$Ba_4(tmhd)_8$	colorless solid	194–197 >305, dec.	190–210, 10^{-3} torr	s = f, j	177–180
$Ba_5(thd)_9(H_2O)_3(OH)$	solid	158–165	230, 5 torr	s = b, f	152, 157
$Ba_5(thd)_9Cl(H_2O)_7$	colorless solid	*	heat, 10^{-2} torr, dec.[h]	s = g	181
$Ba_5(thd)_6(pivalate)$	colorless solid	*	[i]	s = f	178
$[HNEt_3]_2[Ba_6(thd)_{10}(H_2O)_4(OH)_2(O_2)]$	colorless solid	*	*	s = k	182
$Ba_6(thd)_{10}(H_2O)_6(O_2)$	colorless solid	*	[j]	s = l	183
$[Ba_6(thd)_4(pivalate)_8(py)_4 \cdot 2py$	colorless solid	*	*	s = b	181
$Na_2Ba_7(thd)_2(pivalate)_{14}(py)_4(H_2O)_2$	colorless solid	*	*	s = m	181
$Ba_7(pivalate)_{12}(thd)_2(py)_4(H_2O)_4 \cdot 2py$	colorless solid	*	heat, 10^{-2} torr[k]	s = m	175
$[H_{0.5}Sr(tmhd)_{1.5}(OCH_2CH_2OCH_3)]_n$	colorless solid	>170, dec.	*	s = l	184
$[H_{0.5}Ba(tmhd)_{1.5}(OCH_2CH_2OCH_3)]_n$	colorless solid	106–110	*	s = l	184
$Ba(thd)(OCH_2CH_2OCH_3)$	colorless solid	88–92 >175, dec.	*	s = n, o	185
$[Sr(tmhd)(OEt)(EtOH)_{0.66}]_n \cdot$ (toluene)$_{0.66}$	colorless solid	72, desolvates	*	s = j, l	184
$Ba(tmhd)_{1.5}(OEt)_{0.5}(EtOH)_{0.33} \cdot$ (toluene)$_{0.33}$	colorless solid	246–250	*	s = j, l	184
$[Sr(tmhd)(OCH_2CH_2NMe_2)]_n \cdot$ (toluene)	colorless solid	>225, dec.		s = l	184
$[Ca(tmhd)_{1.5}(OCH_2CH_2O\text{-}i\text{-}Pr)_{0.5}]_n$	white solid	*	*	s = l	184
$[Ba(thd)_2(OHCH(CH_3)CH_2NMe_2)]_2$	white solid	168	170–200, 10^{-3} torr	s = h, j	186

Compound	Appearance	Melting point (°C)	Sublimation conditions	Solvent	References
$H_2Ba(tmhd)_6(OCH_2CH_2O\text{-}i\text{-}Pr)_4$	white solid	108–111	95–110, 5×10^{-3} torr[l]	s = l	184
$Ca_4(tmhd)_4(OEt)_4(EtOH)_4$	colorless solid	67, desolvates	170–200, 5×10^{-3} torr	s = j, l	184
$Ca_4(tmhd)_6(OCH_2CH_2NMe_2)_2$	colorless solid	>280, dec.	185–220, 5×10^{-3} torr[m]	s = l	184
$Sr_5(thd)_3(OCH_2CH_2OCH_3)_7$	colorless solid	84–86, >95, dec.	*	s = n, o	185
$Ba_5(OH)(thd)_5(OCH(CH_3)CH_2NMe_2)_4$	white solid	*	260, 5×10^{-3} torr	s = e, h	186
$Ca_2(thd)_6(OCH_2CH_2OCH_3)_8$	colorless solid	>232, dec.	*	s = i, o	185
$Ca(hfac)_2(triglyme)$	colorless solid	120–122	100, 2×10^{-2} torr	s = f, h	187
$Ca(hfac)_2(tetraglyme)$	white solid	94–97, 82–85	90, 10^{-2} torr; 130–150, 10^{-2} torr	s = h, l	188, 189
$Sr(hfac)_2(tetraglyme)$	colorless solid	137–139	115, 1×10^{-2} torr	s = f, h	187, 189
$Ba(hfac)_2(tetraglyme)$	white solid	152–155	145, 10^{-2} torr	s = f, h	155, 187, 189–191
$Ba(hfac)_2(hexaglyme)$	colorless solid	59–62	145, 2×10^{-2} torr	s = f, h	187
$Ca(hfac)_2(18\text{-crown-6})$	white solid	*	150, 10^{-1} torr	s = e, h	192, 193
$Sr(hfac)_2(18\text{-crown-6})$	white solid	*	*	s = e, h	193
$Ba(hfac)_2(18\text{-crown-6})$	white solid	*	150–200, 10^{-3} torr; 160–180, 10^{-2} torr	s = e, h	155, 168, 193, 194
$Ba(hfac)_2(CH_3O(CH_2CH_2O)C_2H_5)$	colorless solid	109–110	100, 5×10^{-2} torr	s = h	195
$Ba(hfac)_2(CH_3O(CH_2CH_2O)_6\text{-}n\text{-}C_4H_9)$	colorless solid	52–54	*	s = h	195
$Ca(hfac)_2(diaza\text{-}18\text{-crown-6})$	white solid	*	*	s = e, h	193
$Sr(hfac)_2(diaza\text{-}18\text{-crown-6})$	white solid	*	*	s = e, h	193
$Ba(hfac)_2(diaza\text{-}18\text{-crown-6})$	white solid	*	180–200, 10^{-2} torr; 170, 1.3 Pa	s = e, h	155, 193
$Ba(tdfnd)_2(tetraglyme)$	white solid	70, 61–63	160, 760 torr; 90, 10^{-2} torr	s = j	162, 196
$Ba(tfac)_2(18\text{-crown-6})$	off-white solid	190–192 dec.	150–170, 1×10^{-2}–5×10^{-2} torr, dec.[n]	s = a, b	168
$Sr(tfac)_2(hmtt)$[o]	white solid	69–72	150, 10^{-3} torr, dec.[p]	s = n	197
$Ba(tfac)_2(hmtt)$[o]	white solid	143–147	105, 10^{-3} torr, dec.[p]	s = h, j	198

(continued)

TABLE V (continued)

Compound	Color/state	m.p. (°C)	Sublim. temp. (°C)	Solubility[b]	Ref.
Sr(tfac)$_2$(triglyme)	pale orange solid	73–75	140–160, 10^{-3} torr	s = l, n	199
Ba(tfac)$_2$(tetraglyme)	white solid	133–136	150, 10^{-1} torr	s = l, n	190
Ba(acac)$_2$(18-crown-6) · 0.5H$_2$O	off-white solid	>140, dec.	120–140, 1 × 10^{-2}–5 × 10^{-2} torr dec.^{n}	s = b	168
Sr(Ph$_2$acac)$_2$(tetraglyme) · 0.5(toluene)	solid	96–98	100, 10^{-3} torr^p	s = i, l	199
Ba(tclac)$_2$(18-crown-6)	off-white solid	158–160, dec.	does not sublime	s = b, n	168
Ba(thd)$_2$(18-crown-6) · H$_2$O	off-white solid	130–135	110–130, 1 × 10^{-2}–5 × 10^{-2} torr dec.	s = l	168
Ba(thd)$_2$(pmdt)^q	colorless solid	*	120–150, 5 × 10^{-2} torr, dec.^{r}	s = f, h	200
Ba(thd)$_2$(hmtt)^o	colorless solid	*	95–120, 5 × 10^{-2} torr, dec.^{r}	s = h, j	200
Ba(thd)$_2${NH(CH$_2$CH$_2$NH$_2$)$_2$}$_2$	colorless solid	*	140, 5 × 10^{-2} torr^s	*	201
Ba(thd)$_2${Me(OCH$_2$CH$_2$)$_2$OH}	colorless solid	125–130	160, 7 × 10^{-2} torr^s	s = f, n	201
Ba(thd)$_2${Me(OCH$_2$CH$_2$)$_3$OH}	colorless solid	116–120	116, 8 × 10^{-2} torr^n	*	201
Ca(thd)$_2$(triglyme)	white solid	*	100, 10^{-3} torr, dec.^{p}	s = h, j	188
Sr(thd)$_2$(triglyme)	solid	87–89	150, 10^{-3} torr^p	s = h, j	197
Ba(thd)$_2$(triglyme)	white solid	*	100, 10^{-3} torr, dec.^{p}	s = h, j	188
Sr(thd)$_2$(tetraglyme)	colorless solid	55–60	100, 10^{-3} torr^p	s = h, j	199
Ba(thd)$_2$(tetraglyme)	colorless solid	90–92	190–210, 0.01 torr^p	s = f, h	179, 190, 199, 202, 203
Sr$_2$(thd)$_4$(μ-H$_2$O)(diglyme)$_2$	solid	96–98	150, 10^{-3} torr^p	s = h, j	197
Ba$_2$(thd)$_4$(μ-H$_2$O)(diglyme)$_2$	colorless solid	*	190–210, 1 × 10^{-2} torr^p	s = f, h	179, 202
Ba{MeC(O)CHC(O)(CH$_2$)$_3$OCH$_2$CH$_2$OMe}$_2$	orange oil	*	involatile	s = l	204
Ba{MeC(O)CC(O)(CH$_2$)$_3$(OCH$_2$CH$_2$)$_2$OMe}$_2$	*	*	involatile	s = l	204
Ba{PhC(O)CHC(O)(CH$_2$)$_3$OCH$_2$CH$_2$OMe}$_2$	*	*	involatile	s = l	204

270

Ba(t-BuC(O)CHC(O)(CH₂)₃OCH₂CH₂OMe)₂	*	*	involatile	s = l	204
Ba(2,4,6-Me₃C₆H₂C(O)CHC(O)(CH₂)₃OCH₂CH₂OMe)₂	*	*	involatile	s = l	204
[Ba(miki)₂]₂	white solid	138–140	150–200, 10^{-3} torr[f]	s = f, h	205
Ba(diki)₂	white solid	170–171	150–200, 10^{-3} torr[f]	s = h	205, 206
Ba(triki)₂	white solid	80–82	150–200, 10^{-3} torr[f]	s = h	205, 206
[Ba(dpmiki)₂]₂	white solid	165–168	150–200, 10^{-3} torr[f]	s = h	205
Ba(dpdiki)₂	white solid	137–139	150–200, 10^{-3} torr[f]	s = h	205
Ba(dptriki)₂	white solid	127–128	150–200, 10^{-3} torr[f]	s = h	205

[a] * = information not available.

[b] s = soluble; a = diethyl ether, b = alcohols, c = ether : petroleum (40–60°C), d = alcohol : water, e = tetrahydrofuran, f = pentane, g = water, h = most organic solvents, i = chloroform, j = hexane, k = triethylamine, l = toluene, m = pyridine, n = benzene, o = methoxyethanol.

[c] Light-sensitive.

[d] Partial melting followed by complete melting (yellow solution).

[e] Residue (14%) Ba₂(thd)₄(NH₃)₂.

[f] Sublimate is [Ba(dpm)₂(HOC₆H₃-t-Bu₂-3,5)]ₙ.

[g] Decomposes by loss of bipy and sublimation of Ba₄(thd)₈.

[h] Decomposes by loss of water (20–30°C) and sublimation of Ba₄(thd)₈; BaCl₂ present in residue.

[i] Decomposition, sublimate ''Ba(thd)₂.''

[j] Compound obtained from the sublimate of [Ba(μ₄, μ₂-dpm)(dpm)(HOC₆H₃-t-Bu₃-3,5)₂(THF)]₂.

[k] Decomposes to Ba₄(thd)₈ and Ba(pivalate)₂.

[l] Sublimate is Ba(tmhd)₁.₅(OR)₀.₅.

[m] Sublimes in low yield.

[n] Dissociation with evaporation of polydentate ligand.

[o] hmtt is 1,1,4,7,10,10-hexamethyltriethylenetetramine.

[p] Sublimate is [M(β-diketonate)₂]ₙ.

[q] pmdt is N,N,N',N',N''-pentamethyldiethylenetriamine.

[r] Some polyamine dissociation occurs.

[s] Sublimes with some loss of polydentate ligand.

[t] Sublimes with some decomposition.

271

TABLE VI

β-Diketones and β-Ketoimines Used as Ligands, Their Formulas, and Commonly Used Abbreviations for Their Anions

β-Diketone or β-Diketoimine	Formula	Abbreviation of anion
Pentane-2,4-dione	$CH_3C(O)CH_2C(O)CH_3$	acac
1,1,1-Trifluoropentane-2,4-dione	$CF_3C(O)CH_2C(O)CH_3$	tfa, tfac
1,1,1,5,5,5-Hexafluoropentane-2,4-dione	$CF_3C(O)CH_2C(O)CF_3$	HFa, hfac
1,1,1-Trifluoro-5,5-dimethylhexane-2,4-dione	$CF_3C(O)CH_2C(O)C(CH_3)_3$	tpm
1,1,1,5,5,6,6,6-Octafluorohexane-2,4-dione	$CF_3C(O)CH_2C(O)CF_2CF_3$	ofhd
2,2,6,6-Tetramethylheptane-3,5-dione	$(CH_3)_3CC(O)CH_2C(O)C(CH_3)_3$	dpm, tmhd, thd
1,1,1,2,2-Pentafluoro-6,6-dimethylheptane-3,5-dione	$(CH_3)_3CC(O)CH_2C(O)CF_2CF_3$	ppm
1,1,1,5,5,6,6,7,7,7-Decafluoroheptane-2,4-dione	$CF_3C(O)CH_2C(O)CF_2CF_2CF_3$	dfhd
1,1,1,2,2,3,3-Heptafluoro-7,7-dimethyloctane-4,6-dione	$(CH_3)_3CC(O)CH_2C(O)CF_2CF_2CF_3$	hpm, fod
1,1,1,2,3,3,7,7,8,8,9,9,9-Tetradecafluorononane-4,6-dione	$CF_3CF_2C(O)CH_2C(O)CF_2CF_2CF_3$	tdfn, tdfhd
1,1 Dimethyl-8-methoxyoctane-3,5-dione	$(CH_3)_3CC(O)CH_2C(O)CH_2CH_2CH_2OCH_3$	dmmod
1,1,1-Trichloropentane-2,4-dione	$CCl_3C(O)CH_2C(O)CH_3$	tclac
1,3-Diphenylpropane-1,3-dione	$PhC(O)CH_2C(O)Ph$	dpp, Ph$_2$acac
2,2-Dimethyl-5-N-(2-methoxyethylimino)-3-hexanone	$(CH_3)_3CC(O)CH_2C(N(CH_2)_2OCH_3)CH_3$	miki
2,2-Dimethyl-5-N-(2-(2-methoxy)ethoxy-ethylimino)-3-hexanone	$(CH_3)_3CC(O)CH_2C(N(CH_2CH_2O_2CH_3)CH_3$	diki
2,2-Dimethyl-5-N-(2-(2-ethoxy)ethoxy)-ethoxyethylimino)-3-hexanone	$(CH_3)_3CC(O)CH_2C(N(CH_2CH_2O)_3CH_3)CH_3$	triki
5-N-(2-Methoxyethylimino)-2,2,6,6-tetramethyl-3-heptanone	$(CH_3)_3CC(O)CH_2C(N(CH_2)_2OCH_3)C(CH_3)_3$	dpmiki
5-N-(2-(2-Methoxy)ethoxyethylimino)-2,2,6,6-tetramethyl-3-heptanone	$(CH_3)_3CC(O)CH_2C(N(CH_2CH_2O)_2CH_3)C(CH_3)_3$	dpdiki
5-N-(2-(2-(2-Ethoxy)ethoxy)ethoxyethylimino)-2,2,6,6-tetramethyl-3-heptanone	$(CH_3)_3CC(O)CH_2C(N(CH_2CH_2O)_3CH_3)C(CH_3)_3$	dptriki

β-diketonate complexes of the Group 2 elements, those methodologies that have been encountered in previous sections are discussed only briefly, while more emphasis is given to methods that have not yet been described. For additional information on β-diketonate compounds see Refs. 19, 158, and 207–209. Heterometallic β-diketonate complexes are not reviewed here, however, because many of these species also contain alkoxide ligands; see Refs. 26, 75, 117, 118, 185, and 210, and for those with other anionic ligands, see Refs. 211 and 212, and references therein.

1. *β-Diketonates: Syntheses of M(β-Diketonate)$_2$L$_n$ (M = Ca, Sr or Ba; L = Mono- or Bidentate Base; n = 0–3) Compounds*

Belcher *et al.* reported the synthesis of the strontium derivatives of tpm, ppm, and hpm by addition of the β-diketone to an aqueous solution of the metal chloride (see later discussion for other method using MCl_2 salts). After adjustment of the solution pH to 7 with ammonia, the metal bis(β-diketonate) compounds precipitate as white solids.[151] The same authors have also reported the synthesis of the calcium derivatives of these three ligands by addition of the ligand to calcium hydroxide suspended in methanol. The compounds are precipitated by addition of water. The barium derivatives, however, were prepared directly by addition of the β-diketone to an aqueous solution of barium hydroxide. The $M^{II}(L)_2 \cdot 2H_2O$ (M = Ca, Sr, or Ba; L = tpm, ppm, or hpm) compounds are all similar in composition. Metal hydroxide methods have also been used for the preparation of $M(fod)_2$,[152] $M(hfac)_2(H_2O)_n$,[153] $(enH_2)_{1.5}[Ba(hfac)_5] \cdot C_2H_5OH$,[157,158] $Ba(dfhd)_2$,[213] $Ba(ofhd)_2$,[213] $Ba(acac)_2$,[164] $Ba(thd)_2(CH_3OH)_3 \cdot CH_3OH$,[171,172] $Ba_5(thd)_9(H_2O)_3(OH)$,[157] and $Ba(tclac)_2 \cdot 0.5H_2O$.[168] The general reaction as given in Eq. (16) is often conducted with the octa-aquo hydrate of the metal hydroxide, $M(OH)_2 \cdot 8H_2O$, which is slightly soluble in alcohols.[214] The acid–base reaction between MCO_3 and β-diketone has also been used in the synthesis of hydrated $M(tfac)_2 \cdot (H_2O)_n$ compounds (e.g., $Ba(tfac) \cdot 0.5H_2O$).[168] Yields by these methods are often moderate to good (40–80%).

$$M(OH)_2 + 2 \; \beta\text{-diketone} \longrightarrow M(\beta\text{-diketonate})_2 + 2 \; H_2O \qquad (16)$$

Anhydrous $M(hfac)_2$, $M(tfac)_2$ (M = Ca, Sr or Ba), "$Ba(thd)_2$,"[169] and $Ba(dmmod)_2$[167] compounds have been prepared by reaction of the metal with the β-diketone as both ligand and solvent [Eq. (17)].[154] Any excess ligand can be removed *in vacuo* or by washing of the product. Troyanov *et al.* have prepared $[Ba_2(hfac)_4(Et_2O)]_\infty$,[156] $Ba_4(thd)_8$,[178] $Ba_2(thd)_4(bipy)_2$,[175] $Ba_5(thd)_9(pivalate)$,[178] and $Ba_6(thd)_4(pivalate)_8(py)_4$[178] by a similar method, but with solvents other than the free β-diketones.

$$M^0 + 2 \ \beta\text{-diketone} \longrightarrow M(\beta\text{-diketonate})_2 + H_2$$

$$\text{(solvent} = \beta\text{-diketone, hydrocarbon, or ether)} \qquad (17)$$

Troyanov *et al.* have also reported the synthesis of $Ba_5(thd)_9Cl(H_2O)_7$[181] and $Na_2Ba_7(thd)_2(pivalate)_{14}(py)_4(H_2O)_2$[181] by reaction of $BaCl_2 \cdot 2H_2O$ with the sodium derivative of the ligands. The alkaline-earth metal bromides have been used in the synthesis of $M(tdfnd)_2 \cdot H_2O$ and $M(dfhd)_2 \cdot H_2O$.[160] The generalized reaction is shown in Eq. (18). One drawback of these reactions is that they usually yield hydrated compounds. However, it has been shown that tdfnd compounds can be dehydrated by sublimation.[161]

$$BaX_2 \cdot 2H_2O + 2 \ Na(\beta\text{-diketonate}) \longrightarrow Ba(\beta\text{-diketonate})_2 \cdot nH_2O + 2 \ NaX$$

$$\text{(X = Cl or Br)} \qquad (18)$$

Reaction of the metal ethoxide $[M(OEt)_2(HOEt)_4]$ with the desired β-diketone has been used as an anhydrous method to Group 2 β-diketonates and mixed ligand β-diketonate–alkoxide compounds. Presumably, the liberated HOR has a lower binding ability compared to liberated H_2O. Drake *et al.* have prepared $[Sr_3(tmhd)_6(Htmhd)]$ and $[Ba_4(tmhd)_8]$ by the reaction of Eq. (19).[177] Equation (20) gives the generalized reaction for formation of mixed ligand compounds of formula $[M(X)_n(OR)_{2-n}]_m$.[184,185]

$$[M(OEt)_2(HOEt)_4] + 2 \ HX \longrightarrow [M(X)_2]_n + 6 \ EtOH$$

$$\text{(M = Sr or Ba; X = tmhd)} \qquad (19)$$

$$M(OR)_2 + n \ HX \longrightarrow [M(X)_n(OR)_{2-n}]_m + n \ ROH$$

(M = Ca, Sr, Ba; X = tmhd; R = Me, Et, CH_2CH_2OMe, $CH_2CH_2O\text{-}i\text{-}Pr$, and $CH_2CH_2NMe_2$, $i\text{-}Pr$; solvent = hexane, ROH) \qquad (20)

The reactivity of Group 2 β-diketonates with two-electron Lewis bases has been demonstrated in the reaction between anhydrous "$Ba(tmhd)_2$" and ammonia gas.[174] The compound "$Ba(tmhd)_2$" was dissolved in C_6H_{14}, and a 5 : 1 mixture of N_2/dry ammonia was bubbled through this solution to give a precipitate. Single-crystal X-ray diffraction studies determined that the resulting compound was $[Ba(tmhd)_2(NH_3)_2]_2$. In a similar reaction, dissolution of oligomeric "$Ba(tmhd)_2$," prepared under anhydrous conditions, in Et_2O gave the dimer $[Ba(tmhd)_2(Et_2O)]_2$.[169]

Other less general synthetic methods include the preparation of the β-diketonate peroxo compound $[HNEt_3]_2[Ba_6(thd)_{10}(H_2O)_4(OH)_2(O_2)]$[182] by evaporation of a solution of "$Ba(thd)_2$" in NEt_3; the preparation of $Ba_6(thd)_{10}(H_2O)_6(O_2)$[183] by crystallizing the sublimate of $[Ba(\mu_1,\mu_2\text{-dpm})$ (dpm) $(HOC_6H_3Bu_2^t\text{-}3, 5)_2(THF)]_2$ from toluene; and the formation of $[Ba_4(tmhd)_8]$ by sublimation of $Ba(thd)_2(CH_3OH)_3 \cdot 2CH_3OH$.[172]

In summary, this brief review of the synthesis of Group 2 β-diketonate compounds shows how sensitive the outcome of the reaction is to the nature of the reaction conditions. As a result, it is often difficult to determine exactly which species have been used in CVD experiments when, for example, the species "Ba(dpm)$_2$" has been used and obtained from a commercial source. Indeed, the CVD literature is often misleading, with figures showing schematic diagrams of the "molecular structure" of monomeric "Ba(dpm)$_2$" as the source used for CVD experiments.[215]

2. β-Diketonates: Chemical and Physical Properties of M(β-Diketonate)$_2$L$_n$ (M = Ca, Sr or Ba; L = Mono- or Bidentate Base; n = 0–3) Compounds

Many of the β-diketones listed in Table V are fluorinated to varying degrees and have fluoroalkyl chain lengths of between five and nine carbon atoms. Although the presence of fluorine tends to improve volatility, it causes problems with MF_2 contamination of the thin films when compounds with these ligands are used for CVD (see Section IV). Within a given series of compounds, the calcium member tends to exhibit a lower sublimation temperature and presumably is more volatile than the heavier barium member. In the series ML_2 (M = Ca, Sr, and Ba; L = hfac and tfac), where hfac is a β-diketonate substituted with six fluorines and tfac is a β-diketonate substituted with three fluorines, both having carbon chain lengths of five, the more fluorinated hfac compounds are reported to sublime at lower temperatures than the less fluorinated tfac complexes.[154,216] For this same series of compounds, M(hfac)$_2$ (M = Ca, Sr, and Ba), the Ba(hfac)$_2$ complex possesses the highest sublimation temperature (see Table V).[154] The hfac complexes all reportedly decompose upon melting. The tdfnd compounds show the best volatility with least residue of any member of the M(β-diketonate)$_2$L$_n$ (L = two-electron donor Lewis base; n = 0, 1, 2) series. Ba(tdfnd)$_2$ sublimes at 220°C at atmospheric pressure with only 0.5–1% residue.[161] Similar trends have been claimed in fluorinated β-diketonate derivatives of other metals, such as Cu, used as precursors for CVD. Although the reason for the superior volatility of tdfnd compounds is not well understood, it is safe to say that the C_3F_7 units of Ba(RC(O)CH$_2$C(O)R)$_2$ must be quite effective in reducing intermolecular interactions.

In many instances, the products obtained from stoichiometrically similar reactions differ significantly in composition and structure depending on reaction conditions. For example, the correct formulation of the nominal species "Ba(thd)$_2$," over which a great deal of controversy has occurred, is dependent on whether its preparation is conducted by a hydrous or anhydrous method. The product from the reaction of the metal hydroxide octahydrate and two equivalents of Hthd is not a homoleptic β-diketonate compound "$[Ba(thd)_2]_n$," but rather $Ba_5(thd)_9(H_2O)_3(OH)$ (Fig. 31).[152,157] The structure of this compound consists of five barium atoms arranged in a roughly square pyramidal geometry with oxygen atoms from the β-diketonate, water, and hydroxy ligands bridging the metal centers. An alternative formulation of $Ba_5(thd)_9(H_2O)_3(OH)$, $Ba_5(thd)_5(Hthd)_4$ $(O)(OH)_3$, constructed by rearrangement of the crystallographically unobserved hydrogen atoms, has been proposed based on ligand geometry and literature precedent.[170] It has been noted that freshly prepared "$Ba(thd)_2$" (prepared by a hydrous method) loses volatility over time when stored in air. The authors note that this is in part due to hydrolysis and "carbonization" processes.[179]

When anhydrous preparative methods are employed, such as direct reaction between the metal and Hthd, the tetramer $Ba_4(thd)_8$, which has been shown to exist in two slightly different crystalline forms, is obtained (see Fig. 32).[177,178,180] Under similar reaction conditions, strontium forms the trimeric compound $Sr_3(tmhd)_6(Htmhd)$ (Fig. 33).[177] Both compounds contain chelating and bridging β-diketonate ligands. The bonding modes of these ligands vary to achieve metal coordination numbers of 6–8. The

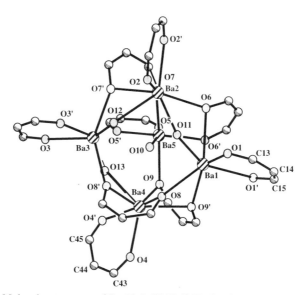

FIG. 31.   Molecular structure of $Ba_5(thd)_9(H_2O)_3(OH)$ showing the pyramidal Ba_5 unit (t-Bu groups have been omitted for clarity). In this formulation $O(11)$–$O(13)$ are assigned as H_2O ligands and $O(10)$ as an OH ligand. Assignment of hydrogen atom locations was based on bond length comparisons. (Redrawn from Ref. 157.)

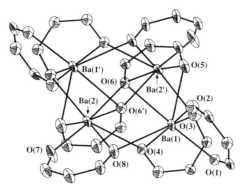

FIG. 32.   The structure of $Ba_4(tmhd)_8$ (*t*-Bu groups omitted for clarity). The barium atoms occupy the vertices of a fairly regular rhombohedron with an average Ba···Ba distance of 4.152(5) Å. The β-diketonate ligands bridge and chelate the metal centers in four distinct bonding modes. (Redrawn from Ref. 177.)

analogous calcium compound, $Ca_3(tmhd)_6$, has only been mentioned recently.[184] It is unclear whether this compound and the compound listed in Table V as $Ca(dpm)_2$[176] (made by a hydrous method) have the same composition. The same may be said for the relationship between $Sr_3(tmhd)_6$ (Htmhd) and $Sr(dpm)_2$.[176]

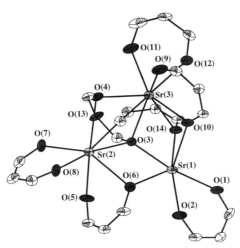

FIG. 33.   Molecular structure of $Sr_3(tmhd)_6(Htmhd)$ (*t*-Bu groups omitted for clarity). The Sr atoms are arranged in a triangular array capped on one face by a μ_3-oxygen atom of a tmhd ligand with the other ketonate oxygen atoms bridging μ_2 along an edge. The remaining coordination sites are filled by five tmhd ligands in various bonding modes and one chelating Htmhd ligand on Sr(3). (Redrawn from Ref. 177.)

Compounds such as $[Ba(thd)_2(Et_2O)]_2$[169] (Fig. 34), $[Ba_2(hfac)_4 \cdot Et_2O]_x$[156] (Fig. 35), $Ba(thd)_2(CH_3OH)_3$[171,172] (Fig. 36), $[Ca(hfac)_2 (H_2O)_2]_2$,[153] and $[Ba(tmhd)_2(NH_3)_2]_2$[174] can be formed from the anhydrous compounds by addition of the respective bases Et_2O, CH_3OH, H_2O, and NH_3. The species $[Ba(thd)_2(Et_2O)]_2$ exists as a dimer in the solid state, with two thd ligands bridging the barium centers through both ketonate oxygen atoms (i.e., two β-diketonates lie roughly perpendicular to the Ba–Ba vector; see Fig. 34) and two others chelating the metal centers. In addition, one ether group is terminally bonded to each Ba center to give each metal a coordination number of 7. $[Ba(tmhd)_2(2NH_3)_2]_2$, $Ba_2(thd)_4$ $(bipy)_2$,[175] and $[Ba(\mu_1, \mu_2\text{-dpm})(dpm)(HOC_6H_3\text{-}t\text{-Bu}_2\text{-}3,5)_2(THF)]_2$[131] have similar structures, but with one additional terminal ligand per metal. The structure of the etherate, $[Ba_2(hfac)_4(Et_2O)]_x$ (Fig. 35), has also been described.[156] This compound is polymeric with infinite zigzag chains of barium atoms. Interestingly, fluorine atoms from the C—F bonds interact with the metal centers. The Ba—F distances are in the range 2.97–3.09 Å. The weak basicity of the hfac ligand and the coordinative unsaturation

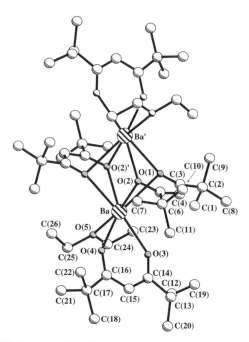

FIG. 34.   ORTEP diagram of $[Ba(thd)_2(Et_2O)]_2$ showing the μ_2-bridging oxygen atoms of the thd ligands and their roughly perpendicular arrangement to the Ba$\cdots$Ba vector. The remainder of the coordination sphere around each barium is then completed by a chelating thd ligand and a terminally bonded ether ligand. (Redrawn from Ref. 169.)

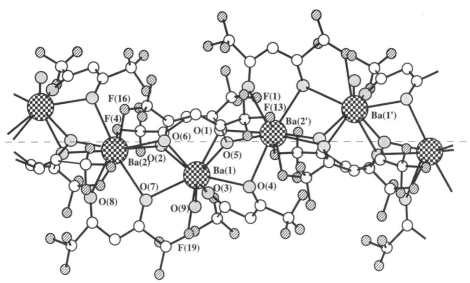

FIG. 35. ORTEP diagram of $[Ba_2(hfac)_4(Et_2O)]_x$ projected on the O_{yz} plane (Et groups of the ether ligand have been omitted). The structure consists of infinite zigzag chains of Ba(2)Ba(1)(2')Ba(1') atoms bridged and chelated by the hfac ligands. An exceptional feature of the structure is the Ba–F(F$_2$C) interactions (2.97–3.09 Å), most easily seen for Ba(2) and Ba(2'). (Redrawn from Ref. 156.)

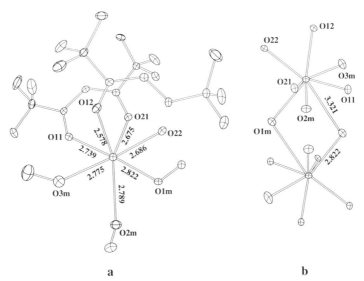

a   b

FIG. 36. (a) ORTEP diagram of $Ba(thd)_2(CH_3OH)_3$. The coordination sphere comprises two chelating thd ligands and three methanol ligands. The O1m atoms of methanol ligands of neighboring molecules, related to each other by an inversion center, form a very long double bridge as shown in diagram (b). (Redrawn from Ref. 171.)

of the barium centers may make these types of interactions energetically favorable. These interactions may also aid in the formation of metal fluorides on thermal decomposition of the $M(\beta\text{-diketonate}^F)_2$ compounds.

The species $Ba(thd)_2(TMEDA)_2$,[170] $Ba(thd)_2(H_2O)_2(CH_3OH)_2$,[172] and $(enH_2)_{1.5}[Ba(hfac)_5] \cdot C_2H_5OH$[157,158] (Fig. 37) are the only monomeric β-diketonate compounds containing mono- or bidentate bases that have been structurally characterized, although $Ba(thd)_2(CH_3OH)_3$ (Fig. 36) is only weakly associated through hydrogen bonds and long methanol bridges in the solid state. The structure of $Ba(thd)_2(CH_3OH)_3$ is comprised of three terminally ligated methanol groups and two chelating thd ligands. The barium center has a coordination number of 7. One methanol ligand from each monomer has a long-distance interaction (3.32 Å) with a neighboring metal center. The anionic chelate $(enH_2)_{1.5}[Ba(hfac)_5] \cdot C_2H_5OH$ (Fig. 37) has been claimed to have high volatility (in fact, it was claimed as the most volatile barium-containing compound at the time based on TGA data).[158] The structure consists of one dangling and four chelating hfac ligands, giving the barium atom a coordination number of 9. The ethanol molecule is not coordinated to the metal center. The surrounding $[enH_2]^{2+}$

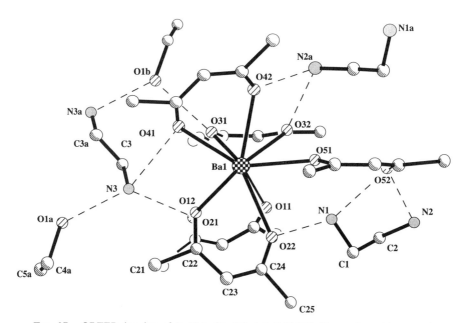

FIG. 37.   ORTEP drawing of $(enH_2)_{1.5}[Ba(hfac)_5] \cdot C_2H_5OH$. The amine cations are involved in hydrogen bonding interactions with the oxygen atoms of the hfac ligands and the ethanol solvent molecule. (Redrawn from Refs. 157 and 158.)

ions are hydrogen bonded to hfac ligands of the same and adjacent [Ba (hfac)$_5$]$^{3-}$ ions. Only one type of hfac ligand is observed by NMR (in D$_2$O), suggesting that rapid exchange of the terminal and chelating hfac ligands occurs in this solvent. (enH$_2$)$_{1.5}$[Ba(hfac)$_5$] · C$_2$H$_5$OH has good thermal stability (>250°C) and sublimes with only 1–4% residue.[157] The volatility of this salt may result from a dissociative process that would give "Ba (hfac)$_2$," Hhfac, and en (or [enH$_2$][hfac]$_2$) on heating, as shown in Eq. (21) (consistent with TGA data). This equilibrium would be driven to the right by higher temperatures and removal of the volatile products by the carrier gas. A similar process has been proposed to account for the volatility observed for (C$_2$H$_5$)$_3$NH[Eu(hfac)$_4$].[217]

$$2\ (enH_2)_{1.5}[Ba(hfac)_5] \cdot C_2H_5OH \rightleftharpoons 2\ \text{``Ba(hfac)}_2\text{''} + 6\ Hhfac + 3\ en + 2\ C_2H_5OH \quad (21)$$

The compound [Ca(hfac)$_2$(H$_2$O)$_2$]$_2$[153] (Fig. 38) is a centrosymmetric dimer in the solid state, while the analogous barium complex, [Ba(hfac)$_2$ (H$_2$O)]$_\infty$,[153] exists as a polymer. Both compounds have M–F interactions similar to those described for [Ba$_2$(hfac)$_4$(Et$_2$O)]$_\infty$. The smaller size of calcium is probably responsible for the structural differences between these two water adducts. The volatility of these fluorinated complexes is certainly decreased by the M–F interactions, as well as by the overall oligomeric structure of the compounds. What relationship metal ligand interactions of this type play in the incorporation of BaF$_2$ in films prepared from these compounds has not yet been elucidated. Interestingly, no base-free fluorinated β-diketonate compounds have been structurally characterized.

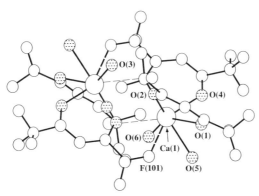

FIG. 38.   ORTEP diagram of [Ca(hfac)$_2$(H$_2$O)$_2$]$_2$ showing that each calcium atom is bonded to one chelating hfac ligand, two water molecules, and a second hfac ligand that is both chelating and bridging. The coordination sphere around each metal is completed by a Ca–F(F$_2$C) interaction (2.92–2.97 Å). (Redrawn from Ref. 153.)

Slow evaporation of a solution of "Ba(thd)$_2$" (not scrupulously anhydrous or anaerobic) in triethylamine resulted in the formation of crystals of the previously mentioned peroxo compound [Ba$_6$(thd)$_{10}$(H$_2$O)$_4$(OH)$_2$(O$_2$)][HNEt$_3$]$_2$,[182] shown in Fig. 39. Until recently, this was the only molecular peroxide compound of barium known.[183] In this compound, the six barium atoms are arranged in an octahedral array with each of the oxygen atoms of the O$_2^-$ ligand bonded to four metals (i.e., μ_4). The Ba$_6$ core is further supported by bridging hydroxides and thd ligands. Each barium atom retains one chelating thd ligand to give a coordination number of 7.

Mass spectral studies have been conducted on many of the β-diketonate compounds discussed in this review. Mass spectra of the compounds in the series ML$_2$ (M = Ca, Sr, and Ba; L = tpm, ppm, and hpm) have shown that in the gas phase, from a source temperature of 200°C, many oligomeric species, such as [Ca$_4$(tpm)$_7$]$^+$ and [Ca$_5$(tpm)$_9$]$^+$, are present.[151] The molecular ion peak corresponding to ML$_2^+$ was reported to be very small or nonexistent for these compounds. The mass spectral studies of anhydrous "Ba(thd)$_2$" (presumably [Ba$_4$(tmhd)$_8$]) have shown that several oligomeric species also exist; the ions [Ba$_4$(thd)$_7$]$^+$, [Ba$_3$(thd)$_5$]$^+$, [Ba$_2$(thd)$_3$]$^+$, and [Ba(thd)]$^+$ were detected in the mass spectrum.[152] The complexes ML$_2$ (M = Ca, Sr, and Ba; L = hfac and tfac) have been

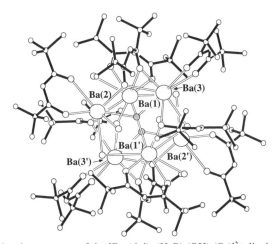

FIG. 39.   Molecular structure of the [Ba$_6$(thd)$_{10}$(H$_2$O)$_4$(OH)$_2$(O$_2$)]$^{2-}$ dianion. The peroxide ligand is represented by the two small spheres that are discernible through the triangular face of the Ba$_6$ octahedron. Ba—O$_{peroxide}$ bond distances have a range of 2.68(1)–2.76(1) Å; the O—O bond distance of 1.48(1) Å in [Ba$_6$(thd)$_{10}$(H$_2$O)$_4$(OH)$_2$(O$_2$)]$^{2-}$ compares closely with the 1.51(2) Å distance for the peroxide ligand in Ba$_6$(thd)$_{10}$(H$_2$O)$_6$(O$_2$),[183] a compound with the same core structure. (Redrawn from Ref. 182.)

shown to sublime with some decomposition.[154] The decomposition of these complexes involves the loss of the free ligand, as determined by mass spectroscopy. The authors note that it is not known whether ligand loss was due to heating or from electron impact. Peaks assigned as ML^+, $M_2F_2L^+$, $M_2L_2F^+$, and $M_2L_3^+$ were found in the mass spectra of the hfac and tdfnd complexes.[154,160] Similar results have been obtained for the $M(fod)_2$ series.[152] However, the mass spectra of $Ba(ofhd)_2$ and $Ba(dfhd)_2$ have their most prominent peaks corresponding to the molecular ion ML_2^+ and the fragment ML^+ when very mild ionization methods are used.[159] This suggests that these compounds may exist as aggregates with only weak intermolecular interactions in the solid state. In general, these data suggest that although many complexes are oligomeric in the solid state, they are less so in the gas phase.

A number of authors have reported significant increases in the volatility of "$Ba(tmhd)_2$" (prepared by both hydrous and anhydrous methods) and fluorinated β-diketonate compounds when they are heated in the presence of bases such as thf, NH_3, pyridine, and Htmhd.[164,218,219] The increase in volatility has been attributed to deoligomerization of the $[Ba_5(thd)_9(H_2O)_3(OH)]$, $[Ba_4(tmhd)_8]$, and $[Ba(\beta\text{-diketonate}^F)_2]_n$ aggregates by these bases. This is supported by isolation of compounds such as $[Ba(thd)_2 \cdot Et_2O]_2$, $[Ca(hfac)_2(OH_2)_2]_2$, and $[Ba(tmhd)_2(NH_3)_2]_2$.[153,169,174]

F. $M(\beta\text{-Diketonate})_2$ (Polydentate) Compounds

This section describes the synthesis as well as chemical and physical properties of Group 2 metal β-diketonate compounds that also contain coordinated polydentate ligands to prevent oligomerization.

1. $M(\beta\text{-Diketonate})_2$ (Polydentate) Compounds: Synthesis of Ca, Sr, and Ba β-Diketonate and Ketoiminate Compounds with Polydentate Bases

Most compounds of general formula $M(\beta\text{-diketonate})_2L_n$, L = a polydentate ligand that can be separate or appended to the β-diketonate and $n = 1$ or 2, have been synthesized by one of the following methods: method A, addition of the polydentate ligand to a solution of $[M(\beta\text{-diketonate})_2]_n$ [Eq. (22)] or method B, in situ preparation of $M(\beta\text{-diketonate})_2$, by reactions similar to those described for the $M(\beta\text{-diketonate})_2L_n$ (L = mono- or bidentate base; $n = 0\text{--}3$) compounds [Eqs. (16–19)], in the presence of the polydentate ligand. By method A, $Ba(acac)_2(18\text{-crown-}6) \cdot 0.5H_2O$,[168] $Ba(thd)_2(pmdt)$,[200,201] $Ba(thd)_2(hmtt)$,[200,201] $Sr(thd)_2(triglyme)$,[197] $Sr_2(thd)_4(\mu\text{-H}_2O)(diglyme)_2$,[197] $Sr(tfac)_2(hmtt)$,[197] $Ba(hfac)_2(tetraglyme)$,[187,190] $Ba(hfac)_2(hexaglyme)$,[187] $Sr(hfac)_2(tetraglyme)$,[187]

Ca(hfac)$_2$(triglyme),[187] Ba(hfac)$_2$(18-crown-6),[168] Ba(tclac)$_2$ (18-crown-6),[168] Ba(tfac)$_2$(18-crown-6),[168] Ba(thd)$_2$(18-crown-6)H$_2$O,[168] Ba(thd)$_2$ (NH(CH$_2$CH$_2$NH$_2$)$_2$)$_2$,[201] Ba(thd)$_2$(Me(OCH$_2$CH$_2$)$_2$OH),[201] and Ba(thd)$_2$ (Me(OCH$_2$CH$_2$)$_3$OH)[201] have been prepared. Yields by this method are usually on the order of 60–80%.

$$[M(\beta\text{-diketonate})_2]_n + L \longrightarrow M(\beta\text{-diketonate})_2L$$

(M = Ca, Sr, Ba; L = glymes, crown ethers, and polyamines; solvent = aliphatic or aromatic hydrocarbons)   (22)

By method B, Eq. (16): Ba(hfac)$_2$(tetraglyme),[189] Sr(hfac)$_2$(tetra-glyme),[189] Ba(thd)$_2$(tetraglyme)[203]; Eq. (17): Ba$_2$(thd)$_4$(μ-H$_2$O)(di-glyme)$_2$,[202] Ba(thd)$_2$(tetraglyme)[190,202]; Eq. (19): Ca(thd)$_2$(triglyme),[188] Ba(thd)$_2$(triglyme),[188] Ca(hfac)$_2$(tetraglyme),[188] Ba(tfac)$_2$(hmtt),[198] Sr (thd)$_2$(tetraglyme),[199] Ba(thd)$_2$(tetraglyme),[199] Sr(tfac)$_2$(triglyme),[199] and Sr(Ph$_2$acac)$_2$(tetraglyme)[199] have been synthesized. A few compounds— M(hfac)$_2$(18-crown-6),[155,193] M(hfac)$_2$(diaza-18-crown-6),[193,220] Ba(tdfnd)$_2$ (tetraglyme),[161] Ba(hfac)$_2$(tetraglyme),[155] Ba(tfac)$_2$(tetraglyme),[190] and the β-diketonate and ketoiminate lariats, Ba(RC(O)CH$_2$C(O)(CH$_2$)$_3$(OCH$_2$ CH$_2$)$_n$OMe)$_2$,[204] [Ba(miki)$_2$]$_2$, Ba(diki)$_2$, Ba(triki)$_2$, Ba(dpdiki)$_2$, [Ba (dpmiki)$_2$]$_2$, and Ba(dptriki)$_2$,[206,221] have been prepared by reaction of MH$_2$ with the protonated form of the ligand in the presence of polyether and solvent [Eq. (23)].

$$MH_2 + 2\ \beta\text{-diketone} + L \longrightarrow M(\beta\text{-diketonate})_2L + H_2(g) \qquad (23)$$

Metal nitrates were reacted with the sodium salt of the β-diketonate and polyether in the preparation of Ca(hfac)$_2 \cdot$ (18-crown-6),[192] Ba(hfac)$_2 \cdot$ (CH$_3$O(CH$_2$CH$_2$O)$_5$C$_2$H$_5$),[195] and Ba(hfac)$_2 \cdot$ (CH$_3$O(CH$_2$CH$_2$O)$_6$-n-C$_4$H$_9$)[195] [Eq. (24)], and CaO reacts with Hhfac and tetraglyme to give Ca(hfac)$_2 \cdot$ (tetraglyme)[189] according to Eq. (25).

$$M(NO_3)_2 + 2\ Na(\beta\text{-diketonate}) + L \longrightarrow M(\beta\text{-diketonate})_2L + 2\ NaNO_3(s) \quad (24)$$

$$CaO + 2\ \beta\text{-diketone} + L \longrightarrow Ca(\beta\text{-diketonate})_2L + H_2O \qquad (25)$$

2. *M(β-Diketonate)$_2$ (Polydentate) Compounds: Chemical and Physical Properties of β-Diketonate and Ketoiminate Compounds with Polydentate Bases*

The M(β-diketonate)$_2$(polyether) compounds are normally white or colorless solids with good solubility in many organic solvents. As a group, they have the lowest melting points and sublimation temperatures, and therefore the best apparent volatilities, of any of the compounds discussed in this review.[216] As noted previously, passing nitrogen-containing bases over oligomeric base-free β-diketonate compounds such as M(hfac)$_2$ and Ba$_4$(tmhd)$_8$ enhances their volatility. This behavior is thought to arise from deoligomerization of these species by coordination of the base. One would, therefore, expect that by incorporation of polydentate bases, which have the ability to occupy numerous coordination sites and thereby greatly reduce oligomerization, the volatility of the resulting compounds should be greatly enhanced over that of the parent compound. This is, in fact, observed in at least some M(β-diketonate)$_2$(polydentate) systems. For example, all the M(hfac)$_2$(polyether) compounds for which data have been reported (see Table V) sublime with minimal decomposition at temperatures and pressures below that of their base-free counterparts M(hfac)$_2$. They also melt at lower temperatures and without decomposition, unlike the parent compounds. Other perfluorinated β-diketonate compounds behave similarly. However, all the tfac compounds, with the exception of Sr(tfac)$_2$(triglyme) and Ba(tfac)$_2$(tetraglyme), lose some or all their polydentate ligand on sublimation to give sublimates of composition [M(β-diketonate^F)$_2$]$_n$. The nonfluorinated M(β-diketonate)$_2$L compounds undergo similar polyether loss. Nevertheless, these compounds are still useful sources of volatile [M(β-diketonate)$_2$]$_n$ species. The Ba(tdfnd)$_2$ (tetraglyme) compound has a melting point 96°C *below*, and a sublimation temperature 60°C *below*, that reported for the base-free compound Ba(tdfnd)$_2$.[161,196] In fact, Ba(tdfnd)$_2$(tetraglyme) appears to be the most volatile barium compound reported to date.[196] Although volatility increases have been attained in many cases by formation of polyether and polyamine adducts such as the ones described earlier, these compounds behave similarly to their base-free progenitors in one unfortunate respect: films deposited from these precursors can contain significant MF$_2$ contamination.[16,190,222]

Loss of the polyether ligand through sublimation of partially fluorinated and nonfluorinated I β-diketonate compounds may be a consequence of the stronger basicity of the β-diketonate ligands in these compounds. One would expect that as basicity of the β-diketonate ligands increases, the bonds between the metal and ether oxygen atoms become weaker. A comparison of bond lengths in fluorinated and nonfluorinated compounds seems to bear this out.[199] One obvious way to prevent this type of behavior is by connecting the ether and β-diketonate ligands together through a chemical bond. Unfortunately, the M(β-diketonate)$_2$ compounds with lariat ether appendages that have been synthesized to date are involatile.[204] This is probably due to intermolecular binding of the polyether side chains. However, related M(β-ketoiminate)$_2$ compounds with lariat ether appendages sublime, but with some decomposition. The major volatile decomposition product is the free ligand (e.g., Hdiki for Ba(diki)$_2$).[206,221] These compounds have either monomeric or dimeric structures, depending on the particular ketoiminate employed.

The majority of M(β-diketonate)$_2$ polyether and polyamine adducts are monomeric in the solid state. This is, of course, to be expected, because many of these compounds have metal coordination numbers of 9 and 10. The structures of Ba(hfac)$_2$(18-crown-6),[193] Ba(hfac)$_2$(tetraglyme),[191] Ca(hfac)$_2$(tetraglyme),[188] Ba(tfac)$_2$(hmtt),[198] and Ba(triki)$_2$[221] are presented in Figs. 40–44, respectively. The structures of Ba(hfac)$_2$(tetraglyme) and Ba(hfac)$_2$(18-crown-6) are quite similar, in that both consist of a polyether ligand wrapped around the center of the barium atom and sandwiched between chelating hfac ligands. In fact, similar structures are found for all the structurally characterized barium and strontium β-diketonate compounds containing tetraglyme, triglyme, and 18-crown-6 ethers. The metal coordination number in these compounds range from 8 to 10, varying with the number of oxygen atoms of the ether. When smaller glymes are used, such as diglyme, greater aggregation occurs because of coordinative unsaturation of the metal centers (e.g., M$_2$(thd)$_4$(μ-H$_2$O)(diglyme)$_2$).[197,202] If smaller metal centers are matched with large glymes, structures similar to that of Ca(hfac)$_2$(tetraglyme) are observed. In this compound the calcium atom has a coordination number of 7. One terminal MeOCH$_2$–ether linkage remains uncoordinated; the rest of the structure is similar to that of Ba(hfac)$_2$(tetraglyme). In general, the structures of barium and strontium compounds are more similar than those of barium and calcium. This is one might expect based on a size comparison of barium, strontium, and calcium. The difference in the ionic

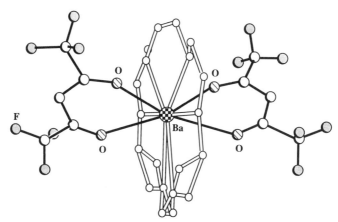

Fig. 40.   Molecular structure of Ba(hfac)$_2$(18-crown-6) showing the disorder in the 18-crown-6 ether. The average Ba–O hfac bond distance is 2.82 Å. (Redrawn from Ref. 193.)

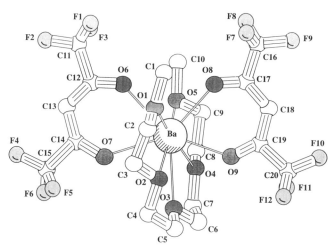

FIG. 41.   PLUTON drawing of the molecular structure of Ba(hfac)$_2$(tetraglyme). The Ba—O ether bond distances have a range of 2.809(2)–2.893(1) Å, and the average Ba—O$_{hfac}$ bond distance is 2.703(14) Å. (Redrawn from Ref. 191.)

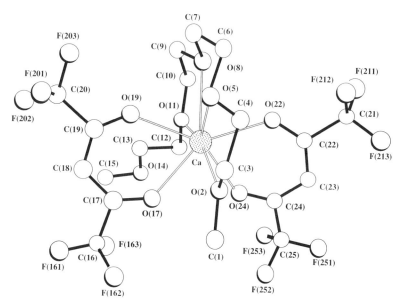

FIG. 42.   Molecular structure of Ca(hfac)$_2$(tetraglyme) showing the nonbonded terminal ether group (O(14)). The Ca—O ether and β-diketonate bond distances have a range of 2.434(3)–2.497(3) Å and 2.395(3)–2.410(3) Å, respectively. (Redrawn from Ref. 188.)

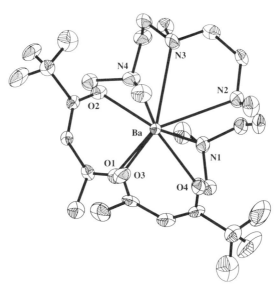

FIG. 43. Molecular structure of Ba(tfac)$_2$(hmtt) showing the square prismatic geometry around the barium atom. The average Ba—O$_{tfac}$ and Ba—N bond distances are 2.677(9) Å and 2.983(1) Å, respectively. (Redrawn from Ref. 198.)

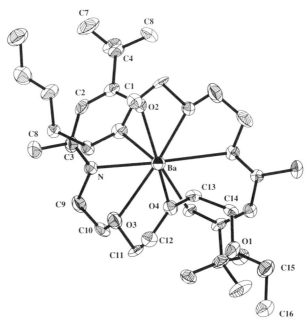

FIG. 44. Molecular structure (50% probability ellipsoids) of Ba(triki)$_2$ showing the chelating ketoiminate ligands and the nonbonded terminal ethoxy group of the lariat ether. (Redrawn from Ref. 221.)

radii of barium and strontium is only 0.17 Å, while the difference between barium and calcium is double this at 0.35 Å.[51] Whereas tetraglyme is a more suitable acyclic ether to achieve coordinative saturation for barium and strontium, triglyme seems to be a better match for calcium. More compact cyclic ethers such as 18-crown-6 are able to coordinate all their oxygen atoms to the metal center, as found in the M(β-diketonate)$_2$(18-crown-6) (M = Ca, Sr, Ba) compounds.[192–194]

M(β-diketonate)$_2$ compounds containing polyamine ligands[198,200] such as pmdt and hmtt, which are analogous to the triglyme and tetraglyme polyethers in number of coordinating sites, have much the same arrangement of ligands as found in the M(hfac)$_2$(tetraglyme) and M(hfac)$_2$(triglyme) compounds. However, the pyramidal geometry associated with the nitrogen atoms of the hmtt amine in Ba(tfac)$_2$(hmtt), shown in Fig. 43, leads to a more side-on bonding of the polyamine in this compound than found for the polyethers in the M(β-diketonate)$_2$(polyether) compounds. This behavior is also observed in compounds of type M(hfac)$_2$(di-aza-18-crown-6), where diaza-18-crown-6 sits to one side of the metal.[220]

Barium compounds with β-ketoiminate ligands containing lariat ether appendages are either monomeric or dimeric in the solid state, depending on the length of the ether appendage. Compounds with $-CH_2CH_2OMe$ (miki and dpmiki) appendages on the nitrogen atom are dimeric, whereas compounds with $-CH_2CH_2OCH_2CH_2OMe$ (diki and dpdiki) and $-CH_2CH_2OCH_2CH_2OCH_2CH_2OMe$ (triki and dptriki) appendages on the nitrogen atoms were found to be monomeric.[206,221] The now-familiar relationship between coordinative saturation and degree of oligomerization seems to be at work here. The miki and dpmiki compounds can only achieve barium coordination numbers in excess of 6 by aggregation, while those with diki, dpdiki, triki, and dptriki can attain coordination numbers of 8–10 while retaining a monomeric form. However, even as a dimer, the barium centers in [Ba(dpmiki)$_2$]$_2$ have only coordination number 6 because two nitrogen atoms of the β-ketoiminate ligands are nonbonding. The authors attribute this to constraints imposed by the sterically demanding t-Bu groups of the β-ketoiminate ligand.[221] It is unclear whether this type of coordination also occurs in the less crowded [Ba(miki)$_2$]$_2$ compound. The structure of Ba(triki)$_2$ is shown in Fig. 44. The terminal ethoxy groups of the lariat remain nonbonding to give the compound a coordination number of 8. The coordination geometry is best described as distorted trigonal dodecahedral.

The mass spectra of M(β-diketonate)$_2$(polydentate) and M(β-diketonate^F)$_2 \cdot$ polydentate compounds are quite different from each other. The mass spectra of nonfluorinated compounds typically have intense peaks for oligomeric ions, such as [M$_2$(β-diketonate)$_4$]$^+$, implying that dissociation of the Lewis base and formation of base-free $[M(\beta\text{-diketonate})_2]_n$ species occurs either before or during vaporization.[188,199] The fluorinated compounds have their highest-mass peaks corresponding to the molecular ion or the $[M(\beta\text{-diketonate}^F)\,(\text{polydentate})]^+$ ion, indicating that these species remain monomeric in the gas phase and do not easily lose their polyether.[155,168,187,189,195,196,198] These data are entirely consistent with the apparent volatility data previously discussed.

G. *Other Compounds with Mono-, Bi-, or Polydentate Ligands*

The final part of the synthesis and properties section of this review addresses other alkaline-earth complexes containing mono-, bi-, or polydentate ligands and anionic ligands such as carboxylates, halides, thiocyanates, or nitrates, where the low degree of oligomerization may result in sufficient volatility for CVD. We have not attempted to be comprehensive in this section, but rather to present examples of the different types of compounds formed by these anionic ligands. More comprehensive coverage of this area can be found elsewhere.[74,170,223,224]

1. *Other Compounds with Mono- or Bidentate Ligands: Synthesis of Ca, Sr, and Ba Compounds of Type* MX_mL_n ($X = O_2SOMe$, O_3SCF_3, $\{OCS\}OMe$, *COT, Carborane, Cp, Alkyl,* RCO_2, *RCSO,* $SeSi(SiMe_3)_3$, $TeSi(SiMe_3)_3$, Te_4, $SbSe_2$, *SCN,* NO_3, ClO_4, ClO_3, *Halide;* $L = Mono\text{-} or$ *Bidentate Ligand;* $n = 0\text{--}6$; $m = 1$ or 2)

Perhaps the most interesting compounds from this group of anionic ligands, from a materials point of view, are the selenolates and tellurolates: $M(SeSi(SiMe_3)_3)_2(TMEDA)_2$, $M(TeSi(SiMe_3)_3)_2(THF)_4$, and $M(TeSi(SiMe_3)_3)_2(\text{pyridine})_n$ ($n = 4$, $M = $ Ca, Sr; $n = 5$, $M = $ Ba).[225,226] These Lewis base adducts have been prepared from solutions of the metal bis(trimethylsilyl) amide adducts, $M(N(SiMe_3)_2)_2(THF)_2$, by addition of HTeSi$(SiMe_3)_3$ or HSeSi$(SiMe_3)_3$ and base (TMEDA, THF, or pyridine) according to Eq. (26).

$$M(N(SiMe_3)_2)_2(THF)_2 + 2\ HESi(SiMe_3)_3 + nL \longrightarrow M(ESi(SiMe_3)_3)_2L_n + 2HN(SiMe_3)_2$$

$$(M = Ca, Sr, Ba;\ E = Se, Te;\ L = TMEDA, THF, \text{pyridine}) \qquad (26)$$

The strontium selenolate $Sr(Se(2,4,6\text{-}t\text{-}Bu_3C_6H_2))_2(THF)_4$ has been synthesized by reaction of SrI_2 with $KSe(2,4,6\text{-}t\text{-}Bu_3C_6H_2)$ in THF. The compound is low melting ($120\text{--}125°C$) and soluble in THF.[227] Tellurium- and selenium-containing salts, $Ca(DMF)_6(Te_4)$[228] and $Ba(en)_4(SbSe_2)_2$;[229] with mono- and bidentate bases such as dimethylformamide and ethylenediamine have been synthesized by reaction of the metal with the Group 16 element in the presence of base or by addition of the base to a base-free salt (i.e., $Ba_4Sb_4Se_{11}$). Interestingly, although numerous alkaline-earth compounds with oxygen containing anionic ligands have been described, only a few compounds with ligated sulfur atoms have been reported (*vide infra*, thiocarboxylates). Such species could be important as precursors in the formation of metal sulfide phases with interesting electrical and optical properties. The thiolate compound $Sr(S(2,4,6\text{-}t\text{-}Bu_3C_6H_2))_2(THF)_4$ was reported as the product of the reaction between $KS(2,4,6,\text{-}t\text{-}Bu_3C_6H_2)$ and SrI_2.[227] The heterocyclic ligand 5-(1-naphthylamino)-1,2,3,4-thiatriazole (naphthyl-NHC̄NNNS̄) rearranges on reaction with $Ba(OH)_2$ and HMPA in toluene to give a mercaptide–tetrazole ligand, which then chelates to the barium center through both sulfur and nitrogen atoms to give Ba(naphthyl-N̄NNNC̄(S))$_2 \cdot$3HMPA.[230] Related compounds, $M(\overline{C_6H_4OC(S)N})_2 \cdot L_n$ (M = Ca, Sr, Ba; L = H_2O, HMPA), can be prepared by reaction of two equivalents of the protonated ligand $\overline{C_6H_4OC(S)N}H$ with either the metal or MH_2 and base (anhydrous method).[231]

Carboxylate (RCO_2),[232–236] thiocarboxylate $(RCSO)$,[237–239] sulfonate (triflate) (CF_3SO_3),[71,168] and sulfito (O_2SOMe)[240] compounds have been prepared, but most are oligomeric or polymeric with monodentate (alcohols, water, THF) bases (see later discussion on polydentate bases). The $M(RCO_2)_2L$ (carboxylate), $M(RCSO)_2L$ (thiocarboxylate), $M(O_3SCF_3)_2L$ (sulfonate), and $M(RCSO)(RCO_2) \cdot L$ (thiocarboxylate–carboxylate) compounds are obtained by reaction of the acids with MCO_3 (also MO or $M(OH)_2$ in the case of the triflate) (hydrous method) or MH_2 (anhydrous method) in the presence of the base. The sulfito compound $[Ca(O_2SOMe)_2(MeOH)_2]_\infty$[240] was prepared in a novel way, by reaction of $[Ca(OMe)_2]_n$ with SO_2 gas in methanol. The reaction is thought to proceed either by a direct insertion of SO_2 into the metal–alkoxide bond or by alcohol-catalyzed formation of the ^-O_2SOR ion followed by nucleophilic substitution on the metal–alkoxide bond. Carbonyl sulfide (COS) was found to behave similarly, inserting into the alkoxide bonds of $[Ca(OMe)_2]_n$ to give $[Ca(\{OCS\}OMe)_2(MeOH)_3]_2$,[241] and CO_2 also appears to undergo insertion reactions with metal alkoxides.[242,243]

The diacetamide compounds, $MX_2(CH_3C(O)NHC(O)CH_3)_n \cdot mH_2O$ (M = Ca, X = Br (n = 4), ClO_4 (n = 5), ClO_3 (n = 4, m = 1)[244–246]; M = Sr, Ba, X = ClO_4 (n = 4 or 5)),[247,248] were obtained by reaction of the anhydrous or hydrated salt $(MX_2 \cdot mH_2O)$ with diacetamide. MX_2L (M = Ca, Sr; X = Cl or Br; L = inositol, galactose, lactose, alcohols),[233,249] $M(NCS)_2(2,2'\text{-bipyridine})_2 \cdot mH_2O$,[250] $M(ClO_4)_2(1,10\text{-phenanthroline})_n \cdot mH_2O$,[251] and the Schiff base complex $CaX_2(H_2salpd)_n \cdot mH_2O$ (X = Cl or NO_3; H_2salpd = N,N'-propane-1,3-diylbis(salicylideneimine);

$n = 1$ or 2) and derivatives were synthesized in an analogous fashion.[252] In contrast, $Sr(NCS)_2(HMPA)$ and $SrI_2(HMPA)$ were not prepared by dissolution of their respective anhydrous salts by base, but rather by reaction of NH_4X (X = SCN or I) with Sr metal in the presence of base.[253] The main advantage of this method is that it yields anhydrous compounds. In addition, this method is useful in the preparation of salt adducts when the dissolution method fails for kinetic reasons (i.e., when the metal coordination sites of $[MX_2]_n$ are inaccessible to the base).

The majority of the developments in the organometallic chemistry of the alkaline-earth metals have been confined to the lighter members of the group (Be, Mg), where covalency and stability are greater.[224] Only relatively few organometallic compounds of Ca, Sr, and Ba, other than metallocenes, have been reported.[74,94,224,254] Many of these are of type $MR_2(L)_n$ (R = Me, Et, Ph, allyl, alkenyl, and alkynyl) and $MXR(L)_n$ (R = Bu, Ph, tolyl; X = halide) and exhibit poor thermal stability.[94,224] $MR_2(L)_n$ and $MXR(L)_n$ compounds are typically prepared by either transmetallation reactions between the metal and HgR_2 or direct reaction of the metal with RX. For example, $Ca(CH(SiMe_3)_2)_2(1,4\text{-dioxane})_2$ was synthesized by condensation of the metal, $BrCH(SiMe_3)_2$, and THF at 77 K, followed by warming to room temperature and crystallization of the resulting solid from dioxane.[57] The reaction is thought to proceed by oxidative addition of $BrCH(SiMe_3)_2$ to the metal followed by a Schlenk equilibrium to give $Ca(CH(SiMe_3)_2)_2(THF)_3$ (product which precedes crystallization from dioxane) and $CaBr_2(THF)_n$. The $Cp^*Ca(CH(SiMe_3)_2)(THF)_3$ compound was prepared by reaction of $CaCp_2^*(THF)_2$ with $LiCH(SiMe_3)_2$, $LiCp^*$ precipitates and is easily separated from the Ca alkyl product.[60] The analogous barium reaction gives $Li[Cp_2^*BaCH(SiMe_3)_2](THF)_2$.[60] Both reactions are similar to the amide reactions described in Section II,A,1. Many other, less general methods have also been employed in preparation of these types of compounds.[224]

Monocyclopentadienyl compounds such as $[Cp^*Ca(I)(THF)_2]$ and $[Cp^{4i}Ca(I)(THF)_{(1,2)}]$ have been prepared from the metal diiodides and potassium decamethylcyclopentadienide by Eq. (27).[75,255] An alternative method of preparation is given in Eq. (28).

$$CaI_2 + KCp^{(*,4i)}(\text{excess}) \longrightarrow [Cp^{(*,4i)}Ca(I)(THF)_2] + KI(s) \qquad (27)$$

$$CaCp_2^{(*,4i)} + CaI_2 \longrightarrow 2\ [Cp^{(*,4i)}Ca(I)(THF)_2] \qquad (28)$$

$$(\text{solvent} = THF)$$

$[Cp^*Ca(I)(THF)_2]$ crystallizes from toluene as an asymmetrically iodide-bridged dimer $[Cp^*Ca(\mu\text{-}I)(THF)_2]_2$,[255] which disproportionates to the homoleptic compounds in ether or dilute hydrocarbon solvents. [Cp^{4i} Ca(I)(THF)$_2$] crystallizes from toluene as a symmetrically iodide-bridged dimer, [Cp^{4i}Ca(μ-I)(THF)]$_2$, which disproportionates only at elevated temperatures in solution or in the presence of dioxane. The difference in reactivity of these two compounds is thought to arise from the inability to form a THF-solvate of the Ca(Cp^{4i})$_2$ metallocene. Therefore, the THF ligand of Cp^{4i}Ca(μ-I)(THF) would have to be completely removed from the oxophilic calcium atom in order for another Cp^{4i} ring to bind and form the homoleptic compound Ca(Cp^{4i})$_2$. In Cp*Ca(μ-I)(THF)$_2$, however, this need not occur, because solvated calcium metallocenes with Cp* are known to be stable (see Section III,B,2). In other words, there is a kinetic barrier to disproportionation of Cp^{4i}Ca(μ-I)(THF). The K for disproportionation [K_d, right to left in Eq. (28)] in THF for the Cp* compound is 0.25 and has been estimated at 2.5×10^{-5} for the Cp^{4i} compound.[60,61] Other Cp and Cp-type compounds that were not included in the metallocene section include the fluorenyl compound (C$_{13}$H$_9$)Ba(NH$_3$)$_4$,[256] prepared by reaction of barium metal in liquid ammonia with fluorene at $-80°$C, followed by crystallization from THF; the diindenyls prepared by metal–ligand co-condensation reactions[94,257]; and the enolate-bridged complex [Ca{η^5C$_5$H$_3$-1,3-(SiMe$_3$)$_2$}{μ-OC$_5$H$_2$-2,4-(SiMe$_3$)$_2$}]$_2$,[258] prepared by oxidation of Ca(η^5-C$_5$H$_3$-1,3-(SiMe$_3$)$_2$)$_2$ with half an equivalent of dioxygen in toluene. Cyclooctatetraenediyl compounds of the alkaline-earth metals, M(cot)(L)$_n$(L = THF or pyridine), have also been described.[74,259] They have been prepared by co-condensation of metal, cot, and base at $-196°$C. After warming to room temperature, the base adducts are isolated in ca. 80% yield. Calcium and strontium compounds containing the [$nido$-7,9-C$_2$B$_{10}$H$_{12}$]$^{2-}$ carborane ligand have been isolated as products of a halide metathesis reaction between MI$_2$ and Na$_2$[$nido$-7,9-C$_2$B$_{10}$H$_{12}$] in THF.[260,261] Recrystallization of the solids from acetonitrile/diethyl ether gives [$closo$-1,1,1,1-(MeCN)$_4$-1,2,4-CaC$_2$B$_{10}$H$_{12}$] and polymeric [$closo$-1,1,1-(MeCN)$_3$-1,2,4-SrC$_2$B$_{10}$H$_{12}$]$_n$.

2. *Other Compounds with Mono- or Bidentate Ligands: Chemical and Physical Properties of Ca, Sr, and Ba Compounds of Type MX$_m$L$_n$ (X = O$_2$SOMe, O$_3$SCF$_3$, {OCS}OMe, COT, Carborane, Cp, Alkyl, RCO$_2$, RCSO, SeSi(SiMe$_3$)$_3$, TeSi(SiMe$_3$)$_3$, Te$_4$, SbSe$_2$, SCN, NO$_3$, ClO$_4$, ClO$_3$, Halide; L = Mono- or Bidentate Ligand; n = 0–6; m = 1 or 2)*

The physical data for these species are not tabulated because of the paucity of information available. The selenolates, M(SeSi(SiMe$_3$)$_3$)$_2$ (TMEDA)$_2$, are colorless or light yellow solids that are relatively high melting (200–300°C), are sparingly soluble in hydrocarbons, and decompose to the diselenide on exposure to air.[225] The tellurolates, $M(TeSi(SiMe_3)_3)_2L_n$, are high-melting (most >300°C) yellow solids that are soluble in hydrocarbons and decompose on exposure to O_2 or water to the tellurol and ditelluride, respectively.[225] Volatility data have not been reported on these species. All the compounds are believed to be monomeric in the solid state, which has been structurally confirmed for $Sr(SeSi(SiMe_3)_3)_2(TMEDA)_2$ (Fig. 45[225]), $Ca(TeSi(SiMe_3)_3)_2(THF)_4$, Ba$(TeSi(SiMe_3)_3)_2(pyridine)_5$, and $Sr(Se(2,4,6-t-Bu_3C_6H_2))_2(THF)_4$. $Sr(SeSi(SiMe_3)_3)_2(TMEDA)_2$ has octahedral geometry with a trans arrangement of the $SeSi(SiMe_3)_3$ groups. The reactivity of these complexes has not yet been described. However, if they show similar reactivity to the previously described siloxides (see alkoxide section), $M(OSiPh_3)_2(15$-crown-$5)(THF)$, then elimination of selenium and tellurium ethers and formation of metal selenide and telluride clusters might be expected on thermolysis. $Ca(DMF)_6(Te_4)$[228] was obtained as a black solid, and $Ba(en)_4(SbSe_2)_2$[229] as

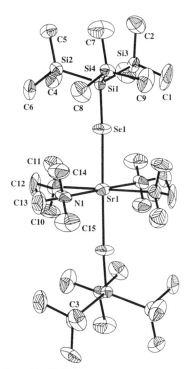

Fig. 45.   ORTEP drawing of $Sr(SeSi(SiMe_3)_3)_2(TMEDA)_2$. The Sr—Se bond distance is 2.946(1) Å, and the Sr–Se–Si bond angle is 167.02(9)°. (Redrawn from Ref. 225.)

an orange solid. Each has good solubility in its respective base. The structure of Ca(DMF)$_6$(Te$_4$) consists of a calcium atom, remote from the bent [Te$_4$]$^{2-}$ ions, in an octahedral environment of six dimethylformamide ligands. In Ba(en)$_4$(SbSe$_2$)$_2$, the barium atom is coordinated by four ethylenediamine ligands and one selenium atom of the infinite [SbSe$_2^-$]$_n$ chains (which are formed from trigonal SbSe$_3$ pyramids connected at common corners). The mercaptotetrazole compound Ba[naphthyl$\overline{NNNNC}$(S)]$_2$ (HMPA)$_3$[230] is a low-melting solid (103–106°C) that has good solubility in aromatic solvents. It is monomeric in the solid state with two chelating mercaptotetrazole ligands and three terminally bonded HMPA ligands. The thiolate Sr(S(2,4,6-t-Bu$_3$C$_6$H$_2$))$_2$(THF)$_4$ (Fig. 46[227]) is a high-melting colorless solid that has good solubility in THF. The compound crystallizes as a monomer with the strontium atom in a distorted octahedral geometry and the thiolate ligands trans to one another.

Ca, Sr, and Ba carboxylate, thiocarboxylate, thiocarboxylate–carboxylate, sulfito, and sulfonate compounds and their monodentate base adducts are white solids, many of which decompose on melting and are only soluble in protic solvents such as alcohols and water.[168,232–240] They are invariably oligomeric or polymeric in the solid state; see, for example, Refs. 235, 239, and 240. On thermolysis, the nonfluorinated carboxylates degrade to the metal carbonates while their fluorinated congeners usually form the metal fluorides.[212,222,262–269] The thiocarboxylates give mixtures of the metal oxides and sulfates on thermolysis in air.[238,239,270] The methylthiocarbonate compound, {Ca[(OCS)OMe]$_2$(MeOH)$_3$}$_2$, is a dimer in the solid state, soluble in polar coordinating solvents and stable under reduced

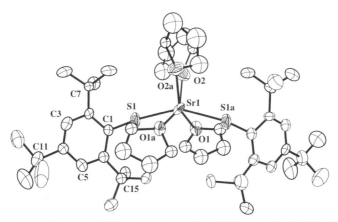

FIG. 46.   ORTEP drawing (35% probability level) of Sr(S(2,4,6-t-Bu$_3$C$_6$H$_2$))$_2$(THF)$_4$. The Sr—S bond distance and S–Sr–S bond angle are 2.951(2) Å and 157.1(1)°, respectively. (Redrawn from Ref. 227.)

pressure up to temperatures of 70°C. The structure of $\{Ca[(OCS)OMe]_2$ $(MeOH)_3\}_2$[241] (Fig. 47) consists of two chelating methylthiocarbonato ligands bonded to the metals through the sulfur and carbonate oxygen atoms, two chelating–bridging methylthiocarbonato ligands bonded through their μ_2-carbonate and methoxo oxygen atoms, and three terminally ligated methanol molecules.

Alkaline-earth compounds containing hard anions, such as NO_3, ClO_4, ClO_3, Cl, and Br, and neutral base donors, such as diacetamide, HMPA, alcohols, bipyridine, and phenanthroline, have physical and chemical properties that corroborate their ionic character.[233,244–249,251,252] Most are white or colorless solids with high melting points and solubility in only protic solvents. In many cases, the cation and anion are remote from one another in the solid. The metal complexes of the very weak bases ClO_4 and ClO_3 almost always adopt such a structure even with large metals such as barium (e.g., $[Ba(CH_3CO)_2NH)_5](ClO_4)_2$[247]), while softer anions such as I and SCN are usually bonded to the cation (e.g., $SrI_2(HMPA)_4$, $SrI_2(THF)_5$, and $CaI_2(THF)_4$).[59,227,253] The structures adopted, of course, are also highly dependent on the size of the cation and coordinating ability of the neutral ligand(s) (see the discussion of polydentates in Section III,G,4).

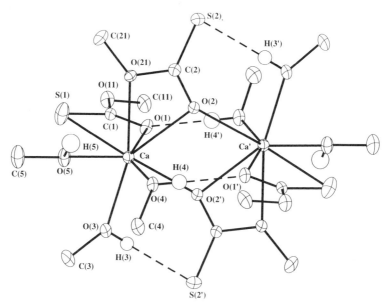

FIG. 47.   ORTEP drawing of $\{Ca[(OCS)OMe]_2(MeOH)_3\}_2$ (dashed lines indicate hydrogen bonding interactions). The Ca···Ca distance is 4.082(2) Å, and the Ca—S bond distance is 2.961(2) Å. (Redrawn from Ref. 241.)

Physical and chemical data on the MR_2L_n (R = Me, Et, Ph, allyl, alkenyl, and alkynyl) and $MXR(L)_n$ (R = Bu, Ph, tolyl; X = halide) compounds are greatly lacking, owing to these compounds' generally poor thermal stability. The aryl $MXR(L)_n$ and alkynyl MR_2L_n compounds constitute the most stable classes.[224] The compounds decrease in stability Ba < Sr < Ca, and $MXR(L)_n$ compounds form most readily for X = I and M = Ca; however, stability with respect to the halide is Cl > Br > I. For example, Ca(Ar)I(1,4-dioxane) can be desolvated at 110–220°C at 10^{-3} torr to give Ca(Ar)I, and the $M(CCPh)_2$ compounds decompose only after heating to 200–300°C.[224,271-274] Although most compounds are written as monomers, their solution behavior suggests, as found for the related Mg compounds, that more complex species may exist.[224] There is some evidence that the $MXR(L)_n$ compounds are in equilibrium with the homoleptic compounds MR_2L_n and MX_2L_n (similar to $Cp^*Ca(I)(THF)_2$; see earlier discussion).[57,271,275] The solid-state structures of these species are wholly unexplored with the exception of $Ca(CH(SiMe_3)_2)_2(1,4\text{-dioxane})_2$ (Fig. 48), which is monomeric with the metal in a distorted tetrahedral ligand environment.[57]

More physical data exist for Cp and Cp-related compounds. The bis(fluorenyl) barium compound $(C_{13}H_9)Ba(NH_3)_4$ is a yellow solid with good solubility in THF. It has a bent structure similar to those of metallocene

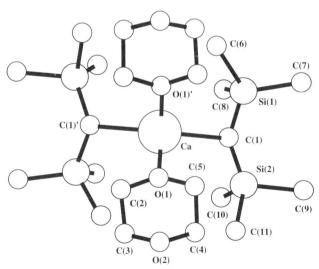

FIG. 48.   ORTEP drawing of $Ca(CH(SiMe_3)_2)_2(1,4\text{-dioxane})_2$. The Ca—C bond distance is 2.483(5) Å, and the internal angles have a range of 79.4(1)–133.7(2)°. (Redrawn from Ref. 57.)

base adducts, MCp_2L_2, but with four bonded ammonia ligands.[256] Interestingly, the ammonia ligands are not displaced by THF, suggesting that the Ba–N interactions are robust and that THF molecules might be too big to access the metal center in the presence of the large fluorenyl ligands. The carborane compounds $[closo\text{-}1,1,1\text{-}(MeCN)_3\text{-}1,2,4\text{-}SrC_2B_{10}H_{12}]_n$[261] and $[closo\text{-}1,1,1,1\text{-}(MeCN)_4\text{-}1,2,4\text{-}CaC_2B_{10}H_{12}]$[260] are colorless solids that have good solubility in DMF and acetonitrile. The calcium compound is monomeric in the solid state with four terminally ligated acetonitrile molecules on one side of the metal in a square pyramidal arrangement, and the carborane on the other side with the calcium atom capping its hexagonal face (Fig. 49). The larger size of strontium, however, results in a polymeric compound where each carborane interacts with two metal centers through hydrogen atoms associated with a triangular side face and its open hexagonal face. The strontium coordination sphere is then completed by three terminally bonded acetonitrile molecules.[261]

3. *Other Compounds with Polydentate Ligands: Synthesis of Ca, Sr, and Ba Compounds of Type MX_2 (Polydentate) ($X = RCO_2$, RCSO, RSO_3, SCN, NO_3, ClO_4, ClO_3, $(RO)_2PO_2$, OH, Te_4, Halide, Tosylate)*

The original interest in these systems stems from the need to selectively extract heavy alkaline-earth metals from various media.[276–280] A number of alkaline-earth compounds with polydentate ligands and anionic ligands

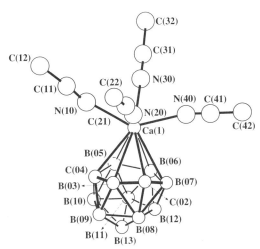

FIG. 49.   ORTEP drawing of $[closo\text{-}1,1,1,1\text{-}(MeCN)_4\text{-}1,2,4\text{-}CaC_2B_{10}H_{12}]$. The average Ca—N bond distance is 2.481(36) Å, and the Ca—C and Ca—B bond distances have ranges of 2.701(5)–2.895(5) Å and 2.649(6)–2.935(6) Å, respectively. (Redrawn from Ref. 260.)

other than alkoxides or β-diketonates have been prepared. They include the thiocyanates: $M(NCS)_2L_n \cdot mH_2O$ ($n = 1$, L = tetraglyme, penta-glyme, 18-crown-6, 12-crown-4, glycols, cryptand, etc.; $n = 2$, M = Ba, L = 12-crown-4)[168,281–290]; the halides: $BaX_2L \cdot mH_2O$ (X = Cl, Br, I, L = 18-crown-6, polyamines; X = Cl, L = cryptand 222B, m = 6)[168,228,281,283,291–294]; MCl_2L (L = ethylene glycols)[295]; the chlorates, ni-trates, and perchlorates; $M(ClO_3)_2L$ (L = diethylenetriamine),[291] $M(NO_3)_2L$,[168,295–297] and $M(ClO_4)_2L$ (L = 18-crown-6 and derivatives, ethylene glycols, polyamines)[223,291,298,299]; the phosphates and tosylates: $M(PO_2(OR)_2)_2$(18-crown-6) $\cdot$ H_2O (M = Sr, Ba; R = n-Bu)[300] and M $(C_7H_7SO_3)_2L_n$ (M = Sr, Ba, n = 1; M = Ca, n = 2; L = hexaglyme)[301]; the hydroxides, selenides, and polytellurides: $Ba(OH)_2L$ (L = 18-crown-6 and derivatives),[168,281] $[Sr(15\text{-crown})\text{-}5_2]Te_4 \cdot H_2O$,[228] and $[Ba(18\text{-crown-}6)(DMF)_4][Cd(Se_4)_2]$[302]; and the sulfonates, carboxylates, and thiocarbox-ylates (Table VII). Excluding the anhydrous compounds in Table VII, most of the other $MX_2 \cdot L$ compounds were prepared by reaction of MX_2 with the polydentate ligand in solvent. The anhydrous compounds were synthesized by reaction of MH_2 with RCOOH, RSO_3H, and RCOS(H) and polyether in ethanol [Eq. (29)]. Yields by these methods are typically excellent ($>80\%$).

$$MH_2 + 2HA + L \longrightarrow MA_2L + 2H_2 \text{ (g)} \tag{29}$$

(M = Ca, Sr, or Ba; A = RCO_2, RCOS, RSO_3; L = polyether; solvent = THF or alcohol)

4. *Other Compounds with Polydentate Ligands: Chemical and Physical Properties of Ca, Sr, and Ba Compounds of Type MX_2(Polydentate) (X = RCO_2, $RCSO$, RSO_3, SCN, NO_3, ClO_4, ClO_3, $(RO)_2PO_2$, OH, Te_4, Halide, Tosylate)*

Most of these compounds are colorless or white solids that melt at high temperatures or in many instances decompose before melting. Although many are monomeric, they are involatile and lose polyether on attempted sublimation. They are insoluble in hydrocarbons, but soluble in coordinat-ing solvents such as alcohols and water. This behavior is accordant with that observed for the base-free and mono- and bidentate base-containing salts and suggests that the polyether adducts are similarly ionic in charac-ter. Structures vary; the $M(NCS)_2L$ (Ca, Sr, or Ba) compounds usually have the polydentate ligand around the center of the metal with the SCN ligands, bonded through nitrogen, either on the same side of the metal with a water molecule coordinated on the other side or on opposite sides when no water or protic groups (–OH) are present.[168,281–290] Polyether ligands coordinate around the center of the metal, except for 12-crown-4, which sits to one side.[288]

TABLE VII

Ca, Sr, and Ba Carboxylate, Thiocarboxylate, and Sulfonate Polyether Adducts and Their Physical Properties[a]

Compound	$\nu_{asym}(CO_2)$ (cm^{-1})	Coord. no.	Solubility[b]	Product of therm. dec. (dry air)	Characterization[c]	Ref.
[Ba(pivalate)(dicyclohexano-18-crown-6)(H$_2$O$_2$)](pivalate)H$_2$O	*	9	s = a:b	*	X-ray	303
Sr(pivalate)$_2$(dicyclohexano-18-crown-6) · H$_2$O	*	10	s = a:b	*	X-ray	303
Ba(pivalate)$_2$(18-crown-6)	*	10	s = c	*	X-ray	304
[Ba(adipate)$_2$(18-crown-6) · 8H$_2$O]$_x$	*	10	s = a; ss = b; is = c	BaCO$_3$	TGA, X-ray, ^{1}H NMR	267
Ba(O$_2$CCH$_3$)$_2$ · (18-crown-6) · 1.5H$_2$O	*	10	s = a:b	*	EA, ^{1}H NMR	168
Ba(O$_2$CCH$_3$)$_2$ · (18-crown-6) · 4H$_2$O	1560	10	s = a; ss = b; is = c	BaCO$_3$	IR, TGA, X-ray, ^{1}H NMR	267
[Sr(3,5-dinitrobenzoate)(H$_2$O)(TEG)][3.5-dinitrobenzoate][d]	*	9	*	*	X-ray	305
Ca(3,5-dinitrobenzoate)$_2$(benzo-15-crown-5) · 3H$_2$O	*	9	s = a:b	*	X-ray	306
[CaCl][SrCl{(PhNHCO)$_2$(COO)$_2$-18-crown-6}]	*	9	s = a:b	*	X-ray	293
[Ca{(PhNHCO)$_2$(COO)$_2$-18-crown-6}] · 3H$_2$O	*	9	s = a:b	*	X-ray	293
Ba(O$_2$CCF$_3$)$_2$(18-crown-6) · H$_2$O	*	*	*	*	EA, ^{1}H NMR	168
Ba(O$_2$CCF$_3$)$_2$(18-crown-6)	1696, 1672	*	s = a, b, e; ss = d	BaF$_2$	EA, IR, TGA, ^{13}C, ^{1}H NMR	150

Compound	IR	Coord.	Solubility	Decomposition product	Methods	T (°C)
Ba(O$_2$CCF$_3$)$_2$(18-crown-6)(py)	1700, 1678	10	s = f	BaF$_2$	EA, IR, TGA, ^{13}C, ^{1}H NMR, X-ray	150
Ba$_2$(O$_2$CCF$_3$)$_4$(15-crown-5)$_2$	1717	9	s = a, b; ss = d; is = c	BaF$_2$	EA, IR, TGA, ^{13}C, ^{1}H NMR, X-ray	150
Ba$_2$(O$_2$CCF$_3$)$_4$(12-crown-4)$_2$	1690	8–9	s = a, b; ss = d; is = c	BaF$_2$	EA, IR, TGA, ^{13}C, ^{1}H NMR, X-ray	150
Ba$_2$(O$_2$CCF$_3$)$_2$(18-crown-6'OH)^e	1695, 1669	10	s = a, b	BaF$_2$ and BaCO$_3$	EA, IR, TGA, ^{13}C, ^{1}H NMR, X-ray	150
Ba$_2$(O$_2$CCF$_3$)$_4$(15-crown-5'OH)$_2$^{e}	1707, 1676	9	s = a, b	BaF$_2$	EA, IR, TGA, ^{13}C, ^{1}H NMR, X-ray	150
Ba(O$_2$CCF$_3$)$_2$(cryptand(222))	1694	10	s = b, c	BaF$_2$	EA, IR, TGA, ^{13}C, ^{1}H NMR, X-ray	266
[Ba$_2$(O$_2$CCF$_3$)$_4$(tetraglyme)]$_x$	1686, 1655	9	s = b, d; ss = c	BaF$_2$	EA, IR, TGA, ^{13}C, ^{1}H NMR, X-ray	150
[Sr$_2$(O$_2$CCF$_3$)$_4$(tetraglyme)]$_x$	1697, 1658	9	s = b; ss = d; is = c	SrF$_2$	EA, IR, TGA	150
Ba$_2$(O$_2$CCF$_3$)$_4$(triglyme)	1693	8–9	s = a, b; ss = d; is = c	BaF$_2$	EA, IR, TGA, ^{13}C, ^{1}H NMR, X-ray	150
Ba(O$_3$SCF$_3$)$_2$(18-crown-6) · H$_2$O	*	*	*	*	EA, ^{1}H NMR	168
Ba(O$_3$SCF$_3$)$_2$(tetraglyme)	1288, 1256 (s = 0)	*	s = b	BaF$_2$	EA, IR, TGA, ^{13}C, ^{1}H NMR	150
Ca(SOCCH$_3$)$_2$(15-crown-5)	1534, 1499	8	s = b	CaS (in N$_2$)	EA, IR, TGA, ^{13}C, ^{1}H NMR, X-ray	270

^{a} * = information not available.

^{b} s = soluble, ss = slightly soluble, is = insoluble; a = water, b = alcohols, c = toluene, d = tetrahydrofuran, e = chloroform, g = pyridine.

^{c} EA (elemental analysis), IR (infrared), NMR (nuclear magnetic resonance), TGA (thermogravimetric analysis), X-ray (single crystal X-ray diffraction).

^{d} TEG is tetraethylene glycol.

^{e} ^ = methylene group (—CH$_2$—).

The barium atom in the cryptand compound [Ba(222B)(H$_2$O)$_2$] (Cl)$_2 \cdot$ 4H$_2$O is coordinated by two water molecules and the two nitrogen atoms and six oxygen atoms of the cryptand to give a coordination number of 9 (Fig. 50[294]). The chloride ions are not coordinated to the metal center, which is typical for halide compounds with polydentate ligands. The structures for the chlorate, nitrate, and perchlorate compounds cannot be so easily grouped.[223,291,295–299] In some cases both anions are bonded to the metal, and in others one or none are bonded to the metal. The particular structural type is mostly dependent on the polyether used and the degree of hydration. Water molecules and alcohol groups (glycols) seem to preferentially bond to the metal centers over halide and perchlorate ligands, which in many instances are relegated to the outer coordination sphere.

The carboxylate compounds show the greatest structural diversity and have perhaps the most covalent metal–ligand bonding of this group of adducts. The structurally characterized trifluoroacetate compounds in Table VII have five distinct structural types. The structures of three of these, Ba$_2$(O$_2$CCF$_3$)$_4$(15-crown-5)$_2$, [Ba$_2$(O$_2$CCF$_3$)$_4$(tetraglyme)]$_x$, and Ba(O$_2$CCF$_3$)$_2$(18-crown-6 OH), are shown in Figs. 51–53.[150]

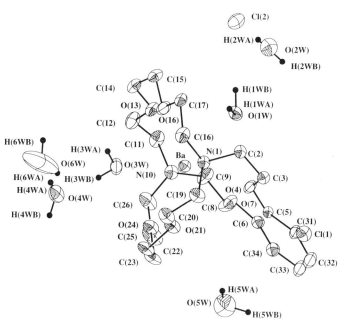

FIG. 50.   ORTEP drawing (25% probability level) of [Ba(222B)(H$_2$O)$_2$](Cl)$_2 \cdot$ 4H$_2$O. (Redrawn from Ref. 294.)

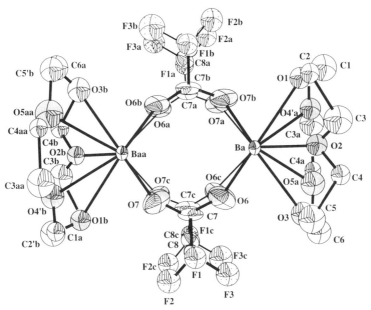

Fig. 51. ORTEP drawing of $Ba_2(O_2CCF_3)_4(15\text{-crown-5})_2$. The average Ba—O ether and carboxylate bond distances are 2.991 Å and 2.644 Å, respectively, and the Ba···Ba distance is 4.735(1) Å. (Redrawn from Ref. 150.)

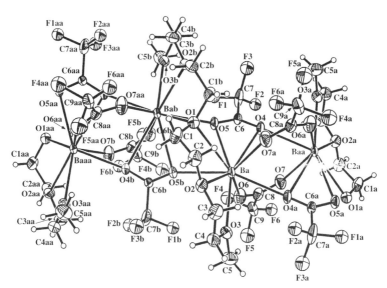

Fig. 52. ORTEP drawing of $[Ba_2(O_2CCF_3)_4(\text{tetraglyme})]_\infty$. The Ba—$O_{\text{ether}}$ (bridging) and average Ba—O_{ether} (chelating) bond distances are 3.080(5) Å and 2.791(15) Å, respectively. The average Ba—O_{carbox} bridging (two types) and chelating bond distances are 2.674(9) Å, 2.758(21) Å, and 2.903(70) Å, respectively. (Redrawn from Ref. 150.)

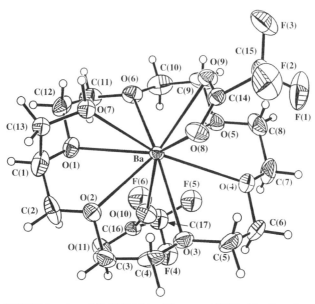

FIG. 53.   ORTEP drawing of Ba(O₂CCF₃)₂ · (18-crown-6ˆOH). The Ba—$O_{alcohol}$ and average Ba—O_{ether} bond distances are 2.856(7) Å and 2.828(29) Å, respectively. The average Ba—O_{carbox} chelating and dangling bond distances are 2.836(13) Å and 2.677(6) Å, respectively. (Redrawn from Ref. 150.)

Ba₂(O₂CCF₃)₄(15-crown-5)₂, and related species with similar structures such as Ba₂(O₂CCF₃)₄(triglyme)₂ and Ba₂(O₂CCF₃)₄(12-crown-4)₂, has four bridging trifluoroacetates with side-on bonding of the polyether ligands. The three compounds are quite similar in that all have small polyethers that prefer to sit to one side of large metal centers. This, coupled with the proclivity of CF₃CO₂ to bridge metal centers, gives the structure shown in Fig. 51. These centrosymmetric compounds have one $\nu_{asym}(CO_2)$ stretch in the range 1690–1720 cm⁻¹.[150] The polymeric compound [Ba₂(O₂CCF₃)₄ · (tetraglyme)]∞ has long zigzag chains of barium atoms in the solid state. The barium atoms are held together by two types of trifluoroacetate coordinations. One set of trifluoroacetate ligands bridges the zig metal centers to the zag metal centers, and the other set acts in the dual capacity of chelating and bridging zig metal centers (Fig. 52). The tetraglyme ligand is shared between each two-metal center unit, with the center ether oxygen atom μ_2-bridging the metals. Two $\nu_{asym}(CO_2)$ stretches are observed for both the Ba and Sr compounds between 1650 and 1700 cm⁻¹.[150] The compounds are soluble in THF, which suggests that some deoligomerization occurs in coordinating solvents. The alcohol-functionalized 18-crown-

6 compounds, $Ba(O_2CCF_3)_2$(18-crown-6^OH) and $Ba_2(O_2CCF_3)_4$(15-crown-5^OH)$_2$, have different structures. The species $Ba(O_2CCF_3)_2$ (18-crown-6^OH) is monomeric and similar in structure to $Ba(O_2CCF_3)_2$ (18-crown-6)(py), with one chelating and one dangling trifluoroacetate group and the 18-crown-6 ligand bonded to the metal center through its one alcohol and six ether oxygen atoms (Fig. 53). The trifluoroacetate ligands lie on opposite sides of the crown ether. Apparently, a Ba–O$_{alcohol}$ interaction is more favorable than the chelating of both trifluoroacetate ligands. The intermolecular hydrogen bond between the nonbonded carboxylate oxygen atom and an alcohol group of a neighboring 18-crown-6^OH ligand would seem to make this statement precarious; however, a similar situation occurs in $Ba(O_2CCF_3)_2$(18-crown-6)(py), where the Ba—N$_{pyridine}$ bond results in a dangling trifluoroacetate ligand, and no hydrogen bonding exists in this compound. In both compounds the metal coordination number is 10, which seems to be the upper bound for these types of species. Both compounds have two $\nu_{asym}(CO_2)$ stretches between 1630 and 1700 cm^{-1}. In $Ba_2(O_2CCF_3)_4$(15-crown-5^OH)$_2$, the alcohol group also bonds to the metal, but side-on bonding of the ether opens up one side of the metal center, and two trifluoroacetate groups bridge to give a dinuclear structure (Fig. 54). The chelating trifluoroacetate ligand has two distinct Ba—O bond distances, a long bond distance of 3.266(20) Å for the oxygen atom that is involved in a intramolecular hydrogen bond to the alcohol group of the 15-crown-5^OH ligand, and a short bond distance of 2.183(10) Å for the oxygen atom bonded only to the barium atom. Again, two $\nu_{asym}(CO_2)$ stretches are observed for this compound.

The structure of $Ba(O_2CCF_3)_2$(cryptand(222)) is perhaps the most interesting in that each trifluoroacetate group is bonded to the barium center through only one oxygen atom (Fig. 55). This is probably a consequence of both the encapsulation of the barium atom in the cryptand and the chelate effect of coordinating eight donor atoms of the cryptand to the metal. This compound has good solubility in alcohols and THF and is moderately soluble in toluene.[150]

The structures of most of the nonfluorinated carboxylates show striking similarity to those of the M(β-diketonate)$_2$(18-crown-6) compounds. The carboxylates chelate on opposite sides of the metal center with the ethers between. When a dicarboxylate ligand is employed, such as adipate, a polymeric structure with chelating carboxylate groups and bridging adipate ligands results.[267] A few nonfluorinated carboxylate compounds have other structural types. In [Ba(pivalate)·(dicyclohexano-18-crown-6)(H$_2$O)$_2$](pivalate)·H$_2$O, only one of the carboxylate groups is bonded to the barium center.[303] The analogous strontium compound has both carboxylates chelating the metal center. The authors attribute this differ-

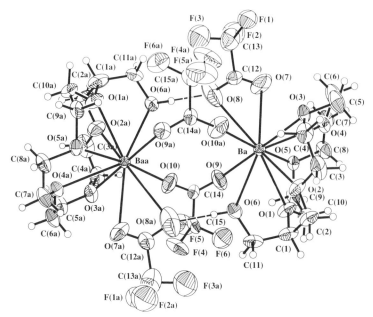

FIG. 54. ORTEP drawing of $Ba_2(O_2CCF_3)(15\text{-crown-}5\char`\^OH)_2$ (dashed lines indicate a probable hydrogen bond). The $Ba—O_{ether}$ bond distances have a range of 2.762(6)–2.962(6) Å. The $Ba—O_{alcohol}$ and average $Ba—O_{carboxylate}$ bridging bond distances are 2.844(6) Å and 2.684(59) Å, respectively. (Redrawn from Ref. 150.)

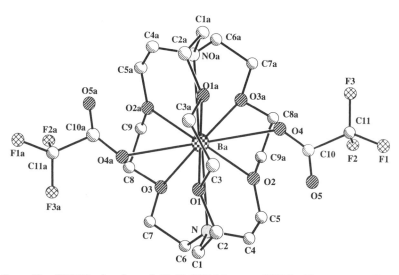

FIG. 55. ORTEP drawing of $Ba(O_2CCF_3)_2[\text{cryptand}(222)]$. The average Ba—N, $Ba—O_{ether}$, and $Ba—O_{carboxylate}$ bond distances are 2.963(4), 2.831(28), and 2.778(5) Å, respectively. The C—O bond distances in the trifluoroacetate ligand are similar: 1.194(9) Å and 1.221(8) Å (oxygen atom bonded to metal). (Redrawn from Ref. 150.)

ence to factors relating to ion size.[303] The anhydrous 18-crown-6 compound, Ba(pivalate)$_2$(18-crown-6), has both carboxylate groups chelating to the metal and has good solubility in toluene.[304] The nitrobenzoates usually have the carboxylate groups bonded to the metal through both oxygen atoms, and in some instances the –NO$_2$ groups also coordinate.[305,306]

The thiocarboxylate Ca(SOCMe)$_2$(15-crown-5), Fig. 56, is monomeric, with the 15-crown-5 ligand on one side of the metal center with a chelating and a dangling thioacetate ligand on the other side.[270] The dangling ligand is bonded through oxygen to the calcium center. This structure is in contrast to that of Ba$_2$(O$_2$CCF$_3$)$_4$(15-crown-5)$_2$,[150] which is dimeric with bridging carboxylates. Apparently, the thiocarboxylate ligand is less inclined to assume a bridging position than a trifluoroacetate ligand. Also, the smaller size of calcium and the coordinating ability of sulfur are undoubtedly important factors in determining the structure. The structure of Ca(SOCMe)$_2$(15-crown-5) is also in contrast to that of the corresponding Sr compound, Sr(SOCMe)$_2$(15-crown-5),[307] Fig. 57, which is similar, but now with both of the thioacetate ligands chelating. Space-filling models are consistent with the explanation that the larger size of Sr compared to Ca enables the second thioacetate to chelate rather than dangle. The barium compound, Ba(SOCMe)$_2$(18-crown-6),[307] Fig. 58, exhibits the now-

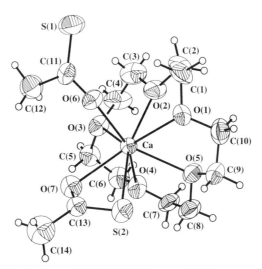

FIG. 56.   ORTEP drawing of Ca(SOCMe)$_2$(15-crown-5). The average Ca—O$_{ether}$ bond distance is 2.493(26) Å. The Ca—O$_{thioacetate}$ dangling bond distance is 2.282(3) Å, and the average Ca—O and Ca—S bond distances for the chelating thioacetate, which is disordered over two positions, are 2.459(44) Å and 2.917(52) Å, respectively. (Redrawn from Ref. 270.)

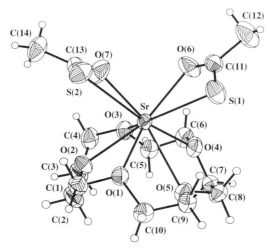

FIG. 57. ORTEP drawing of Sr(SOCMe)₂(15-crown-5). The average Sr—O$_{ether}$, Sr—O$_{thioacetate}$, and Sr—S bond distances are 2.717(50) Å, 2.566(21) Å, and 3.132(71) Å, respectively. (Redrawn from Ref. 307.)

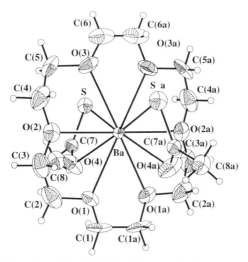

FIG. 58. ORTEP drawing of Ba(SOCMe)₂(18-crown-6). The average Ba—O$_{ether}$, Ba—O$_{thioacetate}$, and Ba—S bond distances are 2.802(17) Å, 2.875(7) Å, and 3.325(71) Å, respectively. (Redrawn from Ref. 307.)

familiar structure with the Ba in the coordination plane of the 18-crown-6 ligand with the remaining anionic ligands, thioacetates in this case, above and below this coordination plane.

$Ca(SOCMe)_2$(15-crown-5) gives a mixture of metal oxide and sulfate on thermolysis in air. However, thermolysis under scrupulously anaerobic conditions yields CaS. The compound is slightly volatile and has been successfully employed as a precursor to CaS thin films by CVD (*vide infra*).[270]

The polyether carboxylate and sulfonate adducts lose crown ether on attempted sublimation. The fluorinated carboxylates and sulfonates form MF_2 when thermally decomposed in dry air, whereas the nonfluorinated carboxylates form MCO_3.[136,267] Their dissociation on sublimation makes them unattractive for scaleable CVD applications, but they show promise in $MTiO_3$ film formation by solution methods. This is discussed further in the CVD section.

IV

CHEMICAL VAPOR DEPOSITION OF FILMS CONTAINING GROUP 2 ELEMENTS

In this section the results of deposition experiments using some of the Group 2 compounds discussed in the previous sections are described. This discussion is focused on the characteristics and observations of the various Group 2 compounds for film deposition, rather than on a detailed description of the deposition experiments and properties of the films, because the latter aspects have been reviewed elsewhere. Tables VIII–X provide an overview of representative examples of materials, precursors, and experimental conditions that have been used for the deposition of Group 2 metal-containing films. These tables are organized according to the materials deposited, namely, MF_2 (Table VIII), perovskite-phase $MTiO_3$ (Table IX), and ceramic superconductors (Table X). The deposition of Group 2 element-containing films is described in the same order as the ligand chemistry described in Section III.

A. *Group 2 Metal Amides and Phosphides*

Although a number of Group 2 metal amides and phosphides have been prepared that are potentially suitable as precursors for CVD, to our knowledge they have not been utilized for this purpose.

TABLE VIII

Summary of Conditions and Results of CVD of Metal Fluoride Films

Entry no.	Group 2 precursor[a]	Other precursor	Carrier gas/other reagents	Deposition temp. (°C)	Substrate	Deposit	Growth rate	Comments	Ref.
1	Ba(dfhd)$_2$ (200–210)	Pr(ppm)$_3$ (160–180)	O$_2$/Ar	750	CaF$_2$ (111)	Pr-doped BaF$_2$	200–400 nm/hr		308
2	Ba(dfhd)$_2$ (200–230)	—	O$_2$/Ar	750	CaF$_2$ (111)	BaF$_2$	100–500 nm/hr		309
3	Ba(ofhd)$_2$ (210)	—	O$_2$/Ar	700	CaF$_2$ (111)	BaF$_2$	50–300 nm/hr		159, 310
4	M(hfac)$_2$ (180–255)	—	O$_2$	500–525	Si	CaF$_2$ SrF$_2$ BaF$_2$	Up to 500 nm/hr		154
5	BaX$_2$(tetraglyme) (175) X = hfac, tfac, thd	—		650	Quartz	BaF$_2$	0.2 μm/hr		190
6	M(hfac)$_2$ (tetraglyme) M = Ca, Sr, Ba (90–130)	—		450–550	Si(111)	MF$_2$	—		189
7	Ca(thd)$_2$ (195)	—	N$_2$/HF	260–400	Glass	CaF$_2$ SrF$_2$	—	Atomic layer deposition	311

[a] Numbers in parentheses are the temperatures to which the precursor was heated.

B. *Group 2 Metal Cyclopentadienyl Compounds*

There have been few reports of the use of Group 2 cyclopentadienyl compounds as precursors for CVD. The derivatives containing unsubstituted cyclopentadienyl rings are oligomeric, but some of the substituted derivatives are monomeric and therefore potentially suitable. Tasaki *et al.*[102] have reported that Ba and Sr pentamethylcyclopentadienyl compounds vaporized at much lower temperatures than the dpm complexes. They assessed the suitability of these reagents for depositing Group 2 metal-containing films by transporting them to an MgO (100) substrate heated to 800°C in the presence of N_2O and O_2 as oxidants. They observed that the lowest temperature to which the precursor container needed to be heated to produce a film was 130 and 100°C for $M(C_5Me_5)(THF)_2$, where M = Ba and Sr, respectively, and 190 and 150°C for $M(dpm)_2$, M = Ba and Sr, respectively (with presumably some loss of THF; see Section III). However, some of the disadvantages of using this class of precursors were also noted in this work, especially their high reactivity toward small molecules. For example, Tasaki *et al.* observed with both O_2 and N_2O that under certain conditions, increasing the concentration of these oxidizing agents reduced the deposition rate. This is consistent with gas-phase reactions between the reagents, which decrease the concentration of the Group 2 precursor in the vapor phase that reaches the surface.

The only other reports of the use of this class of compounds as precursors for CVD we have found are in Japanese patent applications and general descriptions of work in progress in review articles. Sato and Sugawaran[341] have reported the use of Ba alkenyl, benzyl, and cyclopentadienyl derivatives in the CVD of various ceramic superconductors, while Yamada and Morimoto[342] have claimed the use of Group 2 cyclopentadienyl reagents such as $Ba(C_5Me_5)_2$, together with $Cu(dpm)_2$ and $Y(dpm)_3$, for the deposition of $YBa_2Cu_3O_y$ superconducting films by CVD; they also report the use of barium porphryn compounds,[343] but no details are available.

The species $Ba[1,2,3,4-(i-Pr)_4C_5H]_2$ has been reported to have similar volatility as $Ba(dpm)_2$ and $Ba(fod)_2$, but the use of this compound for CVD of Ba-containing films has not been reported. Schulz and Marks[18] have reported the growth of $BaPbO_3$ films using the more sterically encumbered species $Ba[(i-Bu)_3C_5H_2]_2$ as a Ba source. The donor functionalized derivatives $M[C_5H_4\{(CH_2)_nER\}]_2$, where M = Ca, Sr, Ba; ER = OMe, OEt, and NMe_2, have also been prepared but not yet successfully used to transport Ba for the deposition of ceramic superconductor films.[19]

TABLE IX

SUMMARY OF CONDITIONS AND RESULTS OF CVD OF PEROVSKITE-PHASE METAL OXIDE FILMS

Entry no.	Group 2 precursor[a]	Other precursor	Carrier gas/other reagents	Deposition temp. (°C)	Substrate	Deposit	Growth rate	Comments	Ref.
1	Ba(thd)$_2$ (240)	Ti(O-i-Pr)$_4$	O$_2$	800–1000	Pt (100) MgO (100)	BaTiO$_3$	1.2 μm/hr	BaTiO$_3$ (200)	312
2	Sr(dpm)$_2$ (215)	Ti(O-i-Pr)$_4$	O$_2$	750	Si(100) Mg (100)	SrTiO$_3$	10–15 nm/hr	reduced pressure	323
3	Ba(tmhd)$_2$ (250)	Ti(O-i-Pr)$_4$	N$_2$O	280–400	p-InP (100)	BaTiO$_3$	~/nm/min		314
4	Ba(tmhd)$_2$ (230)	Ti(O-i-Pr)$_4$	O$_2$	600	quartz	BaTiO$_3$	~0.3 μm/hr	reduced pressure	315
5	Ba(thd)$_2$·(H$_2$O)$_n$ (230)	Ti(O-i-Pr)$_4$	O$_2$	600	MgO (100)	BaTiO$_3$	190 nm/hr	—	316
6	Ca(dpm) (220–250)	Ti(O-i-Pr)$_4$ Fe(acac)$_3$ (115)	N$_2$/O$_2$	650–900	SiO$_2$	Ca(Ti, Fe)O$_3$	not reported	—	317
7	Ba(dpm)$_2$ (250) Sr(dpm)$_2$ (250)	Ti(O-i-Pr)$_4$	O$_2$/N$_2$O thf	650	Pt	Ba$_{1-x}$Sr$_x$TiO$_3$	—	liquid delivery	318
8	Ba(thd)$_2$ (245)	Ti(O-i-Pr)$_4$	Ar/O$_2$	600	Si (100) Al$_2$O$_3$ ($\bar{1}$102) Pt NaCl (100)	BaTiO$_3$	<0.5 μm/hr	—	30
9	Sr(dpm)$_2$ (200–230)	Ti(O-i-Pr)$_4$	O$_2$/H$_2$O	600–850	sapphire (0001)		0.3–4.5 μm/hr	reduced pressure	31
10	Ba(dpm)$_2$ (325)	Fe(Cp)$_2$ (65)	O$_2$	775	MgO ZnGa$_2$O$_4$	BaFe$_{12}$O$_{19}$	—	reduced pressure	319
11	M(dpm)$_2$ (150–190) M(C$_5$Me$_5$)$_2$ (100–130) M = Ba, Sr	—	O$_2$/N$_2$O	800	MgO (100)	MO	80 Å/hr	—	102

No.	Precursor	Second precursor	Oxidant/reactant	Temp (°C)	Substrate	Product	Growth rate	Notes	Ref.
12	Sr(thd) (200–340)	—	H₂O, H₂S	200–420	Ag	SrS SrO	—	ALE	320
13	Ba(diethylhexanoate)₂	Ti(O-*i*-Pr)₄(acac)₂	N₂/O₂/ *n*-BuOH (solvent)	450–600	*p*-Si (100)	BaTiO₃	~200 Å/min	aerosol-assisted CVD	38
14	Sr(dpm)·L (100–150)	Ti(O-*i*-Pr)₄	O₂	450–650	Pt/Ta/SiO₂	SrTiO₃	0.5 nm/min	L = ~50 different adducts studied	321
15	Ba(thd)₂ (tetraglyme)	Ti(O-*i*-Pr)₄(thd)₂	—	840	MgO (100)	BaTiO₃	—	—	322
16	Ba(thd)₂(pmdt)	Ti(O-*i*-Pr)₄	HO-*i*-Pr	not reported	not reported	BaTiO₃	not reported	liquid delivery	42
17	Sr(hfac)₂ (tetraglyme) (102–108)	Ti(O-*i*-Pr)₄	O₂	800–810	LaAlO₃ (100)	SrTiO₃	40–120 Å/min	reduced pressure	323, 324
18	Ba(hfac)₂ (tetraglyme) (105–120)	Ti(O-*i*-Pr)₄	O₂/H₂O	600–800	LaAlO₃ (100)	BaTiO₃	0.3–1.5 μm/hr	reduced pressure	325, 326
19	Ba(hfac)₂ (tetraglyme) (111–115) Sr(hfac)₂ (tetraglyme) (108–113)	Nb(OEt)₅ (100–112)	O₂/H₂O	>800°C	MgO (100)	SrₓBa₁₋ₓNb₂O₆	0.2–1.5 μm/hr	reduced pressure	327
20	M(hfac)₂ (tetraglyme) (90–120)	Cu(acac)₂	not reported	450–550°C	Si(111)	MF, BaCaCuOₓ	not reported	TGA/DSC study of Group 2 precursors	189
21	Ba(hfac)₂(poly-ether) (116)	Ti(O-*i*-Pr)₄	O₂/H₂O	775	LaAlO₃	BaTiO₃	not reported	—	328

a Numbers in parentheses are the temperatures to which the precursor was heated.

TABLE X

Summary of Conditions and Results of CVD of Ceramic Superconductor Films

Entry no.	Group 2 precursor[a]	Other precursor	Carrier gas/other reagents	Deposition temp. (°C)	Substrate	Deposit	Growth rate	Comments	Ref.
1	$Sr(dpm)_2$ (145–230), $Ca(dpm)_2$	$BiPh_3$, $Cu(acac)_2$	O_2/H_2O	550	MgO (100)	$Bi_2(Sr, Ca)_3Cu_2O_x$	2–3 μm/hr	—	329
2	$Sr(dpm)_2$, $Ca(dpm)_2$	$BiPh_3$, $Cu(acac)_2$	Ar/N_2O	625–700	Ag	Various BSCCO	25 nm/hr	—	330, 331
3	$Ba(dpm)_2$	$Y(dpm)_3$, $Cu(dpm)_2$	Ar/N_2O	740	Ag	$YBa_2Cu_3O_{7-x}$	0.2–2 μm/hr	PE-MOCVD	332, 333
4	$Ba(tmhd)_2$	$Cu(tmhd)_2$, $Y(tmhd)_3$	He/O_2	700–710	$LaAlO_3$	YBa_2CuO_7	50–450 Å/min	powders mixed and evaporated	334
5	$Ba(dpm)_2$ (250)	$Y(dpm)_3$ (100), $Cu(acac)_2$ (150)	O_2/N_2O	610	YSZ, $SrTiO_3$ (100)	$YBa_2Cu_3O_{7-x}$	10–30 nm/min	low-pressure PE-CVD	335
6	$Ba(dpm)_2$ (220)	$Y(dpm)_3$ (120), $Cu(acac)_2$ (180)	Ar/N_2O	700	$SrTiO_3$	$YBa_2Cu_3O_{7-x}$	10–30 nm/min	low pressure	336
7	$Ba(thd)_2$ (180–250), $Ba(thd)_3$(tetraglyme)	$(Y(thd)_3)$		500	sapphire	$BaCO_3$	not reported	atm. pressure reactor	203
8	$Ba(fod)_2$ (160–210), $Ca(dpm)_2$	$Cu(acac)_2$, TlCp	O_2/H_2O	600	YSZ	Tl–Ba–Ca–Cu–O	—	—	337
9	$Ba(tdfnd)_2 \cdot H_2O$ (80–107)	$Cu(tmhd)_2$ (101), $Y(tmhd)_3$ (108)	O_2/N_2	660	Si (100), MgO	BaF_2, $Y_3Ba_3Cu_3O_7$	~5 nm/min for BaF_2	no details of YBC deposition	162
10	$Ba(hfac)_2$(18-crown-6) (189)	$Cu(thd)_2$ (126), $Y(thd)_3$ (128)	O_2/H_2O	690	MgO (001), $LaAlO_3$ (001)	$YBa_2Cu_3O_{7-x}$	0.02 μm/hr	—	338
11	$Ba(hfac)_2$ (tetraglyme) (111–175)	$Cu(thd)_2$, $Y(thd)_3$ (181–173)	O_2/H_2O	800–900	MgO (100)	$YBa_2Cu_3O_7$, Y_2BaCuO_5 + BaO	13 nm/min	—	339
12	$Ca(hfac)_2(triglyme)$ (90), $Sr(hfac)_2$ (tetraglyme) (120)	$BiPh_3$ (125), $Cu(acac)_2$ (132)	Ar/N_2O	650	MgO	Bi–Sr₁–Ca–Cu–O	10 nm/min	low pressure	340

[a] Numbers in parentheses are the temperatures to which the precursor was heated.

C. Group 2 Metal Poly(pyrazolyl)borates

To our knowledge, Group 2 poly(pyrazolyl)borates have not been used as precursors to deposit Group 2 element-containing films by CVD.

D. Group 2 Metal Alkoxides

Group 2 metal alkoxide compounds are potentially suitable as CVD precursors, especially because hydrolysis and subsequent thermally induced dehydration are likely to lead to complete removal of the supporting ligands. There are examples of Group 2 metal alkoxide compounds that are claimed to have suitable volatility for CVD, but to our knowledge these have not been used successfully. However, Group 2 metal alkoxide compounds have found widespread application in solution routes to metal oxide materials, an area that has been reviewed.[1]

E. Group 2 Metal β-Diketonates

The Group 2 β-diketonates have been used extensively in CVD experiments, and representative examples are reviewed here. The Group 2 β-diketonates can be divided into a number of different classes. Here we discuss the attributes of these compounds as precursors for CVD by classifying them according to the presence or absence of polydentate donor ligands. In the literature, species with empirical formula $M(\beta\text{-diketonate})_2$ have been termed "first-generation precursors," while those that possess additional polydentate ligands to avoid oligomerization have been termed "second-generation precursors."[20] In addition, there are a number of different subclasses of precursors, according to whether or not they possess fluorinated β-diketonate ligands, are β-ketoiminates, or have the polydentate ligands bonded to the β-diketonate ring (so-called lariats).

1. Adduct-free M(β-Diketonate)₂

Unsubstituted Group 2 bis-β-diketonate compounds have been used extensively as precursors for the deposition of Group 2 metal-containing films. This section is organized according the nature of the material deposited. Metal fluoride films are described first, followed by metal oxide films.

   a. Fluorinated M(β-Diketonate)₂ as Precursors for MF₂ Films.   Metal fluorides have been deposited using fluorinated β-diketonate as precursors, taking advantage of the presence of fluorine in the precursor as the source of fluoride in the film. Films deposited in the absence of an oxidizing agent also contain significant levels of carbon impurities.[154,190] This problem is solved by addition of O_2 and/or N_2O, which oxidizes carbon at typical deposition temperatures (500–750°C), but does not lead to oxide impurities because of the thermodynamic stability of the MF_2 phases.[213,308,309,344] The process of depositing metal fluorides from a "single-source" precursor (a precursor that contains all the elements required in the final film) is inherently less complex than deposition of ternary materials: only a single metal-containing reagent is involved, and so the process is less sensitive to changes in precursor stability or volatility. As a result, fluorinated β-diketonates have been used successfully, with the observations that tfac derivatives are generally less volatile than their hfac analogues, and that the order of decreasing volatility for a given series, M(β-diketonate)$_2$, is Ca > Sr > Ba.[154] Sato and Sagawara observed that the derivatives, Ba(ofhd)$_2$ and Ba(dfhd)$_2$, sublime almost completely based on TGA data and are insensitive to hydrolysis, making them the optimum precursors for CVD of metal fluorides.[213] The observed evaporation rates as a function of temperature are plotted in Fig. 59. Films deposited from these precursors are generally high-quality fluorides—e.g., BaF_2, which is highly preferentially (111) oriented—and have C and O impurity levels at the limits of detection of Rutherford back-scattering (RBS) and secondary ionization mass spectroscopy (SIMS).

Other nonfluorinated β-diketonates such as Ba(dpm)$_2$ have been used in conjunction with HF, in one case by sequential reaction using atomic layer deposition (ALD).[311] This method also gave high-quality fluoride films and enabled deposition of multilayer stacks for optical filtering applications.

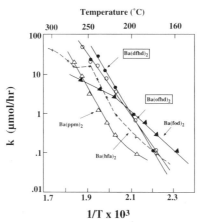

FIG. 59.   Graph of "relative volatility" (evaporation) rates vs reciprocal temperature for Ba(dfhd)$_2$, Ba(ofhd)$_2$, Ba(fod)$_2$, Ba(hfac)$_2$, and Ba(ppm)$_2$. (Redrawn from Ref. 213.)

b. *M(β-Diketonate)₂ as Precursors to Metal Oxide Films.*   Fluorinated and nonfluorinated Group 2 β-diketonates have been used extensively for the deposition of metal oxide films. Fluorinated precursors can lead to films with unacceptable levels of fluoride contamination, which is a problem particularly for films used for electronic applications. However, fluorinated β-diketonates generally exhibit the highest volatilities, so there has been a great deal of interest in the development of volatile nonfluorinated precursors. This section reviews CVD of metal oxide films using nonfluorinated and then fluorinated β-diketonates.

The species "M(tmhd)₂" have been used extensively for the deposition of metal oxide films containing M = Sr and Ba. However, the literature describing deposition experiments using these compounds is plagued with comments on irreproducibility and problems associated with transport of the precursors. The origin of these problems lies in the high temperatures, often above 200°C, required to transport these precursors, which leads to thermally induced reactions during long exposures. In addition, there has been some ambiguity associated with the nature of some of these precursor species, especially in the case of "Ba(tmhd)₂."[215] As mentioned earlier, the material described as "Ba(tmhd)₂" is either $Ba_4(tmhd)_8$ or $Ba_5(tmhd)_9(OH)(H_2O)_3$, depending on the method by which it was prepared.

The irreproducibility in transport of Ba compounds has been verified by attempts to measure the vapor pressures of such species. The compound "Ba(dpm)₂" decomposed during these experiments, but "Ca(dpm)₂" remained intact and corresponded to a Clausius–Clapeyron fit of log P (torr) = $-3760/T + 7.06$ (T = temperature in kelvins). This gives vapor pressure values of approximately 1 mtorr at 100°C and 1 torr at 250°C.[345] Although vapor pressure data for "Ba(dpm)₂" do not appear to exist, the evaporation rate has been quantified. The steady-state evaporation rate, $n_0(T)$ in mol·h⁻¹, can be expressed as a function of T, $n_0(T) = 7.17 \times 10^5 \exp(-24000/RT)$.[346] This equation also gives a value for the enthalpy of vaporization (24 kcal·mol⁻¹), close to the value previously obtained by Schmaderer *et al.*[347] Gas chromatographic data provide supporting evidence that M(thd)₂ species are oligomeric in the vapor phase.[348] Nevertheless, these species have been used to deposit metal oxide films, including BaTiO₃ and ceramic superconductor films. Attempts to deposit metal oxide films, MO, from M(dpm)₂ resulted in extensive contamination with carbon in the form of the metal carbonate and amorphous C. In the same report, fluorinated derivatives, M(fod)₂, gave the corresponding fluorides.[349]

Representative examples of the deposition of perovskite-phase MTiO₃ films are summarized in Table IX. The films generally have been deposited using the following conditions: Ti source = Ti(O-i-Pr)$_4$; ~200–250°C; other reagents, H_2O and/or O_2, and/or N_2O in reduced pressure, cold-wall CVD reactors, although there are some examples of liquid delivery. Representative examples of ceramic superconductor films are summarized in Table X. The films have been deposited under similar general conditions to the MTiO$_3$ films, but with metal β-diketonates such as Y(tmhd)$_3$, Cu(acac)$_2$ or Cu(tmhd)$_2$, and TlCp as the sources of the other metals. Most deposition experiments using "Ba(tmhd)$_2$" have been carried out using fresh samples of the precursor for each run as a result of the irreproducibility in transport rates at the source temperature necessary to establish sufficient vapor pressure. The Sr and Ca analogues are usually more volatile at lower temperatures, and so are less prone to thermal decomposition, although heating "Sr(dpm)$_2$" results in degradation of its properties,[313] and the plot of deposition rate of CaO as a function of "Ca(dpm)$_2$" source temperature shows a reduction with increasing temperature above 230°C (Fig. 60).[317] The first example of an *in situ* MOCVD-derived BaTiO$_3$ film was claimed as recently as 1991,[30] although attempts to grow epitaxial films were unsuccessful. The first report of epitaxial BaTiO$_3$ films grown by CVD was claimed in 1992 using Ba(hfac)$_2$(tetraglyme) as a Ba source.[326] Attempts to use other reagents such as Sr(acac)$_2$, Sr(OAc)$_2$, Sr(O-i-Pr)$_2$, and Sr (cyclohexanebutyrate)$_2$ were unsuccessful, either because these compounds had insufficient vapor pressures, or because they decomposed

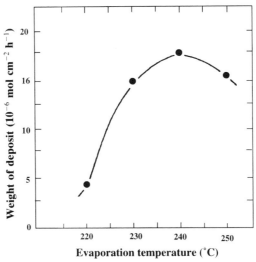

FIG. 60. Weight of CaO deposited plotted against the evaporation temperature of Ca(dpm)$_2$. (Redrawn from Ref. 317.)

upon heating.[31] Yoon *et al.*[350,351] deposited crystalline $BaTiO_3$ at the unusually low temperature of 370°C using $Ba(tmhd)_2$, $Ti(O\text{-}i\text{-}Pr)_4$, and N_2O as reagents.

Group 2 metal sulfide films have been deposited by ALD and ALE using "$M(thd)_2$" compounds. For example, SrS was deposited from "$Sr(thd)_2$" and H_2S at a substrate temperature of 200–340°C.[320]

Some of the problems of the irreproducibility in the delivery of "$M(dpm)_2$" compounds have been circumvented using liquid delivery approaches as described earlier. Films of the solid solution $Ba_{1-x}Sr_xTiO_3$, BST, were deposited using a mixture of "$M(tmhd)_2$" compounds, M = Ba or Sr, dissolved in THF, a coordinating solvent.[38] This method led to BST films with excellent properties and step coverage, as shown in Fig. 61.[318]

Fluorinated Ba β-diketonates have been seldom used for the deposition of metal oxide films as a result of the potential problem of fluoride incorporation and the poor volatility of the base-free compounds. Films deposited from fluorinated precursors generally require a subsequent anneal under an oxidizing ambient to remove metal fluoride impurities.

2. *Lewis Base Adducts, $M(\beta\text{-}Diketonate)_2L_n$*

The other solution to the problem of low precursor volatility and instability that has been used extensively in the deposition of Group 2 metal-containing films is the incorporation of various donor ligands into the coordination sphere of the metal ion, which prevents oligomerization.

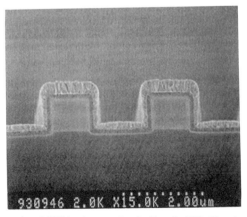

FIG. 61.   Cross-sectional SEM micrograph of a $Ba_{1-x}Sr_xTiO_3$ film showing step coverage using $Ba(dpm)_2$, $Sr(dpm)_2$, and $Ti(O\text{-}i\text{-}Pr)_4$ at 753 K. (Redrawn from Ref. 318.)

This approach also has some drawbacks, which are associated with the dissociation of the neutral ligands; this is discussed as it is encountered.

a. *M(β-Diketonate)$_2$L$_n$, Where L = a Monodentate Donor.* In order to improve the volatility of Ba(β-diketonate)$_2$ compounds, various donor species have been added to reduce or prevent oligomerization. For example, it was noted that the addition of Htmhd to the carrier gas used for evaporation of "Ba(tmhd)$_2$" resulted in a constant vapor pressure over several hours.[352] A dramatic increase in the stability and volatility of "Ba(tmhd)$_2$" was reported in the presence of nitrogen-based donors, which resulted in a lowering of the melting point (70–100°C) and evaporation temperature (130–230°C) of the Ba(tmhd)$_2$ source.[164,218,353] Nitrogen-free barium oxide films were deposited at substrate temperatures of 500°C under ambient pressure with NEt$_3$ added to the N$_2$/O$_2$ carrier gas stream. The presence of the amine was proposed to dissociate the Ba$_4$(tmhd)$_8$[180] tetramer to explain the improved transport under these conditions. Subsequent to this study, the compound [Ba(tmhd)$_2$(NH$_3$)$_2$]$_2$ was isolated and structurally characterized in the solid state by single-crystal X-ray diffraction.[174] However, for this species the amine ligand is not retained on vapor transport, which accounts for the need for a large excess of donor in the system. The vapor phase transport of Ba(thd)$_2$ was studied by dissolving it in NEt$_3$ and injecting this solution into a flowing Ar gas stream. This evaporative precursor transport (EPT) method led to greatly improved reproducibility in reagent transport. The method was used to deposit Ba-containing films in the presence of O$_2$ at substrate temperatures of 750°C on sapphire (−102) or Si (100). Barium carbonate films were formed with growth rates of up to 1 μm·h^{-1}. The residue deposited in the thermal evaporator during the course of these experiments, when conducted without the rigorous exclusion of moisture, led to the identification of the interesting Ba cluster [HNEt$_3$]$_2$[Ba$_6$(thd)$_{10}$(H$_2$O)$_4$(OH)$_2$(O$_2$)].[182]

Other donor adducts of monodentate ligands, such as MeOH and H$_2$O, are known and have been structurally characterized.[172] These are typically lower-nuclearity clusters compared to the parent Ba(β-diketonate)$_2$, which probably accounts for many observations associated with the improvement of precursor transport in the presence of these additives.

b. *M(β-Diketonate)$_2$L$_n$, Where L = a Polydentate Donor.* The monodentate donor adducts of the Group 2 β-diketonates typically dissociate the neutral donor ligand when heated, which requires that an excess of the donor be used during vapor phase transport. A potential alternative solution to adding excess ligand is to use a polydentate donor ligand. A polydentate ligand can potentially solve this problem by having a higher entropic barrier to dissociation, and can also lead to further deoligomerization of the parent Group 2 β-diketonate because the polydentate ligand is better able to satisfy the desired coordination number of the central metal. A number of cyclic and acyclic polyether and polyamine neutral donor compounds have been prepared and structurally characterized, as described earlier, and a number of these derivatives have been used in the deposition of Group 2 metal-containing films. Here, the fluorinated derivatives are described, followed by the nonfluorinated derivatives.

The deposition characteristics using polydentate donor ligand derivatives of $M(hfac)_2$ are summarized in Tables IX and X. The deposition of $SrTiO_3$ films using $Sr(hfac)_2(tetraglyme)$ showed a number of advantages in terms of precursor transport compared to unligated precursors. In one report,[323] the source temperature for $Sr(hfac)_2(tetraglyme)$ (105°C) was 100°C lower than that typically used for "$Sr(dpm)_2$" and, in addition, $Sr(hfac)_2(tetraglyme)$ was stable over several months. $Ba(hfac)_2(tetraglyme)$ exhibited significantly improved volatility and vapor pressure stability in comparison to $Ba(fod)_2$ and $Ba(dpm)_2$. Again, the ligated precursor, $Ba(hfac)_2(tetraglyme)$, could be transported (105–120°C) at temperatures significantly lower than unligated analogues.

A number of thermogravimetric analyses of $M(hfac)_2L$ adducts have been carried out. The studies generally show that the adducts sublime to leave a residue of about 5–10 wt.% of the original sample. For example, the sublimation of the compounds $M(hfac)_2(tetraglyme)$, where M = Ca, Sr, Ba, occurred in the ranges 100–300°C, 140–300°C, and 160–350°C, respectively, as a single step.[189] In contrast, the parent compound, $Ba(hfac)_2 \cdot H_2O$, showed two well-defined steps consistent with the initial loss of coordinated water.[354] Deposition experiments using $M(hfac)_2$ (tetraglyme) compounds as sources resulted in deposition of crystalline MF_2 films on Si(111) substrates at 450–550°C in the presence of O_2.[189]

A variety of studies that employ polydentate ligand donor adducts of fluorinated Group 2 β-diketonates have shown that crystallographically phase-pure metal oxide films can be deposited. However, there are some problems associated with fluoride incorporation. The presence of fluoride in ceramic superconductors is less of a problem compared to BST films, which require a very low (ppm) level of undesirable charge carriers. Strontium titanate films deposited at 810°C from $Sr(hfac)_2(tetraglyme)$ in the presence of water vapor showed no evidence for fluoride contamination using both X-ray diffraction and Auger electron spectroscopy.[323,324] Epitaxial films were grown, as illustrated by the X-ray diffraction data given in Fig. 62. However, it is likely that fluoride contamination is present at lower levels that are still detrimental to certain film properties. As a result, polydentate donor ligand adducts of nonfluorinated Group 2 β-diketonates

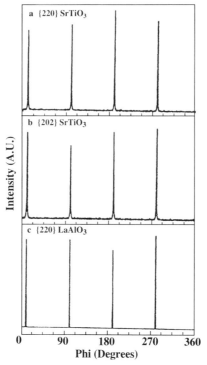

FIG. 62. Phi scan X-ray diffraction patterns of (a) the SrTiO₃ {220} reflections, (b) the SrTiO₃ {202} reflections, and (c) the LaAlO₃ substrate {220} reflections, showing the epitaxial relationship between the SrTiO₃ film (deposited from Sr(hfac)₂(tetraglyme) and Ti(O-i-Pr)₄) and the (100) LaAlO₃ substrate. (Redrawn from Ref. 323.)

have been investigated as precursors, but these species have proven less useful. The problem frequently encountered during the transport of these precursors is thermal dissociation of the polydentate ligand prior to evaporation. This problem arises because the metal center is more electron-rich when it possesses β-diketonate ligands substituted with electron-donating (e.g., t-Bu), rather than electron-withdrawing (e.g., CF_3), ligands; as a result, the metal center binds the neutral donor less strongly in the case of tmhd than that of hfac. A comparison of the relative transport properties of "Ba(thd)₂" and Ba(thd)₂(tetraglyme) revealed many of the problems associated with transporting "Ba(thd)₂" described earlier, and also that Ba(thd)₂(tetraglyme) exhibited significant precursor transport at vaporization temperatures of 220–250°C.[203] However, dissociation of the tetraglyme ligand was observed at "moderate" temperatures, and the amount of barium transported was lower when compared to Ba(thd)₂.

These authors suggest that while $Ba(thd)_2(tetraglyme)$ is not suitable for deposition of Ba-containing metal oxide films using conventional precursor delivery, it might be suitable for use in nontraditional liquid delivery methods.

Liquid delivery of polydentate donor adducts of nonfluorinated β-diketonates has been reported. The species $Ba(thd)_2(tetraglyme)$, $Ba(thd)_2$ (pmdt), and $Ba(thd)_2(hmtt)$ have been used as precursors to deposit $BaTiO_3$ films.[42,355] A comparison of the sublimation behavior of these compounds reveals that $Ba(thd)_2(tetraglyme)$ dissociates tetraglyme without transporting $Ba(thd)_2$, while $Ba(thd)_2(pmdt)$ and $Ba(thd)_2(hmtt)$ dissociate their respective polyamine ligands during sublimation, but permitting transport of $Ba(thd)_2$. It was concluded that $Ba(thd)_2(pmdt)$ is the slightly more thermally stable of the two polyamine compounds. Deposition of $BaTiO_3$ and $Ba_{1-x}Sr_xTiO_3$, using liquid precursor delivery with flash evaporation, revealed that the pmdt compound is more suitable because it allowed a larger temperature window for effective vapor transport and was less sensitive to changes in reactor wall temperatures.

A precursor design strategy has been described that avoids the problem of fluoride contamination resulting from fluorinated ligands and, at the same time, the problem of precursor transport associated with the dissociation of the neutral polydentate donor ligand. This design involved the isoelectronic replacement of one of the β-diketonate oxygen atoms with an imino group, $=NR$, where the R group contained a polydentate donor as shown in Section III.[205,206] When compared to "$Ba(dpm)_2$," these complexes exhibited greater volatilities and comparable residues after sublimation. The deposition of $BaPbO_3$ was demonstrated using one of these compounds as precursor. A similar strategy was developed by Rees *et al.*, in the substitution of one of the alkyl groups of the β-diketonate ligand for a polydentate donor.[356] The latter species was also a liquid Ba compound under ambient conditions, which is a valuable property and can lead to better control over precursor delivery as described earlier.

Other strategies to volatile Ba compounds that are also liquids under the transport conditions include the use of polydentate ether ligands where the number of ether oxygens is greater than can coordinate to a Ba center, which results in the presence of uncoordinated, dangling polyether groups that disrupt crystal packing forces. The species $Ba(hfac)_2(MeO(CH_2 CH_2O)_6$-$n$-Bu and $Ba(hfac)_2(MeO(CH_2CH_2O)_6$-$C_2H_5$ are typical examples. They exhibit similar sublimation behavior to $Ba(hfac)_2(tetraglyme)$, with $Ba(hfac)_2(MeO(CH_2CH_2O)_6$-$C_2H_5$ evaporating approximately three times faster than $Ba(hfac)_2(tetraglyme)$ under typical $BaTiO_3$ delivery conditions.[20,328] The liquid precursors could be used reproducibly over a period of several months without detectable degradation of vapor pressure or transport characteristics. This is in contrast to solid $Ba(hfac)_2$(tetraglyme) samples, which slowly lose volatility with time and require pulverization to restore volatility.

F. Group 2 Metal Carboxylates

In order to build on the results described earlier and also to address some of the outstanding issues, a number of polydentate ligand Group 2 metal carboxylate compounds have been prepared as precursors for the deposition of Group 2 metal-containing films. Group 2 metal carboxylates themselves are expected to be oligomers or infinite polymers, and therefore completely involatile. However, by employing an analogous strategy to that proposed earlier, polydentate ligand adducts could have a much lower degree of aggregation and may exhibit sufficient volatility to allow deposition of CVD-derived films. The potential advantage of carboxylate precursors over their β-diketonate counterparts is that the ligands might be eliminated at a significantly lower temperature, by ester elimination, than is possible through thermal decomposition of β-diketonate ligands in the presence of an additional oxide ion source such as H_2O or O_2. It has been demonstrated that the reaction between a metal alkoxide and a metal carboxylate can lead to complete elimination of organic supporting ligands, with concomitant formation of an ester according to Eqs. (30) and (31).[357,358]

$$M(O_2CR)_n + M(OR')_n \longrightarrow M(O_2CR)_{n-x}(OR)_{n-x} + x\ RCO_2R' \tag{30}$$

$$M(O_2CR)_{n-x}(OR)_{n-x} \longrightarrow MO_y + (n-x)RCO_2R' \tag{31}$$

As a result, it may be possible to achieve a surface initiated ester elimination reaction in which a neutral polyether adduct of a Group 2 carboxylate and a titanium alkoxide are delivered to the substrate surface. Because the Group 2 carboxylate is coordinatively saturated, it is unlikely to react with the metal alkoxide at ambient temperatures. However, when the Group 2 carboxylate adsorbs on the heated substrate surface, the neutral donor ligand can dissociate, releasing a highly reactive, coordinately unsaturated monomer $M(O_2CR)_2$ that can react with the $Ti(OR')_4$ present to eliminate two equivalents of ester. The remaining alkoxide ligands can be removed through β-hydride elimination processes that have been observed previously. Because ligand oxidation (e.g., by O_2 or N_2O) is not required, because the dissociation of neutral polydentate donor ligands is a low-energy process, and because it has been shown that ester elimination can also occur at low temperatures ($<100°C$), this process should be capable of depositing Group 2 metal-containing films at low temperatures relative to existing processes. Furthermore, if the Group 2 element ligands are completely eliminated by ester elimination at relatively low temperatures, lower than their thermal decomposition temperature, there is a greater possibility of low impurity levels, especially using fluorinated ligands and avoiding the detrimental side effect of fluoride contamination. Finally, fluorinated carboxylate ligands may render the Group 2 metal center more electron deficient than fluorinated β-diketonate ligands. This feature may help to avoid premature dissociation of the neutral polydentate ligand.

A number of examples of compounds with the empirical formula $[M(O_2CR)_2]_m(L)_n$ have been prepared and characterized in the solid state as described earlier. Attempts to deposit $MTiO_3$ films from these species using AACVD has resulted in low deposition rates, mainly as a result of their low volatility and their propensity to dissociate the polydentate donor ligand. In control experiments, in the absence of a Ti source, Group 2 element-containing films have been deposited, but their deposition rates are extremely low. To provide further insight, the deposition of Group 2 element-containing films has been studied by liquid-phase deposition techniques using representative examples of this class of compounds. The reaction of these polyether barium carboxylate compounds with $Ti(O-i-Pr)_4$ was studied by spin coating ethanol solutions of these species onto Si substrates, followed by thermal treatments to different temperatures. X-ray diffraction data of films prepared from $Ba_2(O_2CCF_3)_4$ (15-crown-5)$_2$ showed only the presence of crystalline BaF_2 at 350 and 450°C and a mixture of crystalline $BaTiO_3$ and BaF_2 at 550 and 650°C. In contrast, X-ray diffraction of films deposited from solutions of M_2 $(O_2CCF_3)_4$(tetraglyme) and $Ti(i-OPr)_4$ showed that amorphous films was formed in the temperature range 250–500°C and crystalline $MTiO_3$ (M = Sr, Ba, $Ba_{0.7}Sr_{0.3}$) at 575°C and above. Auger electron spectroscopy of amorphous films prepared at 350°C showed that the films had an elemental composition consistent with the presence of amorphous $MTiO_3$ (M = Sr, Ba, $Ba_{0.7}Sr_{0.3}$) and that after further heating to 575°C, the $MTiO_3$ (M = Sr, Ba, $Ba_{0.7}Sr_{0.3}$) composition was retained. This is consistent with the crystallographic data, which reveal the presence of $MTiO_3$ as the only crystalline phase. The crystalline films showed no fluorine contamination by Auger electron spectroscopy.

Liquid-phase metal-organic decomposition methods to prepare Group 2 element-containing films have employed carboxylate derivatives of the Group 2 elements quite extensively. However, we can find very few examples of CVD using Group 2 metal carboxylates. Kim et al. used an ultrasonic spray technique to deposit (110) preferential orientation $BaTiO_3$ films using barium hexanoate and $Ti(acac)(O-i-Pr)_2$, but it is unclear whether this was by CVD or a solution route.

G. Group 2 Metal Thiocarboxylates

The Group 2 thioacetate compounds, $M(SOCMe)_2L$, have been studied as single-source precursors for the formation of binary Group 2 metal sulfides, MS. The rationale to investigate these compounds is that if the reaction pathway involves thermally induced elimination of thioacetic anhydride according to Eq. (32), then the metal sulfide film should be relatively pure.

$$M(SOCMe)_2L \longrightarrow MS + (MeCO)_2S + L \qquad (32)$$

The thermal decomposition behavior of these compounds was studied by thermogravimetric analysis (TGA). In general, thermal decomposition in air resulted in loss of the polyether ligand and reaction to form mixtures of the corresponding oxides and sulfates, as was observed previously for the thermal decomposition of $M(SOCMe)(O_2CMe)$, M = Ca and Sr However, thermal decomposition in dinitrogen resulted in loss of the polyether ligand, reaction of the thioacetate ligands, and formation of the corresponding crystalline metal sulfide, MS, by 300°C.

Metal sulfide films were also deposited by aerosol-assisted (AA)CVD, as exemplified by the results for $Ca(SOCMe)_2(15\text{-crown-}5)$ described here.

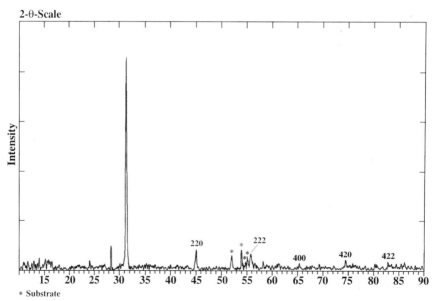

FIG. 63. X-ray diffraction pattern from CaS formed from CVD of $Ca(SOCMe)_2(15\text{-}$crown-5). (Redrawn from Ref. 270.)

Aerosol delivery of the precursors was chosen because it was expected that although the species were monomeric, their volatility was likely to be low based on the TGA results described earlier, and that the crown ether might dissociate on heating for extended periods. The compound $Ca(SOCMe)_2(15$-crown-5) was dissolved in ethanol and delivered in a nitrogen gas stream to a cold-wall atmospheric pressure CVD reactor where the substrate temperature was varied between 300 and 400°C. Analysis of films deposited at 310°C revealed that crystalline CaS was deposited, approximately 100 nm thick, with a deposition rate of 10 nm/min. The X-ray powder pattern indicates that CaS was formed with preferential (200) orientation (see Fig. 63). The SEM data showed that the films comprised cubic crystallites with dimensions consistent with the linewidth of the peaks observed by X-ray diffraction. The bulk composition corresponded to CaS as determined by AES.

V

SUMMARY, CONCLUSIONS, AND FUTURE DIRECTIONS

A great deal of progress has been made in the past 5 years in the development of Group 2 compounds suitable for the deposition of Group 2 element containing films. It seems most logical to organize these concluding remarks on precursor design and deposition strategies based on the material reviewed in this article according to the categories of precursor design criteria described in the introduction.

Volatility: A large variety of different ligand systems have been used in an attempt to impart volatility to Group 2 compounds. It is difficult quantitatively to compare the volatility of the species described in this review for the reasons stated earlier, but it appears that the fluorinated β-diketonates (e.g., $Ba(tdfnd)_2$) and their polydentate ligand adducts (e.g., $Ba(hfac)_2(18$-crown-6)) are among the most volatile examples. Fluorinated β-diketonate derivatives are generally more volatile than non-fluorinated β-diketonate derivatives, and donor adducts are generally more volatile than base-free compounds.[155,359] However, donor adducts of non-fluorinated metal β-diketonates generally dissociate prior to sublimation. Even though advances have been made in this respect, there has not yet been a significant breakthrough that is likely to make a large impact on industrial applications. Group 2 element precursors that exhibit sufficient volatility to be transported by conventional delivery methods are still desperately needed. Fluorinated ligands have the drawback that fluoride contamination of films is unacceptable in some applications, and as a result nonfluorinated ligand sets are required. However, anionic substituents with highly electronegative substituents are generally required to render the metal center, especially Ba, sufficiently Lewis acidic for coordinated donor ligands to remain bonded to the metal center during sublimation. A better understanding of factors that control intermolecular interactions and lattice energies is required.

Reactivity: The methods by which the supporting organic ligands have been removed during CVD of Group 2 metal-containing films have generally employed either thermally induced oxidation of the ligands or hydrolysis. If significantly lower deposition temperatures are to be achieved (as is necessary for some applications), then alternative strategies are probably needed: complete oxidation of organic ligands (e.g., using O_2) is likely to require temperatures in excess of 500°C, and hydrolysis does not appear to result in complete elimination of ligands, especially fluorinated ligands. In the case of Group 2 precursors with nonfluorinated ligands, hydrolysis (or, in general, the addition of a reagent that will eliminate the ligands and ultimately replace them with the desired atom, i.e., O from H_2O or S from H_2S) might be a good strategy, but will be complicated if another metal-containing reagent is used. For example, in the deposition of $BaTiO_3$, the use of Ba- and Ti-containing reagents that are both hydrolytically sensitive is likely to lead to gas-phase reactions and particle formation unless special precautions are taken. Therefore, a potentially better strategy might involve the design of Group 2-containing compounds that contain ligands that can be completely eliminated by thermal initiation, or that are designed to react with the ligands of an accompanying metal-containing reagent on thermal initiation. Examples of these strategies, which involve ether and ester elimination, were described earlier.

Precursor transport: Although related to the volatility described earlier, transport is mentioned separately here because of potential impact on this area. Liquid delivery methods have demonstrated advantages over conventional delivery methods for thermally unstable and low-volatility precursors, as illustrated by some of the examples described in this review. The development of improved transport techniques might alleviate the necessity for better volatilities and thermal stabilities required by traditional delivery methods of these precursors. However, whether liquid delivery methods will be adopted into Si-based manufacturing processes remains to be seen. It seems more likely that these transport methods will be used in applications where less stringent control over purity is acceptable.

To solve problems of volatility, transport, and reactivity, and simultaneously to obtain high-quality Group 2 metal containing films at relatively low substrate temperatures, is clearly still a significant challenge to chemists. It is reasonable to ask if significant developments, especially in volatility, are likely to be made. We feel that significant improvements in volatility and reproducible precursor transport can be made. There is a large effort in the development of ligands that encapsulate Group 2 elements, which have been studied for biological reasons and could provide insight here.[299,360–371] However, from a technological viewpoint, we feel that the most useful precursors for CVD applications are likely to be those that are simple and inexpensive, and that rely on chemical ingenuity rather than complex ligand design to transport them to a surface.

REFERENCES

(1) Chandler, C. D.; Roger, C.; Hampden-Smith, M. J. *Chem. Rev.* **1993,** *93,* 1205.

(2) Kautek, W. *Vacuum* **1992,** *43,* 403.

(3) Hibst, H. *Angew. Chem., Int. Ed. Engl.* **1982,** *21,* 270.

(4) Hibst, H.; Schwab, E. In *Materials Science and Technology*; Cahn, R. W.; Haasen, P.; Kramer, E. J., Eds.; VCH: Weinheim, 1994.

(5) Braithwaite, N.; Weaver, G. *Electronic Materials*; Butterworth: London, 1990.

(6) Hampden-Smith, M. J.; Kodas, T. T.; Ludviksson, A. In *Materials Chemistry: An Emerging Discipline*; Hampden-Smith, M. J.; Interrante, L., Eds.; VCH: New York, 1995.

(7) Kodas, T. T.; Hampden-Smith, M. J. *The Chemistry of Metal CVD*; VCH: Weinheim, 1994.

(8) Vossen, J. L.; Kern, W., Eds. *Thin Film Processes*; Academic Press: New York, 1978.

(9) Stringfellow, G. B. *Organometallic Vapor-Phase Epitaxy: Theory and Practice*; Academic Press: New York, 1989.

(10) Gladfelter, W. L. *Chem. Mater.* **1993,** *5,* 1372.

(11) Hitchman, M. L.; Jensen, K. F. *Chemical Vapor Deposition: Principles and Applications*; Academic Press: London, 1993.

(12) Hampden-Smith, M. J.; Kodas, T. T. *Adv. Mater.* **1994,** *7,* 8.

(13) Hampden-Smith, M. J.; Kodas, T. T. *Adv. Mater.* **1994,** *7,* 39.

(14) Kodas, T. T.; Hampden-Smith, M. J. In *The Chemistry of Metal CVD*; Kodas, T. T.; Hampden-Smith, M. J., Eds.; VCH: New York, 1994, Chapter 9.

(15) Jaraith, R.; Jain, A.; Tolles, R. D.; Hampden-Smith, M. J.; Kodas, T. T. In *The Chemistry of Metal CVD*; Kodas, T. T.; Hampden-Smith, M. J., Eds.; VCH: New York, 1994, Chapter 1.

(16) Tonge, L. M.; Richeson, D. S.; Marks, T. J.; Zhao, J.; Zhang, J.; Wessels, B. W.; Marcy, H. O.; Kannewurf, C. R. *Adv. Chem. Ser.* **1990,** *226,* 351.

(17) Dahmen, K. H.; Gerfin, T. *Progress in Crystal Growth and Characterization of Materials* **1993,** *27,* 117.

(18) Schulz, D. L.; Marks, T. J. *Adv. Mater.* **1994,** *6,* 719.

(19) Barron, A. R.; Rees, W. S. *Adv. Mater. Opt. Elec.* **1993,** *2,* 271.

(20) Marks, T. J. *Pure Appl. Chem.* **1995,** *67,* 313.

(21) Carlsson, J. O. *Acta Chem. Scand.* **1991,** *45,* 864.

(22) Rees, W. S.; Barron, A. R. *Mater. Sci. Forum* **1993,** *137–139,* 473.

(23) Wessels, B. W. *Annu. Rev. Mater. Sci.* **1995,** *25,* 525.

(24) Rees, W. S. *Mat. Res. Symp. Proc.* **1994,** *335,* 351.

(25) Rees, W. S.; Dippel, K. A., Eds. *New Group 2 Precursors to Metal Oxides*; John Wiley and Sons: New York, 1992.

(26) Hubertpfalzgraf, L. G. *Appl. Organometal. Chem.* **1992**, *6*, 627.

(27) Mehrotra, R. C.; Singh, A.; Sogani, S. *Chem. Soc. Rev.* **1994**, 215.

(28) Vaartstra, B. A.; Huffman, J. C.; Streib, W. E.; Caulton, K. G. *Inorg. Chem.* **1991**, *30*, 121.

(29) Hampden-Smith, M. J.; Kodas, T. T. In *The Chemistry of Metal CVD*; Kodas, T. T.; Hampden-Smith, M. J., Eds.; VCH: New York, 1994, Chapter 8.

(30) Kwak, B. S.; Zhang, K.; Boyd, E. P.; Erbil, A.; Wilkens, B. J. *J. Appl. Phys.* **1991**, *69*, 767.

(31) Feil, W. A.; Wessels, B. W.; Tonge, L. M.; Marks, T. J. *J. Appl. Phys.* **1990**, *67*, 3858.

(32) Xu, C.; Hampden-Smith, M. J.; Kodas, T. T. *Adv. Mater.* **1994**, *6*, 745.

(33) Xu, C.; Kodas, T. T.; Hampden-Smith, M. J. *Chem. Mater.* **1995**, *7*, 1539.

(34) Roger, C.; Corbitt, T. S.; Hampden-Smith, M. J.; Kodas, T. T. *Appl. Phys. Lett.* **1994**, *65*, 1021.

(35) Deschanvres, J. L.; Bochu, B.; Joubert, J. C. *J. Phys. IV* **1993**, *3*, 485.

(36) Deschanvres, J. L.; Joubert, J. C. *J. Phys. IV* **1992**, *2*, 29.

(37) Deschanvres, J. L.; Rey, P.; Delabouglise, G.; Labeau, M.; Joubert, J. C.; Peuzin, J. C. *Sensors and Actuators A—Physical* **1992**, *33*, 43.

(38) Kim, I. T., Lee, C. H.; Park, S. J. *Jpn. J. Appl. Phys. Pt. 1* **1994**, *33*, 5125.

(39) McMillan, L. D.; Huffman, M.; Roberts, T. L.; Scott, M. C.; Paz de Araujo, C. A. *Integr. Ferroelectr.* **1994**, *4*, 319.

(40) Xu, C. Y.; Hampden-Smith, M. J.; Kodas, T. T. *Adv. Mater.* **1994**, *6*, 746.

(41) Nyman, M.; Duesler, E. N.; Hampden-Smith, M. J. *Adv. Mater. Chem.* **1996**, in press.

(42) Gardiner, R. A.; Gordon, D. C.; Stauf, G. T.; Vaartstra, B. A.; Ostrander, R. L.; Rheingold, A. L. *Chem. Mater.* **1994**, *6*, 1967.

(43) Hansen, B. N.; Hybertson, B. M.; Barkley, R. M.; Sievers, R. E. *Chem. Mater.* **1992**, *4*, 749.

(44) Sievers, R. E.; Hanson, B. N. University of Colorado, U.S. Patent 4,970,093, 1990.

(45) Hybertson, B. M.; Hansen, B. H.; Barkley, R. M.; Sievers, R. E. *Mater. Res. Bull.* **1991**, *26*, 1127.

(46) Bradley, D. C.; Mehrotra, R. C.; Gaur, D. P. *Metal Alkoxides*; Academic Press: London, 1978.

(47) Rees, W. S. J.; Dippel, K. A.; Carris, M. W.; Caballero, C. R.; Moreno, D. A.; Hesse, W. *Mater. Res. Soc. Symp. Proc.* **1992**, *271*, 127.

(48) Herrmann, W. A.; Huber, N. W.; Priermeier, T. *Angew. Chem.* **1994**, *106*, 102.

(49) Drake, S. R.; Streib, W. E.; Folting, K.; Chisholm, M. H.; Caulton, K. G. *Inorg. Chem.* **1992**, *31*, 3205.

(50) Zinn, A. In *The Chemistry of Metal CVD*; Kodas, T. T.; Hampden-Smith, M. J., Eds.; VCH: Weinheim, 1994.

(51) Greenwood, N. N.; Earnshaw, E. A. *Chemistry of the Elements*; Pergamon: Oxford, 1984.

(52) Westerhausen, M.; Schwarz, W. *Z. Anorg. Allg. Chem.* **1991**, *604*, 127.

(53) Westerhausen, M. *Inorg. Chem.* **1991**, *30*, 96.

(54) Bradley, D. C.; Hursthouse, M. B.; Ibrahim, A. A.; Abdul Malik, K. M.; Motevalli, M.; Moseler, R.; Powell, H.; Runnacles, J. D.; Sullivan, A. C. *Polyhedron* **1990**, *9*, 2959.

(55) Hitchcock, P. B.; Lappert, M. F.; Lawless, G. A.; Royo, B. *J. Chem. Soc., Chem. Commun.* **1990**, 1141.

(56) Westerhausen, M.; Schwarz, W. *Z. Anorg. Allg. Chem.* **1991**, *606*, 177.

(57) Cloke, F. G. N.; Hitchcock, P. B.; Lappert, M. F.; Lawless, G. A.; Royo, B. *J. Chem. Soc., Chem. Commun.* **1991**, 724.

(58) Boncella, J. M.; Coston, C. J.; Cammack, J. K. *Polyhedron* **1991**, *10*, 769.

(59) Tesh, K. F.; Burkey, D. J.; Hanusa, T. P. *J. Am. Chem. Soc.* **1994**, *116*, 2409.

(60) Sockwell, S. C.; Hanusa, T. P.; Huffman, J. C. *J. Am. Chem. Soc.* **1992**, *114*, 3393.

(61) Burkey, D. J.; Alexander, E. K.; Hanusa, T. P. *Organometallics* **1994**, *13*, 2773.

(62) Mosges, G.; Hampel, F.; Kaupp, M.; Schleyer, P. V. R. *J. Am. Chem. Soc.* **1992**, *114*, 10880.

(63) Shao, P.; Berg, D. J.; Bushnell, G. W. *Can. J. Chem.* **1995**, *73*, 797.

(64) Westerhausen, M.; Hausen, H. D.; Schwarz, W. *Z. Anorg. Allg. Chem.* **1995**, *621*, 877.

(65) Westerhausen, M.; Schwarz, W. *J. Organomet. Chem.* **1993**, *463*, 51.

(66) Westerhausen, M. *J. Organomet. Chem.* **1994**, *479*, 141.

(67) Drake, S. R.; Hall, P.; Lincoln, R. *Polyhedron* **1993**, *12*, 2307.

(68) Westerhausen, M.; Schwarz, W. *Z. Anorg. Allg. Chem.* **1992**, *609*, 39.

(69) Wannagat, U.; Autzen, H.; Wismar, H.-J. *Z. Anorg. Allg. Chem.* **1972**, *394*, 254.

(70) Wannagat, U.; Kuckertz, H. *Angew. Chem.* **1963**, *75*, 95.

(71) Frankland, A. D.; Hitchcock, P. B.; Lappert, M. F.; Lawless, G. A. *J. Chem. Soc., Chem. Commun.* **1994**, 2435.

(72) Drake, S. R.; Otway, D. J. *J. Chem. Soc., Chem. Commun.* **1991**, 517.

(73) Westerhausen, M.; Hildenbrand, T. *J. Organomet. Chem.* **1991**, *411*, 1.

(74) Hanusa, T. P. *Polyhedron* **1990**, *9*, 1345.

(75) Hanusa, T. P. *Chem. Rev.* **1993**, *93*, 1023.

(76) Burkey, D. J.; Williams, R. A.; Hanusa, T. P. *Organometallics* **1993**, *12*, 1331.

(77) Blom, R.; Faegri, K., Jr.; Volden, H. V. *Organometallics* **1990**, *9*, 372.

(78) Williams, R. A.; Hanusa, T. P.; Huffman, J. C. *Organometallics* **1990**, *9*, 1128.

(79) Williams, R. A.; Hanusa, T. P.; Huffman, J. C. *J. Chem. Soc., Chem. Commun.* **1988**, 1045.

(80) Andersen, R. A.; Blom, R.; Burns, C. J.; Volden, H. V. *J. Chem. Soc., Chem. Commun.* **1987**, 768.

(81) Andersen, R. A.; Blom, R.; Boncella, J. M.; Burns, C. J.; Volden, H. V. *Acta Chem. Scand.* **1987**, *A41*, 24.

(82) Andersen, R. A.; Boncella, J. M.; Burns, C. J. *J. Organomet. Chem.* **1986**, *312*, C49.

(83) Burns, C. J.; Andersen, R. A. *J. Organomet. Chem.* **1987**, *325*, 31.

(84) McCormick, M. J.; Williams, R. A.; Levine, L. J.; Hanusa, T. P. *Polyhedron* **1988**, *7*, 725.

(85) Rees, W. S. J.; Lay, U. W.; Dippel, K. A. *J. Organomet. Chem.* **1994**, *483*, 27.

(86) Williams, R. A.; Tesh, K. F.; Hanusa, T. P. *J. Am. Chem. Soc.* **1991**, *113*, 4843.

(87) Coates, G. E.; Glockling, F. *J. Chem. Soc.* **1954**, 22.

(88) Gardiner, M. G.; Raston, C. L.; Kennard, C. H. L. *Organometallics* **1991**, *10*, 3680.

(89) Engelhardt, L. M.; Junk, P. C.; Raston, C. L.; White, A. H. *J. Chem. Soc., Chem. Commun.* **22**, 1500 (1988).

(90) Williams, R. A.; Hanusa, T. P.; Huffman, J. C. *Organometallics* **1990**, *9*, 372.

(91) Zerger, R.; Stucky, G. *J. Organomet. Chem.* **1974**, *80*, 7.

(92) Fischer, E. O.; Stolzle, G. *Chem. Ber.* **1961**, *94*, 2187.

(93) Hammel, A.; Schwarz, W.; Weidlein, J. *J. Organomet. Chem.* **378**, 347 (1989).

(94) Gowenlock, B. G.; Lindsell, W. E.; Singh, B. *J. Chem. Soc., Dalton Trans.* **1978**, 657.

(95) Rieckhoff, M.; Pieper, U.; Stalke, D.; Edelmann, F. T. *Angew. Chem., Int. Ed. Engl.* **1993**, *32*, 1079.

(96) Jutzi, P.; Leffers, W.; Muller, G.; Huber, B. *Chem. Ber.* **1989**, *122*, 879.

(97) Tanner, P. S.; Hanusa, T. P. *Polyhedron* **1994**, *13*, 2417.

(98) Bruce, M. I.; Walton, J. K.; Skelton, B. W.; White, A. H. *J. Chem. Soc., Dalton Trans.* **1982**, *11*, 2221.
(99) Ziegler, K. *Angew. Chem.* **1954**, *66*, 239.
(100) Ziegler, K.; Froitzheim-Kuhlhorn, H.; Hafner, H. *Chem. Ber.* **1956**, *89*, 434.
(101) Tanner, P. S.; Burkey, D. J.; Hanusa, T. P. *Polyhedron* **1995**, *14*, 331.
(102) Tasaki, Y.; Sakamoto, R.; Ogawa, Y.; Yoshizawa, S.; Ishiai, J.; Akase, S. *Jpn. J. Appl. Phys. Pt. 1* **1994**, *33*, 5400.
(103) Williams, R. A.; Hanusa, T. P.; Huffman, J. C. *J. Organomet. Chem.* **1992**, *429*, 143.
(104) Williams, R. A.; Hanusa, T. P.; Huffman, J. C. *J. Am. Chem. Soc.* **1990**, *112*, 2454.
(105) Burkey, D. J.; Hanusa, T. P.; Huffman, J. C. *Adv. Mater. Opt. Electron.* **1994**, *4*, 1.
(106) Kaupp, M.; Schleyer, P. v. R.; Dolg, M.; Stoll, H. *J. Am. Chem. Soc.* **1992**, *114*, 8202.
(107) Green, J. C.; Hohl, D.; Rosch, N. *Organometallics* **1987**, *6*, 712.
(108) Andersen, R. A.; Boncella, J. M.; Burns, C. J.; Green, J. C.; Hohl, D.; Rosch, N. *J. Chem. Soc., Chem Commun.* **1986**, 405.
(109) Ortiz, J. V.; Hoffman, R. *Inorg. Chem.* **1985**, *24*, 2095.
(110) Hollis, T. K.; Burdett, J. K.; Bosnich, B. *Organometallics* **1993**, *12*, 3385.
(111) Belderrain, T. R.; Contreras, L.; Paneque, M.; Carmona, E.; Monge, A.; Ruiz, C. *J. Organomet. Chem.* **1994**, *474*, C5.
(112) Dutremez, S. G.; Leslie, D. B.; Streib, W. E.; Chisholm, M. H.; Caulton, K. G. *J. Organomet. Chem.* **1993**, *462*, C1.
(113) Sohrin, Y.; Kokusen, H.; Kihara, S.; Matsui, M.; Kushi, Y.; Shiro, M. *J. Am. Chem. Soc.* **1993**, *115*, 4128.
(114) Sohrin, Y.; Matsui, M.; Hata, Y.; Hasegawa, H.; Kokusen, H. *Inorg. Chem.* **1994**, *33*, 4376.
(115) Dias, H. V. R.; Lu, H.-L.; Ratcliff, R. E.; Bott, S. G. *Inorg. Chem.* **1995**, *34*, 1975.
(116) Amoroso, A. J.; Jeffery, J. C.; Jones, P. L.; McCleverty, J. A.; Psillakis, E.; Ward, M. D. *J. Chem. Soc., Chem. Commun.* **1995**, 1175.
(117) Hubert-Pfalzgraf, L. G. *Polyhedron* **1994**, *13*, 1181.
(118) Caulton, K. G.; Hubert-Pfalzgraf, L. G. *Chem. Rev.* **1990**, *90*, 969.
(119) Purdy, A. P.; George, C. F. *Inorg. Fluor. Chem.* **1994**, *555*, 405.
(120) Mehrotra, R. C.; Singh, A.; Sogani, S. *Chem. Rev.* **1994**, *94*, 1643.
(121) Purdy, A. P.; George, C. F.; Callahan, J. H. *Inorg. Chem.* **1991**, *30*, 2812.
(122) Caulton, K. G.; Chisholm, M. H.; Drake, S. R.; Folting, K.; Huffman, J. C.; Streib, W. E. *Inorg. Chem.* **1993**, *32*, 1970.
(123) Poncelet, O.; Hubert-Pfalzgraf, L. G.; Toupet, L.; Daran, J. C. *Polyhedron* **1991**, *17*, 2045.
(124) Rees, W. S.; Moreno, D. A. *J. Chem. Soc., Chem. Commun.* **1991**, 1759.
(125) Caulton, K. G.; Chisholm, M. H.; Drake, S. R.; Folting, K.; Huffman, J. C. *Inorg. Chem.* **1993**, *32*, 816.
(126) Caulton, K. G.; Chisholm, M. H.; Drake, S. R.; Huffman, J. C. *J. Chem. Soc., Chem. Commun.* **1990**, 1498.
(127) Caulton, K. G.; Chisholm, M. H.; Drake, S. R.; Folting, K. *J. Chem. Soc., Chem. Commun.* **1990**, 1349.
(128) Drake, S. R.; Streib, W. E.; Chisholm, M. H.; Caulton, K. G. *Inorg. Chem.* **1990**, *29*, 2707.
(129) Goel, S. C.; Matchett, M. A.; Chiang, M. Y.; Buhro, W. E. *J. Am. Chem. Soc.* **1991**, *113*, 1844.
(130) Miele, P.; Foulon, J.-D.; Hovanian, N.; Cot, L. *Polyhedron* **1993**, *12*, 267.
(131) Miele, P.; Foulon, J.-D.; Hovnanian, N. *Polyhedron* **1993**, *12*, 209.

(132) Drake, S. R.; Otway, D. J.; Hursthouse, M. B.; Abdul Malik, K. M. *Polyhedron* **1992,** *11,* 1995.

(133) Caulton, K. G.; Chisholm, M. H.; Drake, S. R.; Streib, W. E. *Angew. Chem., Int. Ed. Engl.* **1990,** *29,* 1483.

(134) Darr, J. A.; Drake, S. R.; Williams, D. J.; Slawin, A. M. Z. *J. Chem. Soc., Chem. Commun.* **1993,** 866.

(135) Caulton, K. G.; Chisholm, M. H.; Drake, S. R.; Folting, K. *Inorg. Chem.* **1991,** *30,* 1500.

(136) Wojtczak, W. A.; Hampden-Smith, M. J.; Duesler, E. N. *Inorg. Chem.,* submitted for publication.

(137) Darr, J. A.; Drake, S. R.; Hursthouse, M. B.; Malik, K. M. A. *Inorg. Chem.* **1993,** *32,* 5704.

(138) Tesh, K. F.; Hanusa, T. P.; Huffman, J. C.; Huffman, C. J. *Inorg. Chem.* **1992,** *31,* 5572.

(139) Cole, L. B.; Holt, E. M. *J. Chem. Soc., Perkin Trans. II* **1986,** 1997.

(140) Kanters, J. A.; Postma, R.; Duisenberg, A. J. M.; Venkatasubramanian, K.; Poonia, N. S. *Acta Cryst.* **1983,** *C39,* 1519.

(141) Singh, T. P.; Reinhardt, R.; Poonia, N. S. *Inorg. Nucl. Chem. Lett.* **1980,** *16,* 293.

(142) Postma, R.; Kanters, J. A.; Duisenberg, A. J. M. *Acta Cryst.* **1983,** *C39,* 1221.

(143) Hughes, D. L.; Wingfield, J. N. *J. Chem. Soc., Chem. Commun.* **1977,** 804.

(144) Tesh, K. F.; Hanusa, T. P. *J. Chem. Soc., Chem. Commun.* **1991,** 879.

(145) Turova, N. Y.; Tureskaya, E. P.; Kessler, V. G.; Yanovsky, A. I.; Struchkov, Y. T. *J. Chem. Soc., Chem. Commun.* **1993,** 21.

(146) Vincent, H.; Labrize, F.; Hubert-Pfalzgraf, L. G. *Polyhedron* **1994,** *13,* 3323.

(147) Borup, B.; Samuels, J. A.; Streib, W. E.; Caulton, K. G. *Inorg. Chem.* **1994,** *33,* 994.

(148) Coan, P. S.; Streib, W. E.; Caulton, K. G. *Inorg. Chem.* **1991,** *30,* 5019.

(149) Gaffney, C.; Harrison, P. G.; King, T. J. *J. Chem. Soc., Chem. Commun.* **1980,** 1251.

(150) Wojtczak, W. A.; Hampden-Smith, M. J.; Duesler, E. **1996,** unpublished results.

(151) Belcher, R.; Cranley, C. R.; Majer, J. R.; Stephen, W. I.; Uden, P. C. *Anal.Chim. Acta* **1972,** *60,* 109.

(152) Turnipseed, S. B.; Barkley, R. M.; Sievers, R. E. *Inorg. Chem.* **1991,** *30,* 1164.

(153) Bradley, D. C.; Hasan, M.; Hursthouse, M. B.; Motevalli, M.; Khan, O. F. Z.; Pritchard, R. G.; Williams, J. O. *J. Chem. Soc., Chem. Commun.* **1992,** 575.

(154) Purdy, A. P.; Berry, A. D.; Holm, R. T.; Fatemi, M.; Gaskill, D. K. *Inorg. Chem.* **1989,** *28,* 2799.

(155) Inerowicz, H. D.; Chandrasekhar, T. M.; Atkinson, G.; White, R. L. *Anal. Chim. Acta* **1994,** *289,* 249.

(156) Drozdov, A.; Troyanov, S. *J. Chem. Soc., Chem. Commun.* **1993,** 1619.

(157) Huang, L.; Turnipseed, S. B.; Haltiwanger, R. C.; Barkley, R. M.; Sievers, R. E. *Inorg. Chem.* **1994,** *33,* 798.

(158) Sievers, R. E.; Turnipseed, S. B.; Huang, L.; Lagalante, A. F. *Coord. Chem. Rev.* **1993,** *128,* 285.

(159) Sato, H.; Sugawara, S. *Inorg. Chem.* **1993,** *32,* 1941.

(160) Thompson, S. C.; Cole-Hamilton, D. J.; Gilliland, D. D.; Hitchman, M. L.; Barnes, J. C. *Adv. Mater. Opt. Electron.* **1992,** *1,* 81.

(161) Shamlian, S. H.; Hitchman, M. L.; Cook, S. L.; Richards, B. C. *J. Mater. Chem.* **1994,** *4,* 81.

(162) Shamlian, S. H.; Hitchman, M. L.; Cook, S. L.; Richards, B. C. *J. Mater. Chem.* **1994,** *4,* 81.

(163) Sahbari, J. J.; Olmstead, M. M. *Acta Cryst.* **1983,** *C39,* 208.
(164) Buriak, J. M.; Cheatham, L. K.; Gordon, R. G.; Graham, J. J.; Barron, A. R. *Eur. J. Solid State Inorg. Chem.* **1992,** *29,* 43.
(165) Hollander, F. J.; Templeton, D. H.; Zalkin, A. *Acta Cryst.* **1973,** *B29,* 1295.
(166) Hollander, F. L.; Templeton, D. H.; Zalkin, A. *Acta Cryst.* **1973,** *B29,* 1303.
(167) Rees, W. S. R. J.; Caballero, C. R.; Hesse, W. *Angew. Chem., Int. Ed. Engl.* **1992,** *31,* 735.
(168) Timmer, K.; Meinema, H. A. *Inorg. Chim. Acta* **1991,** *187,* 99.
(169) Rossetto, G.; Polo, A.; Benetollo, F.; Porchia, M.; Zanella, P. *Polyhedron* **1992,** *2,* 979.
(170) Rees, W. S. In King, R. B., ed. *The Encyclopedia of Inorganic Chemistry;* 1994.
(171) Gleizes, A.; Sans-Lenain, S.; Medus, D.; Morancho, R. *C. R. Acad. Sci. Paris* **1991,** *312,* 983.
(172) Gleizes, A.; Medus, D.; Sans-Lenain, S. *Mat. Res. Soc. Symp. Proc.* **1992,** *271,* 919.
(173) Gleizes, A., Sans-Lenain, S.; Heughebaert, M. *C. R. Acad. Sci., Series II* **1992,** *315,* 299.
(174) Rees, W. S.; Carris, M. W.; Hesse, W. *Inorg. Chem.* **1991,** *30,* 4479.
(175) Drozdov, A.; Troyanov, S. *Polyhedron* **1993,** *12,* 2973.
(176) Berg, E. W.; Herrera, N. M. *Anal. Chim. Acta* **1972,** *60,* 117.
(177) Drake, S. R.; Hursthouse, M. B.; Malik, K. M. A.; Otway, D. J. *J. Chem. Soc., Dalton Trans.* **1993,** 2883.
(178) Drozdov, A. A.; Trojanov, S. I. *Polyhedron* **1992,** *11,* 2877.
(179) Drozdov, A.; Kuzmina, N.; Troyanov, S.; Martynenko, L. *Mater. Sci. Eng. B—Solid State M.* **1993,** *18,* 139.
(180) Gleizes, A.; Sans-Lenain, S.; Medus, D. *Compt. Rendu, Acad. Sci. Paris* **1991,** *313,* 761.
(181) Drozdov, A. A.; Troyanov, S. I.; Pisarevsky, A. P.; Struchkov, Y. T. *Polyhedron* **1994,** *13,* 1445.
(182) Auld, J.; Jones, A. C.; Leese, A. B.; Cockayne, B.; Wright, P. J.; Obrien, P.; Motevalli, M. *J. Mater. Chem.* **1993,** *3,* 1203.
(183) Hovnanian, N.; Galloy, J.; Miele, P. *Polyhedron* **1995,** *14,* 297.
(184) Arunasalam, V. C.; Baxter, I.; Drake, S. R.; Hursthouse, M. B.; Malik, K. M. A.; Otway, D. J. *Inorg. Chem.* **1995,** *34,* 5295.
(185) Bidell, W.; Bosch, H. W.; Veghini, D.; Hund, H. U.; Doering, J.; Berke, H. *Helv. Chim. Acta* **1993,** *76,* 596.
(186) Hubertpfalzgraf, L. G.; Labrize, F. *Polyhedron* **1994,** *13,* 2163.
(187) Timmer, K.; Spee, K. I. M. A.; Mackor, A.; Meinema, H. A.; Spek, A. L.; Van der Sluis, P. *Inorg. Chim. Acta* **1991,** *190,* 109.
(188) Drake, S. R.; Miller, S. A. S.; Williams, D. J. *Inorg. Chem.* **1993,** *32,* 3227.
(189) Malandrino, G.; Castelli, F.; Fragala, I. L. *Inorg. Chim. Acta* **1994,** *224,* 203.
(190) Gardiner, R.; Brown, D. W.; Kirlin, P. S.; Rheingold, A. L. *Chem. Mater.* **1991,** *3,* 1053.
(191) Sluis, P. V. D.; Spek, A. L. *Acta Cryst.* **1990,** *C46,* 1741.
(192) Polyanskaya, T. M.; Furmanova, N. G.; Martynova, T. N. *J. Struct. Chem.—Engl. Tr.* **1993,** *34,* 879.
(193) Norman, J. A. T.; Pez, G. P. *J. Chem. Soc., Chem. Commun.* **1991,** *14,* 971.
(194) Polyanskaya, T. M.; Gatilov, Y. V.; Martynova, T. N.; Nikulina, L. D. *Zh. Struk. Khim.* **1992,** *33,* 190.
(195) Neumayer, D. A.; Studebaker, D. B.; Hinds, B. J.; Stern, C. L.; Marks, T. J. *Chem. Mater.* **1994,** *6,* 878.

(196) Malandrino, G.; Fragala, I. L.; Neumayer, D. A.; Stern, C. L.; Hinds, B. J.; Marks, T. J. *J. Mater. Chem.* **1994,** *4,* 1061.
(197) Drake, S. R.; Hursthouse, M. B.; Malik, K. M. A.; Miller, S. A. S. *J. Chem. Soc., Chem. Commun.* **1993,** 478.
(198) Drake, S. R.; Hursthouse, M. B.; Malik, K. M. A.; Miller, S. A. S.; Otway, D. J. *Inorg. Chem.* **1993,** *32,* 4464.
(199) Drake, S. R.; Miller, S. A. S.; Hursthouse, M. B.; Malik, K. M. A. *Polyhedron* **1993,** *12,* 1621.
(200) Gardiner, R. A.; Gordon, D. C.; Stauf, G. T.; Vaartstra, B. A. *Chem. Mater.* **1994,** *6,* 1967.
(201) Vaartstra, B. A.; Gardiner, R. A.; Gordon, D. C.; Ostrander, R. L.; Rheingold, A. L. *Mater. Res. Soc. Symp. Proc.* **335,** 203 (1994).
(202) Drozdov, A. A.; Troyanov, S. I. *Koord. Khim.* **1994,** *20,* 171.
(203) Watson, I. M.; Atwood, M. P.; Haq, S. *Superconduct. Sci. Technol.* **1994,** *7,* 672.
(204) Nash, J. A. P.; Thompson, S. C.; Foster, D. F.; Cole-Hamilton, D.; Barnes, J. C. *J. Chem. Soc., Dalton Trans.* **1995,** 269.
(205) Schulz, D. L.; Hinds, B. J.; Neumayer, D. A.; Stern, C. L.; Marks, T. J. *Chem. Mater.* **1993,** *5,* 1605.
(206) Schulz, D. L.; Hinds, B. J.; Stern, C. L.; Marks, T. J. *Inorg. Chem.* **1993,** *32,* 249.
(207) Mehrotra, R. C.; Bohra, R.; Gaur, D. P. In *Metal β-Diketonates and Allied Derivatives*; Academic Press: New York, 1978.
(208) Meinema, H. A.; Timmer, K.; Linden, H. L. *Mater. Res. Soc. Symp. Proc.* **1994,** *335,* 193.
(209) Rees, W. S.; Barron, A. R. *Mater. Sci. Forum* **1993,** *137–139,* 473.
(210) Kessler, V. G.; Hubert-Pfalzgraf, L. G.; Daniele, S.; Gleizes, A. *Chem. Mater.* **1994,** *6,* 2336.
(211) Breeze, S. R.; Wang, S. *Inorg. Chem.* **1994,** *33,* 5113.
(212) Krupoder, S. A.; Danilovich, V. S.; Miller, A. O.; Furin, G. G. *Zh. Org. Khim.* **1994,** *30,* 1185.
(213) Sato, H.; Sugawara, S. *Inorg. Chem.* **1993,** *2,* 1941.
(214) *CRC Handbook of Chemistry and Physics*; CRC Press: Boca Raton, FL, 1987.
(215) Zhao, J.; Norris, P. *Mater. Sci. Forum* **1993,** *130–132,* 233.
(216) Inerowicz, H. D.; White, R. L. *J. Therm. Anal.* **1995,** *45,* 653.
(217) Sievers, R. E.; Eisentraut, K. J.; Springer, C. S. *Adv. Chem. Ser.* **1967,** *71,* 141.
(218) Buriak, J. M.; Cheatham, L. K.; Graham, J. J.; Gordon, R. G.; Barron, A. R. *Mater. Res. Soc. Symp. Proc.* **1991,** *204,* 545.
(219) Matsuno, S.; Uchikawa, F.; Yoshizaki, K. *Jpn. J. Appl. Phys.* **1990,** *29,* L947.
(220) Inerowicz, H. D.; Khan, M. A.; Atkinson, G.; White, R. L. *Acta Cryst.* **1994,** *C50,* 688.
(221) Schulz, D. L.; Hinds, B. J.; Neumayer, D. A.; Stern, C. L.; Marks, T. J. *Chem. Mater.* **1993,** *5,* 1605.
(222) Gardiner, R. A.; Kirlin, P. S.; Hussein, G. A. M.; Gates, B. C. *Mater. Res. Soc. Symp. Proc.* **1992,** *271,* 933.
(223) Fenton, D. E.; Parkin, D.; Newton, R. F.; Nowell, I. W.; Walker, P. E. J. *J. Chem. Soc., Dalton Trans.* **1982,** 327.
(224) Lindsell, W. E. In *Comprehensive Organometallic Chemistry*; Wilkinson, G.; Stone, F. G. A.; Abel, E. W., Eds., Vol. 1; Pergamon Press: Oxford, 1982, p. 155.
(225) Gindelberger, D. E.; Arnold, J. *Inorg. Chem.* **1994,** *33,* 6293.
(226) Gindelberger, D. E.; Arnold, J. *J. Am. Chem. Soc.* **1992,** *114,* 6242.
(227) Ruhlandt-Senge, K.; Davis, K.; Dalal, S.; Englich, U.; Senge, M. O. *Inorg. Chem.* **1995,** *34,* 2587.

(228) Fenske, D.; Baum, G.; Wolkers, H.; Schreiner, B.; Weller, F.; Dehnicke, K. Z. Anorg. Allg. Chem. **1993**, *619*, 489.
(229) Konig, V. T.; Eisenmann, B.; Schafer, H. Z. Anorg. Allg. Chem. **1982**, *488*, 126.
(230) Banbury, F. A.; Davidson, M. G.; Martin, A.; Raithby, P. R.; Snaith, R.; Verhorevoort, K. L.; Wright, D. S. J. Chem. Soc., Chem. Commun. **1992**, 1152.
(231) Mikulcik, P.; Raithby, P. R.; Smith, R.; Wright, D. S. Angew. Chem., Int. Ed. Engl. **1991**, *30*, 428.
(232) Hursthouse, M. B. Mol. Struct. Diffr. Methods **1977**, *5*, 416.
(233) Hursthouse, M. B. Mol. Struct. Diffr. Methods **1974**, *2*, 492.
(234) Hodgson, D. J.; Asplund, R. O. Inorg. Chem. **1990**, *29*, 3612.
(235) Gautier-Luneau, I.; Mosset, A. J. Solid State Chem. **1988**, *73*, 473.
(236) Klop, E. A.; Schouten, A.; Sluis, P. V. D.; Spek, A. L. Acta Cryst. **1984**, *C40*, 51.
(237) Mak, T. C. W.; Yip, W.-H.; Smith, G.; O'Reilly, E. J.; Kennard, C. H. Inorg. Chim. Acta **1984**, *88*, 35.
(238) Bernard, M. A.; Borel, M. M.; Ledesert, M. A. Bull. Soc. Chimique de France **1973**, *7*, 2194.
(239) Borel, P. M. M.; Ledesert, M. Acta Cryst. **1976**, *B32*, 2388.
(240) Arunasalam, V.-C.; Baxter, I.; Hursthouse, M. B.; Malik, K. M. A.; Mingos, D. M. P.; Plakatouras, J. C. J. Chem. Soc., Chem. Commun. **1994**, 2695.
(241) Arunasalam, V.-C., Mingos, D. M. P.; Plakatouras, J. C.; Baxter, I.; Hursthouse, M. B.; Malik, K. M. A. Polyhedron **1995**, *14*, 1105.
(242) Stiles, M.; Finkbeiner, H. L. J. Am. Chem. Soc. **1959**, *81*, 505.
(243) Finkbeiner, H. L.; Wagner, G. W. J. Org. Chem. **1963**, *28*, 215.
(244) Gentile, P. S.; Ocampo, A. P. Inorg. Chim. Acta **1978**, *29*, 83.
(245) Roux, J. P.; Boeyens, J. C. A. Acta Cryst. **1970**, *B26*, 526.
(246) Roux, J. P.; Kruger, G. J. Acta Cryst. **1976**, *B32*, 1171.
(247) Gentile, P. S.; White, J.; Haddad, S. Inorg. Chim. Acta **1975**, *13*, 149.
(248) Gentile, P. S.; Dinstein, M. P.; White, J. G. Inorg. Chim. Acta **1976**, *19*, 67.
(249) Wood, R. A.; James, V. J.; Angyal, S. J. Acta Cryst. **1977**, *B33*, 2248.
(250) Watson, W. H.; Grossie, D. A.; Vogtle, F.; Muller, W. M. Acta Cryst. **1983**, *C39*, 720.
(251) Smith, G.; O'Reilly, E. J.; Kennard, C. H. L.; White, A. H. J. Chem. Soc., Dalton Trans. **1977**, 1184.
(252) Bullock, J. I.; Ladd, M. F. C.; Povey, D. C.; Tajmir-Riahi, H.-A. Acta Cryst. **1979**, *B35*, 2013.
(253) Barr, D.; Brooker, A. T.; Doyle, M. J.; Drake, S. R.; Raithby, P. R.; Snaith, R.; Wright, D. S. J. Chem. Soc., Chem. Commun. **1989**, *14*, 893.
(254) Gowenlock, B. G.; Lindsell, W. E. J. Organomet. Chem. Lib. **1977**, *3*, 1.
(255) McCormick, M. J.; Sockwell, S. C.; Davies, C. E. H.; Hanusa, T. P.; Huffman, J. C. Organometallics **1989**, *8*, 2044.
(256) Mosges, G.; Hampel, F.; Schleyer, P. v. R. Organometallics **1992**, *11*, 1769.
(257) Kirilov, M.; Petrov, G.; Angelov, C. J. Organomet. Chem. **1976**, *113*, 225.
(258) Gardiner, M. G.; Hanson, G. R.; Junk, P. C.; Raston, C. L.; Skelton, B. W.; White, A. H. J. Chem. Soc., Chem. Commun. **1992**, 1154.
(259) Hutchings, D. S.; Junk, P. C.; Patalinghug, W. C.; Raston, C. L.; White, A. H. J. Chem. Soc., Chem. Commun. **1989**, *15*, 973.
(260) Khattar, R.; Knobler, C. B.; Hawthorne, M. F. J. Am. Chem. Soc. **1990**, *112*, 4962.
(261) Khattar, R.; Knobler, C. B.; Hawthorne, M. F. Inorg. Chem. **1990**, *29*, 2191.
(262) Ruessel, C. J. Mater. Sci. Lett. **1992**, *11*, 152.
(263) Ruessel, C. J. Non-cryst. Solids **1993**, *152*, 161.

(264) Erin, A. V.; Prozorovskaya, Z. N.; Yarovslavtev, A. B. *Zh. Neorg. Khim.* **1993,** *38*, 618.

(265) Gupta, A.; Jagannathan, R.; Cooper, E. I.; Giess, E. A.; Landman, J. I.; Hussey, B. W. *Appl. Phys. Lett.* **1988,** *52*, 2077.

(266) Wojtczak, W.; Atanasova, P.; Duesler, E.; Hampden-Smith, M. J. *Inorg. Chem.* **1996,** in press.

(267) Archer, L. PhD dissertation, 1996.

(268) Mehrotra, R. C.; Bohra, R. *Metal Carboxylates*; Academic Press: New York, 1983.

(269) Apblett, A. W.; Lei, J.; Georgeva, G. D. *Mater. Res. Soc. Symp. Proc.* **1992,** *271*, 77.

(270) Kunze, K.; Bihry, L.; Atanasova, P.; Hampden-Smith, M. J.; Duesler, E. N. *Adv. Mat. Chem. Vapor Deposition*, **1996,** *2*, 105.

(271) Bogatskii, A. V.; Chumachenko, T. K.; Derkach-Kozhukhova, A. E.; Lyamtseva, L. N.; Suprinovich, E. S. *Zh. Obshch. Khim.* **1977,** *47*, 2297.

(272) Zemlyanichenko, M. A.; Sheverdina, N. I.; Viktorova, I. M.; Barminova, N. P.; Kocheshkov, K. A. *Akad. Nauk SSSR* **1970,** *194*, 95.

(273) Zemlyanichenko, M. A.; Sheverdina, N. I.; Kocheshkov, K. A. *Dokl. Akad. Nauk SSSR* **1972,** *202*, 595.

(274) Coles, M. A.; Hart, F. A. *J. Organomet. Chem.* **1971,** *32*, 279.

(275) Bryce-Smith D.; Skinner, A. C. *J. Chem. Soc.* **1963,** 577.

(276) Rogers, R. D.; Bond, A. H.; Bauer, C. B. *Pure Appl. Chem.* **1993,** *65*, 567.

(277) Kaniansky, D.; Zelensky, L.; Valaskova, I.; Marak, J.; Zelenska, V. *J. Chromatogr.* **1990,** *502*, 143.

(278) Kakiuchi, T. J. *Colloid Interface Sci.* **1993,** *156*, 406.

(279) Kikuchi, Y.; Suzuki, T.; Sawada, K. *Anal. Chim. Acta* **1992,** *264*, 65.

(280) Gloe, K.; Muehl, P.; Ruestig, H.; Beger, J.; *Solvent. Extr. Ion Exch.* **1988,** *6*, 417.

(281) Pedersen, C. J. *J. Am. Chem. Soc.* **1967,** *89*, 7017.

(282) Pedersen, C. J. *J. Am. Chem. Soc.* **1970,** *92*, 386.

(283) Weber, E.; Vogtle, F. *Chem. Ber.* **1976,** *109*, 1803.

(284) Metz, B.; Moras, D.; Weiss, R. *Acta Cryst.* **1973,** *B29*, 1382.

(285) Metz, B.; Moras, D.; Weiss, R. *Acta Cryst.* **1973,** *B29*, 1388.

(286) Wei, Y. Y.; Tinant, B.; Declerq, J. P.; Van Meerssche, M.; Dale, J. *Acta Cryst.* **1988,** *C44*, 77.

(287) Wei, Y. Y.; Tinant, B.; Declerq, J. P.; Van Meerssche, M.; Dale, J. *Acta Crystallogr., Sect. C: Cryst. Struct. Commun.* **1987,** *C43*, 1080.

(288) Wei, Y. Y.; Tinant, B.; Declerq, J. P.; Van Meerssche, M.; Dale, J. *Acta Crystallogr., Sect. C: Cryst. Struct. Commun.* **1988,** *1*, 68.

(289) Yamaguchi, I.; Miki, K.; Yasuoka, N.; Kasai, N. *Bull. Chem. Soc. Jpn.* **1982,** *55*, 1372.

(290) Ohmoto, H.; Kai, Y.; Yasuoka, N.; Kasai, N.; Yanagida, S.; Okahara, M. *Bull. Chem. Soc. Jpn.* **1979,** *52*, 1209.

(291) Gentile, P. S.; Carlotto, J.; Shankoff, T. A. *J. Inorg. Nucl. Chem.* **1967,** *29*, 1427.

(292) Mazzarella, L.; Kovacs, A. L.; Santis, P. D.; Liquori, A. M. *Acta Cryst.* **1967,** *22*, 65.

(293) Behr, J. P.; Lehn, J. M.; Moras, D.; Thierry, J. C. *J. Am. Chem. Soc.* **1981,** *103*, 701.

(294) Pettit, W. A.; Baenziger, N. C. *Acta Cryst.* **1994,** *C50*, 221.

(295) Rogers, R. D.; Jezl, M. L.; Bauer, C. B. *Inorg. Chem.* **1994,** *33*, 5682.

(296) Krasnova, N. F.; Dvorkin, A. A.; Simonov, Y. A.; Abashkin, V. M.; Yakshin, V. V. *Kristallografiya* **1985,** *30*, 86.

(297) Dyer, R. B.; Metcalf, D. H.; Ghirardelli, R. G.; Palmer, R. A.; Holt, E. M. *J. Am. Chem. Soc.* **1986,** *108*, 3621.

(298) Hughes, D. L.; Mortimer, C. L.; Truter, M. R. *Inorg. Chim. Acta* **1978,** *29*, 43.

(299) Drew, M. G. B.; Knox, C. V.; Nelson, S. M. *J. Chem. Soc., Dalton Trans.* **1980,** 942.
(300) Burns, J. H.; Kessler, R. M. *Inorg. Chem.* **1987,** *26,* 1370.
(301) Dale, J.; Kristiansen, P. O. *Acta Chem. Scand.* **1972,** *26,* 1471.
(302) Magul, S.; Dehnicke, K.; Fenske, D. *Z. Anorg. Allg. Chem.* **1992,** *608,* 17.
(303) Burns, J. H.; Bryan, S. A. *Acta Crystallogr., Sect. C: Cryst. Struct. Commun.* **1988,** *C44,* 1742.
(304) Rheingold, A. L.; White, C. B.; Haggerty, B. S.; Kirlin, P.; Gardiner, R. A. *Acta Cryst.* **1993,** *C49,* 808.
(305) Kanters, J. A.; Harder, S.; Poonia, N. S. *Acta Crystallogr., Sect. C: Cryst. Struct. Commun.* **1987,** *C43,* 1042.
(306) Cradwick, P. D.; Poonia, N. S. *Acta Cryst.* **1977,** *B33,* 197.
(307) Kunze, K.; Hampden-Smith, M. J. Unpublished results, 1996.
(308) Sato, H. *Jpn. J. Appl. Phys., Part 2* **1994,** *33,* L371.
(309) Sato, H. *Jpn. J. Appl. Phys., Part 2* **1994,** *33,* L368.
(310) Sato, H.; Sugawara, S. *Jpn. J. Appl. Phys., Part 2* **1993,** *32,* L799.
(311) Ylilammi, M.; Ranta-aho, T. *J. Electrochem. Soc.* **1994,** *141,* 1278.
(312) Nakazawa, H.; Yamane, H.; Hirai, T. *Jpn. J. Appl. Phys., Part 1* **1991,** *30,* 2200.
(313) Kobayashi, I.; Wakao, Y.; Tomigana, K.; Okada, M. *Jpn. J. Appl. Phys.* **1994,** *33,* 4680.
(314) Kim, T. W.; Yom, S. S. *Appl. Phys. Lett.* **1994,** *65,* 1955.
(315) Kaiser, D. L.; Vaudin, M. D.; Gillen, G.; Hwang, C. S.; Robins, L. H.; Rotter, L. D. *J. Cryst. Growth* **1994,** *137,* 136.
(316) Kaiser, D. L.; Vaudin, M. D.; Rotter, L. D.; Wang, Z. L.; Cline, J. P.; Hwang, C. S.; Marinenko, R. B.; Gillen, J. G. *Appl. Phys. Lett.* **1995,** *66,* 2801.
(317) Itoh, H.; Oikawa, I.; Iwahara, H.; Alzawa, M. *J. Mater. Sci.* **1995,** *30,* 2139.
(318) Kawahara, T.; Yamamuka, M.; Makita, T.; Tsutahara, K.; Yuuki, A.; Ono, K.; Matsui, Y. *Jpn. J. Appl. Phys., Part 1* **1994,** *33,* 5897.
(319) Donahue, E. J.; Schleich, D. M. *J. Appl. Phys.* **1992,** *71,* 6013.
(320) Aarik, J.; Aidla, A.; Jaek, A.; Leskela, M.; Niinisto, L. *J. Mater. Chem.* **1994,** *4,* 1239.
(321) Kimura, T.; Yamauchi, H.; Machida, H.; Kokubun, H.; Yamada, M. *Jpn. J. Appl. Phys., Part 1* **1994,** *33,* 5119.
(322) Bihari, B. P.; Kumar, J.; Stauf, G. T.; Vanbuskirk, P. C.; Hwang, C. S. *J. Appl. Phys.* **1994,** *76,* 1169.
(323) Gilbert, S. R.; Wessels, B. W.; Studebaker, D. B.; Marks, T. J. *Appl. Phys. Lett.* **1995,** *66,* 3298.
(324) Feil, W. A.; Wessels, B. W. *J. Appl. Phys.* **1993,** *74,* 3927.
(325) Chen, J.; Wills, L. A.; Wessels, B. W.; Schulz, D. L.; Marks, T. J. *J. Electron. Mater.* **1993,** *22,* 701.
(326) Wills, L. A.; Wessels, B. W.; Richeson, D. S.; Marks, T. J. *Appl. Phys. Lett.* **1992,** *60,* 41.
(327) Nystrom, M. J.; Wessels, B. W.; Lin, W. P.; Wong, G. K.; Neumayer, D. A.; Marks, T. J. *Appl. Phys. Lett.* **1995,** *66,* 1726.
(328) Neumayer, D. A.; Studebaker, D. B.; Hinds, B. J.; Stern, C. L.; Marks, T. J. *Chem. Mater.* **1994,** *6,* 878.
(329) Zhang, J.; Zhao, J.; Marcy, H. O.; Tonge, L. M.; Wessels, B. W.; Marks, T. J.; Kannewurf, C. R. *Appl. Phys. Lett.* **1989,** *54,* 1166.
(330) Zhang, J. M.; Wessels, B. W.; Richeson, D. S.; Marks, T. J. *J. Cryst. Growth* **1991,** *107,* 705.
(331) Zhang, J. M.; Wessels, B. W.; Tonge, L. M.; Marks, T. J. *Appl. Phys. Lett.* **1990,** *56,* 976.

(332) Zhao, J.; Li, Y. Q.; Chern, C. S.; Norris, P.; Gallois, B.; Kear, B.; Wessels, B. W. *Appl. Phys. Lett.* **1991,** *58,* 89.

(333) Zhao, J.; Marcy, H. O.; Tonge, L. M.; Wessels, B. W.; Marks, T. J.; Kannewurf, C. R. *Mater. Res. Soc. Symp. Proc.* **1990,** *169,* 593.

(334) Hiskes, R.; DiCarolis, S. A.; Young, J. L.; Laderman, S. S.; Jacowitz, R. D.; Taber, R. C. *Appl. Phys. Lett.* **1991,** *59,* 606.

(335) Zhao, J.; Marcy, H. O.; Tonge, L. M.; Wessels, B. W.; Marks, T. J.; Kannewurf, C. R. *Solid State Commun.* **1990,** *74,* 1091.

(336) Zhao, J.; Dahmen, K. H.; Marcy, H. O.; Tonge, L. M.; Wessels, B. W.; Marks, T. J.; Kannewurf, C. R. *Solid State Commun.* **1989,** *69,* 187.

(337) Richeson, D. S.; Tonge, L. M.; Zhao, J.; Zhang, J.; Marcy, H. O.; Marks, T. J.; Wessels, B. W.; Kannewurf, C. R. *Appl. Phys. Lett.* **1989,** *54,* 2154.

(338) Watson, I. M.; Atwood, M. P.; Cardwell, D. A.; Cumberbatch, T. J. *J. Mater. Chem.* **1994,** *4,* 1393.

(339) Spee, C. I. M. A.; Vanderzouwenassink, E. A.; Timmer, K.; Mackor, A.; Meinema, H. A. *J. Phys. II* **1991,** *1,* 295.

(340) Zhang, J. M.; Wessels, B. W.; Richeson, D. S.; Marks, T. J.; DeGroot, D. C.; Kannewurf, C. R. *J. Appl. Phys.* **1991,** *69,* 2743.

(341) Sato, K.; Sugawara, S. Japanese Patent 91–221721 910902; Nippon Telegraph and Telephone Corp., Japan, Jpn. Kokai Tokkyo Koho, 1991.

(342) Yamada, Y.; Morimoto, T. Japanese Patent 89–41859 890223; Asahi Glass Co., Ltd., Japan, Jpn. Kokai Tokkyo Koho, 1989.

(343) Yamada, Y.; Morimoto, T. Japanese Patent 89–41860 890223; Asahi Glass Co., Ltd., Japan, Jpn. Kokai Tokkyo Koho, 1989.

(344) Sato, H.; Sugawara, S. *Jpn. J. Appl. Phys. Part 2—Letters* **1993,** *32,* L799.

(345) Yuhya, S.; Kikuchi, K.; Yoshida, M.; Sugawara, K.; Shiohara, Y. *Mol. Cryst. Liq. Cryst.* **1990,** *184,* 231.

(346) Chou, K. S.; Tsai, G. J. *Thermochim. Acta* **1994,** *240,* 129.

(347) Schmaderer, F.; Huber, R.; Oetzmann, H.; Wahl, G. *Appl. Surf. Sci.* **1990,** *46,* 53.

(348) Schwarberg, J. E.; Sievers, R. E.; Moshier, R. W. *Anal. Chem.* **1970,** *42,* 1828.

(349) Donahue, E. J.; Schleich, D. M. *Mater. Res. Bull.* **1991,** *26,* 1119.

(350) Yoon, Y. S.; Yom, S. S.; Kim, H. J.; Kim, T. W.; Jung, M.; Leem, J. Y.; Kang, T. W.; Lee, S. L. *J. Mater. Sci.* **1995,** *30,* 3603.

(351) Yoon, Y. S.; Lee, D. H.; Kim, T. S.; Oh, M. H.; Yom, S. S. *J. Vac. Sci. Technol., A* **1994,** *12,* 751.

(352) Dickinson, P. H.; Geballe, T. H.; Sanjurjo, A.; Hildenbrad, D.; Craig, G.; Zisk, M.; Collman, J.; Banning, S. A.; Sievers, R. E. *J. Appl. Phys.* **1989,** *66,* 444.

(353) Barron, A. R.; Buriak, J. M.; Gordon, R. G. U.S. Patent, 5,139,999, 1992.

(354) Sato, H.; Sugawara, S. *Inorg. Chem.* **1993,** *32,* 1941.

(355) Van Buskirk, P. C.; Gardiner, R. A.; Kirlin, P. S.; Nutt, S. *J. Mater. Res.* **1992,** *7,* 542.

(356) Rees, W. S.; Caballero, C. R.; Hesse, W. *Adv. Mater.* **1992,** *104,* 786.

(357) Caruso, J.; Roger, C.; Schwertfeger, F.; Hampden-Smith, M.; Rheingold, A.; Yap, G. *Inorg. Chem.* **1995,** *34,* 449.

(358) Caruso, J.; Hampden-Smith, M.; Rheingold, A. L.; Yap, G. *J. Chem. Soc., Chem. Commun.* **1995,** *2,* 157.

(359) Inerowicz, H. D.; White, R. L. *J. Therm. Anal.* **1995,** *45,* 653.

(360) Pigot, T.; Duriez, M. C.; Cazaux, L.; Picard, C.; Tisnes, P. *J. Chem. Soc., Perkin Trans.* **1993,** *2,* 221.

(361) Ouchi, M.; Inoue, Y.; Wada, K.; Iketani, S.; Hakushi, T.; Weber, E. *J. Org. Chem.* **1987,** *52,* 2420.

(362) McDowell, D.; Nelson, J.; McKee, V.; *Polyhedron* **1989,** *8*, 1143.
(363) Inoue, Y.; Wada, K.; Liu, Y.; Ouchi, M.; Tai, A.; Hakushi, T. *J. Org. Chem* **1989,** *54*, 5268.
(364) Inoue, Y.; Wada, K.; Ouchi, M.; Tai, A.; Hakushi, T. *Chem. Lett* **1988,** *6*, 1005.
(365) Hay, B. P.; Rustad, J. R. *J. Am. Chem. Soc.* **1994,** *116*, 6316.
(366) Hancock, R. D.; Bhavan, R.; Wade, P. W.; Boeyens, J. C. A.; Dobson, S. M. *Inorg. Chem* **1989,** *28*, 187.
(367) Hancock, R. D.; Shaikjee, M. S.; Dobson, S. M.; Boeyens, J. C. A. *Inorg. Chim. Acta* **1988,** *154*, 229.
(368) Drew, M. G. B.; Esho, F. S.; Nelson, S. M. *J. Chem. Soc., Dalton Trans.* **1983,** *8*, 1653.
(369) Anantanarayan, A.; Fyles, T. M. *Can. J. Chem.* **1990,** *68*, 1338.
(370) Bartlett, J. S.; Costello, J. F.; Mehani, S.; Ramdas, S.; Slawin, A. M. Z.; Stoddart, J. F.; Williams, D. J. *Angew. Chem.* **1990,** *102*, 1463.
(371) Katayama, Y.; Fukuda, R.; Iwasaki, T.; Nita, K.; Takagi, M. *Anal. Chim. Acta* **1988,** *204*, 113.

Index

341

Cumulative List of Contributors
for Volumes 1-36

Abel, E. W., **5,** 1; **8,** 117
Aguiló, A., **5,** 321
Akkerman, O. S., **32,** 147
Albano, V. G., **14,** 285
Alper, H., **19,** 183
Anderson, G. K., **20,** 39; **35,** 1
Angelici, R. J., **27,** 51
Aradi, A. A., **30,** 189
Armitage, D. A., **5,** 1
Armor, J. N., **19,** 1
Ash, C. E., **27,** 1
Ashe, A. J., III, **30,** 77
Atwell, W. H., **4,** 1
Baines, K. M., **25,** 1
Barone, R., **26,** 165
Bassner, S. L., **28,** 1
Behrens, H., **18,** 1
Bennett, M. A., **4,** 353
Bickelhaupt, F., **32,** 147
Birmingham, J., **2,** 365
Blinka, T. A., **23,** 193
Bockman, T. M., **33,** 51
Bogdanović, B., **17,** 105
Bottomley, F., **28,** 339
Bradley, J. S., **22,** 1
Brew, S. A., **35,** 135
Brinckman, F. E., **20,** 313
Brook, A. G., **7,** 95; **25,** 1
Bowser, J. R., **36,** 57
Brown, H. C., **11,** 1
Brown, T. L., **3,** 365
Bruce, M. I., **6,** 273, **10,** 273; **11,** 447; **12,**
  379; **22,** 59
Brunner, H., **18,** 151
Buhro, W. E., **27,** 311
Byers, P. K., **34,** 1
Cais, M., **8,** 211
Calderon, N., **17,** 449
Callahan, K. P., **14,** 145
Canty, A. J., **34,** 1
Cartledge, F. K., **4,** 1
Chalk, A. J., **6,** 119
Chanon, M., **26,** 165

Chatt, J., **12,** 1
Chini, P., **14,** 285
Chisholm, M. H., **26,** 97; **27,** 311
Chiusoli, G. P., **17,** 195
Chojinowski, J., **30,** 243
Churchill, M. R., **5,** 93
Coates, G. E., **9,** 195
Collman, J. P., **7,** 53
Compton, N. A., **31,** 91
Connelly, N. G., **23,** 1; **24,** 87
Connolly, J. W., **19,** 123
Corey, J. Y., **13,** 139
Corriu, R. J. P., **20,** 265
Courtney, A., **16,** 241
Coutts, R. S. P., **9,** 135
Coville, N. J., **36,** 95
Coyle, T. D., **10,** 237
Crabtree, R. H., **28,** 299
Craig, P. J., **11,** 331
Csuk, R., **28,** 85
Cullen, W. R., **4,** 145
Cundy, C. S., **11,** 253
Curtis, M. D., **19,** 213
Darensbourg, D. J., **21,** 113; **22,** 129
Darensbourg, M. Y., **27,** 1
Davies, S. G., **30,** 1
Deacon, G. B., **25,** 237
de Boer, E., **2,** 115
Deeming, A. J., **26,** 1
Dessy, R. E., **4,** 267
Dickson, R. S., **12,** 323
Dixneuf, P. H., **29,** 163
Eisch, J. J., **16,** 67
Ellis, J. E., **31,** 1
Emerson, G. F., **1,** 1
Epstein, P. S., **19,** 213
Erker, G., **24,** 1
Ernst, C. R., **10,** 79
Errington, R. J., **31,** 91
Evans, J., **16,** 319
Evans, W. J., **24,** 131
Faller, J. W., **16,** 211
Farrugia, L. J., **31,** 301

Faulks, S. J., **25**, 237
Fehlner, T. P., **21**, 57; **30**, 189
Fessenden, J. S., **18**, 275
Fessenden, R. J., **18**, 275
Fischer, E. O., **14**, 1
Ford, P. C., **28**, 139
Forniés, J., **28**, 219
Forster, D., **17**, 255
Fraser, P. J., **12**, 323
Friedrich, H., **36**, 229
Friedrich, H. B., **33**, 235
Fritz, H. P., **1**, 239
Fürstner, A., **28**, 85
Furukawa, J., **12**, 83
Fuson, R. C., **1**, 221
Gallop, M. A., **25**, 121
Garrou, P. E., **23**, 95
Geiger, W. E., **23**, 1; **24**, 87
Geoffroy, G. L., **18**, 207; **24**, 249; **28**, 1
Gilman, H., **1**, 89; **4**, 1; **7**, 1
Gladfelter, W. L., **18**, 207; **24**, 41
Gladysz, J. A., **20**, 1
Glänzer, B. I., **28**, 85
Green, M. L. H., **2**, 325
Grey, R. S., **33**, 125
Griffith, W. P., **7**, 211
Grovenstein, E., Jr., **16**, 167
Gubin, S. P., **10**, 347
Guerin, C., **20**, 265
Gysling, H., **9**, 361
Haiduc, I., **15**, 113
Halasa, A. F., **18**, 55
Hamilton, D. G., **28**, 299
Handwerker, H., **36**, 229
Harrod, J. F., **6**, 119
Hart, W. P., **21**, 1
Hartley, F. H., **15**, 189
Hawthorne, M. F., **14**, 145
Heck, R. F., **4**, 243
Heimbach, P., **8**, 29
Helmer, B. J., **23**, 193
Henry, P. M., **13**, 363
Heppert, J. A., **26**, 97
Herberich, G. E., **25**, 199
Herrmann, W. A., **20**, 159
Hieber, W., **8**, 1
Hill, A. F., **36**, 131
Hill, E. A., **16**, 131
Hoff, C., **19**, 123
Hoffmeister, H., **32**, 227

Holzmeier, P., **34**, 67
Honeyman, R. T., **34**, 1
Horwitz, C. P., **23**, 219
Hosmane, N. S., **30**, 99
Housecroft, C. E., **21**, 57; **33**, 1
Huang, Y. Z., **20**, 115
Hughes, R. P., **31**, 183
Ibers, J. A., **14**, 33
Ishikawa, M., **19**, 51
Ittel, S. D., **14**, 33
Jain, L., **27**, 113
Jain, V. K., **27**, 113
James, B. R., **17**, 319
Janiak, C., **33**, 291
Jastrzebski, J. T. B. H., **35**, 241
Jenck, J., **32**, 121
Jolly, P. W., **8**, 29; **19**, 257
Jonas, K., **19**, 97
Jones, M. D., **27**, 279
Jones, P. R., **15**, 273
Jordan, R. F., **32**, 325
Jukes, A. E., **12**, 215
Jutzi, P., **26**, 217
Kaesz, H. D., **3**, 1
Kalck, P., **32**, 121; **34**, 219
Kaminsky, W., **18**, 99
Katz, T. J., **16**, 283
Kawabata, N., **12**, 83
Kemmitt, R. D. W., **27**, 279
Kettle, S. F. A., **10**, 199
Kilner, M., **10**, 115
Kim, H. P., **27**, 51
King, R. B., **2**, 157
Kingston, B. M., **11**, 253
Kisch, H., **34**, 67
Kitching, W., **4**, 267
Kochi, J. K., **33**, 51
Köster, R., **2**, 257
Kreiter, C. G., **26**, 297
Krüger, G., **24**, 1
Kudaroski, R. A., **22**, 129
Kühlein, K., **7**, 241
Kuivila, H. G., **1**, 47
Kumada, M., **6**, 19; **19**, 51
Lappert, M. F., **5**, 225; **9**, 397; **11**, 253; **14**, 345
Lawrence, J. P., **17**, 449
Le Bozec, H., **29**, 163
Lednor, P. W., **14**, 345
Linford, L., **32**, 1

Cumulative Index
for Volumes 37–40

ISBN 0-12-031140-2

90065